T0300012

Computational
Methods in
Plasma Physics

Chapman & Hall/CRC
Computational Science Series

SERIES EDITOR

Horst Simon

Associate Laboratory Director, Computing Sciences
Lawrence Berkeley National Laboratory
Berkeley, California, U.S.A.

AIMS AND SCOPE

This series aims to capture new developments and applications in the field of computational science through the publication of a broad range of textbooks, reference works, and handbooks. Books in this series will provide introductory as well as advanced material on mathematical, statistical, and computational methods and techniques, and will present researchers with the latest theories and experimentation. The scope of the series includes, but is not limited to, titles in the areas of scientific computing, parallel and distributed computing, high performance computing, grid computing, cluster computing, heterogeneous computing, quantum computing, and their applications in scientific disciplines such as astrophysics, aeronautics, biology, chemistry, climate modeling, combustion, cosmology, earthquake prediction, imaging, materials, neuroscience, oil exploration, and weather forecasting.

PUBLISHED TITLES

PETASCALE COMPUTING: Algorithms and Applications
Edited by David A. Bader

PROCESS ALGEBRA FOR PARALLEL AND DISTRIBUTED PROCESSING
Edited by Michael Alexander and William Gardner

GRID COMPUTING: TECHNIQUES AND APPLICATIONS
Barry Wilkinson

INTRODUCTION TO CONCURRENCY IN PROGRAMMING LANGUAGES
Matthew J. Sottile, Timothy G. Mattson, and Craig E Rasmussen

INTRODUCTION TO SCHEDULING
Yves Robert and Frédéric Vivien

SCIENTIFIC DATA MANAGEMENT: CHALLENGES, TECHNOLOGY, AND DEPLOYMENT
Edited by Arie Shoshani and Doron Rotem

COMPUTATIONAL METHODS IN PLASMA PHYSICS
Stephen Jardin

Chapman & Hall/CRC
Computational Science Series

Computational Methods in Plasma Physics

Stephen Jardin

CRC Press
Taylor & Francis Group
Boca Raton   London   New York

CRC Press is an imprint of the
Taylor & Francis Group, an **informa** business

A CHAPMAN & HALL BOOK

On the cover is a computed iso-contour surface for the toroidal current density in the nonlinear phase of an internal instability of a tokamak plasma. Contours on one toroidal plane and one mid-section plane are also shown. (Courtesy of Dr. J. Breslau and the M3D team.)

CRC Press
Taylor & Francis Group
6000 Broken Sound Parkway NW, Suite 300
Boca Raton, FL 33487-2742

© 2010 by Taylor and Francis Group, LLC
CRC Press is an imprint of Taylor & Francis Group, an Informa business

No claim to original U.S. Government works

Printed in the United States of America on acid-free paper
10 9 8 7 6 5 4 3 2 1

International Standard Book Number: 978-1-4398-1021-7 (Hardback)

Library of Congress Cataloging-in-Publication Data

Jardin, Stephen.
    Computational methods in plasma physics / Stephen Jardin.
        p. cm. -- (Chapman & Hall/CRC computational science series)
    Includes bibliographical references and index.
    ISBN 978-1-4398-1021-7 (hardcover : alk. paper)
    1. Plasma (Ionized gases)--Mathematical models. 2. Mathematical physics. I. Title. II. Series.

QC718.J345 2010
530.4'4015118--dc22   2010014068

Visit the Taylor & Francis Web site at
http://www.taylorandfrancis.com

and the CRC Press Web site at
http://www.crcpress.com

to Marilyn

Contents

List of Figures   xiii

List of Tables   xvii

Preface   xix

List of Symbols   xxi

1 Introduction to Magnetohydrodynamic Equations   1
  1.1 Introduction   1
  1.2 Magnetohydrodynamic (MHD) Equations   4
    1.2.1 Two-Fluid MHD   6
    1.2.2 Resistive MHD   8
    1.2.3 Ideal MHD   9
    1.2.4 Other Equation Sets for MHD   10
    1.2.5 Conservation Form   10
    1.2.6 Boundary Conditions   12
  1.3 Characteristics   14
    1.3.1 Characteristics in Ideal MHD   16
    1.3.2 Wave Dispersion Relation in Two-Fluid MHD   23
  1.4 Summary   24

2 Introduction to Finite Difference Equations   27
  2.1 Introduction   27
  2.2 Implicit and Explicit Methods   29
  2.3 Errors   30
  2.4 Consistency, Convergence, and Stability   31
  2.5 Von Neumann Stability Analysis   32
    2.5.1 Relation to Truncation Error   36
    2.5.2 Higher-Order Equations   37
    2.5.3 Multiple Space Dimensions   39
  2.6 Accuracy and Conservative Differencing   39
  2.7 Summary   42

3 Finite Difference Methods for Elliptic Equations   **45**
  3.1   Introduction 45
  3.2   One-Dimensional Poisson's Equation 46
        3.2.1   Boundary Value Problems in One Dimension 46
        3.2.2   Tridiagonal Algorithm 47
  3.3   Two-Dimensional Poisson's Equation 48
        3.3.1   Neumann Boundary Conditions 50
        3.3.2   Gauss Elimination 53
        3.3.3   Block-Tridiagonal Method 56
        3.3.4   General Direct Solvers for Sparse Matrices 57
  3.4   Matrix Iterative Approach 57
        3.4.1   Convergence 59
        3.4.2   Jacobi's Method 60
        3.4.3   Gauss–Seidel Method 60
        3.4.4   Successive Over-Relaxation Method (SOR) 61
        3.4.5   Convergence Rate of Jacobi's Method 61
  3.5   Physical Approach to Deriving Iterative Methods 62
        3.5.1   First-Order Methods 63
        3.5.2   Accelerated Approach: Dynamic Relaxation 65
  3.6   Multigrid Methods 66
  3.7   Krylov Space Methods 70
        3.7.1   Steepest Descent and Conjugate Gradient 72
        3.7.2   Generalized Minimum Residual (GMRES) 76
        3.7.3   Preconditioning 80
  3.8   Finite Fourier Transform 82
        3.8.1   Fast Fourier Transform 83
        3.8.2   Application to 2D Elliptic Equations 86
  3.9   Summary 88

4 Plasma Equilibrium   **93**
  4.1   Introduction 93
  4.2   Derivation of the Grad–Shafranov Equation 93
        4.2.1   Equilibrium with Toroidal Flow 95
        4.2.2   Tensor Pressure Equilibrium 97
  4.3   The Meaning of Ψ 99
  4.4   Exact Solutions 102
        4.4.1   Vacuum Solution 102
        4.4.2   Shafranov–Solovév Solution 104
  4.5   Variational Forms of the Equilibrium Equation 105
  4.6   Free Boundary Grad–Shafranov Equation 106
        4.6.1   Inverting the Elliptic Operator 107
        4.6.2   Iterating on $J_\phi(R, \Psi)$ 107
        4.6.3   Determining Ψ on the Boundary 109
        4.6.4   Von Hagenow's Method 111
        4.6.5   Calculation of the Critical Points 113

	4.6.6	Magnetic Feedback Systems	114
	4.6.7	Summary of Numerical Solution	116
4.7		Experimental Equilibrium Reconstruction	116
4.8		Summary	117

5 Magnetic Flux Coordinates in a Torus — **121**

5.1		Introduction	121
5.2		Preliminaries	121
	5.2.1	Jacobian	122
	5.2.2	Basis Vectors	124
	5.2.3	Grad, Div, Curl	125
	5.2.4	Metric Tensor	127
	5.2.5	Metric Elements	127
5.3		Magnetic Field, Current, and Surface Functions	129
5.4		Constructing Flux Coordinates from $\Psi(R, Z)$	131
	5.4.1	Axisymmetric Straight Field Line Coordinates	133
	5.4.2	Generalized Straight Field Line Coordinates	135
5.5		Inverse Equilibrium Equation	136
	5.5.1	q-Solver	137
	5.5.2	J-Solver	138
	5.5.3	Expansion Solution	139
	5.5.4	Grad–Hirshman Variational Equilibrium	140
	5.5.5	Steepest Descent Method	144
5.6		Summary	147

6 Diffusion and Transport in Axisymmetric Geometry — **149**

6.1		Introduction	149
6.2		Basic Equations and Orderings	149
	6.2.1	Time-Dependent Coordinate Transformation	151
	6.2.2	Evolution Equations in a Moving Frame	153
	6.2.3	Evolution in Toroidal Flux Coordinates	155
	6.2.4	Specifying a Transport Model	158
6.3		Equilibrium Constraint	162
	6.3.1	Circuit Equations	163
	6.3.2	Grad–Hogan Method	163
	6.3.3	Taylor Method (Accelerated)	164
6.4		Time Scales	165
6.5		Summary	168

7 Numerical Methods for Parabolic Equations — **171**

7.1		Introduction	171
7.2		One-Dimensional Diffusion Equations	171
	7.2.1	Scalar Methods	172
	7.2.2	Non-Linear Implicit Methods	175
	7.2.3	Boundary Conditions in One Dimension	179

       7.2.4   Vector Forms 180
   7.3   Multiple Dimensions 183
       7.3.1   Explicit Methods 183
       7.3.2   Fully Implicit Methods 184
       7.3.3   Semi-Implicit Method 184
       7.3.4   Fractional Steps or Splitting 185
       7.3.5   Alternating Direction Implicit (ADI) 186
       7.3.6   Douglas–Gunn Method 187
       7.3.7   Anisotropic Diffusion 188
       7.3.8   Hybrid DuFort–Frankel/Implicit Method 190
   7.4   Summary 192

8   Methods of Ideal MHD Stability Analysis   195
   8.1   Introduction 195
   8.2   Basic Equations 195
       8.2.1   Linearized Equations about Static Equilibrium 195
       8.2.2   Methods of Stability Analysis 199
       8.2.3   Self-Adjointness of $\mathbf{F}$ 200
       8.2.4   Spectral Properties of $\mathbf{F}$ 201
       8.2.5   Linearized Equations with Equilibrium Flow 203
   8.3   Variational Forms 204
       8.3.1   Rayleigh Variational Principle 204
       8.3.2   Energy Principle 205
       8.3.3   Proof of the Energy Principle 206
       8.3.4   Extended Energy Principle 207
       8.3.5   Useful Identities 208
       8.3.6   Physical Significance of Terms in δW_f 210
       8.3.7   Comparison Theorem 211
   8.4   Cylindrical Geometry 213
       8.4.1   Eigenmode Equations and Continuous Spectra 214
       8.4.2   Vacuum Solution 215
       8.4.3   Reduction of δW_f 216
   8.5   Toroidal Geometry 219
       8.5.1   Eigenmode Equations and Continuous Spectra 220
       8.5.2   Vacuum Solution 223
       8.5.3   Global Mode Reduction in Toroidal Geometry 225
       8.5.4   Ballooning Modes 226
   8.6   Summary 232

9   Numerical Methods for Hyperbolic Equations   235
   9.1   Introduction 235
   9.2   Explicit Centered-Space Methods 235
       9.2.1   Lax–Friedrichs Method 236
       9.2.2   Lax–Wendroff Methods 237
       9.2.3   MacCormack Differencing 238

      9.2.4   Leapfrog Method 239

      9.2.5   Trapezoidal Leapfrog 241

  9.3 Explicit Upwind Differencing 242

      9.3.1   Beam–Warming Upwind Method 244

      9.3.2   Upwind Methods for Systems of Equations 245

  9.4 Limiter Methods 247

  9.5 Implicit Methods 249

      9.5.1 θ-Implicit Method 251

      9.5.2   Alternating Direction Implicit (ADI) 252

      9.5.3   Partially Implicit 2D MHD 253

      9.5.4   Reduced MHD 256

      9.5.5   Method of Differential Approximation 257

      9.5.6   Semi-Implicit Method 260

      9.5.7   Jacobian-Free Newton–Krylov Method 262

  9.6 Summary 264

10 Spectral Methods for Initial Value Problems   **267**

  10.1 Introduction 267

      10.1.1 Evolution Equation Example 268

      10.1.2 Classification 269

  10.2 Orthogonal Expansion Functions 269

      10.2.1 Continuous Fourier Expansion 270

      10.2.2 Discrete Fourier Expansion 272

      10.2.3 Chebyshev Polynomials in $(-1, 1)$ 273

      10.2.4 Discrete Chebyshev Series 277

  10.3 Non-Linear Problems 278

      10.3.1 Fourier Galerkin 278

      10.3.2 Fourier Collocation 279

      10.3.3 Chebyshev Tau 280

  10.4 Time Discretization 281

  10.5 Implicit Example: Gyrofluid Magnetic Reconnection 283

  10.6 Summary 286

11 The Finite Element Method   **289**

  11.1 Introduction 289

  11.2 Ritz Method in One Dimension 289

      11.2.1 An Example 290

      11.2.2 Linear Elements 290

      11.2.3 Some Definitions 293

      11.2.4 Error with Ritz Method 294

      11.2.5 Hermite Cubic Elements 295

      11.2.6 Cubic B-Splines 298

  11.3 Galerkin Method in One Dimension 301

  11.4 Finite Elements in Two Dimensions 304

      11.4.1 High-Order Nodal Elements in a Quadrilateral 305

11.4.2 Spectral Elements 309
11.4.3 Triangular Elements with C^1 Continuity 310
11.5 Eigenvalue Problems 316
11.5.1 Spectral Pollution 318
11.5.2 Ideal MHD Stability of a Plasma Column 320
11.5.3 Accuracy of Eigenvalue Solution 322
11.5.4 Matrix Eigenvalue Problem 323
11.6 Summary 323

Bibliography   **325**

Index   **341**

List of Figures

1.1   Gaussian pill box is used to derive jump conditions between two regions. 13

1.2   Characteristic curves $dx/dt = u$. All information is propagated along these lines. 14

1.3   Space is divided into two regions by characteristic curves. . . 15

1.4   Domain of dependence (l) and domain of influence (r). . . . 16

1.5   Characteristics in two spatial dimensions. 17

1.6   Reciprocal normal surface diagram in low-β limit. 20

1.7   Ray surface diagram in low-β limit. 21

1.8   Typical dispersion relation for low-β two-fluid MHD for different angles of propagation relative to the background magnetic field. 24

2.1   Space time discrete points. 28

2.2   Domain of dependence of the point (x_j, t^n) for an (a) explicit and (b) implicit finite difference method. 30

2.3   Amplification factor, r (solid), must lie within the unit circle (dashed) in the complex plane for stability. 35

2.4   Density and velocity variables are defined at staggered locations. 40

2.5   Finite volume method. 41

3.1   Computational grid for Neumann boundary conditions. . . . 52

3.2   Multigrid method sequence of grids. 67

3.3   Basic coarse grid correction. 68

3.4   Full Multigrid V-cycle (FMV). 69

4.1   Cylindrical coordinates (R, ϕ, Z). 94

4.2   Calculate magnetic flux associated with disk in the $z = 0$ plane as shown. 100

4.3   Poloidal magnetic flux contours for a typical tokamak discharge at three times. 101

4.4   For an axisymmetric system, the vacuum toroidal field constant g_0 is proportional to the total current in the toroidal field coils, I_{TF}. 104

4.5   Poloidal flux Ψ on the boundary of the computational domain
       is obtained from a Green's function. External coils are repre-
       sented as discrete circular current loops.   110

4.6   Singularity due to self-field term is resolved by taking the limit
       as ϵ approaches zero from the outside   112

4.7   Poloidal flux at magnetic axis, Ψ_0, is a local minimum. Limiter
       value of flux, Ψ_ℓ, is the minimum of the value of the flux at
       the limiter points and at the saddle points.   113

4.8   A double-nested iteration loop is used to converge the interior
       and boundary values.   116

5.1   Flux coordinates ψ and θ.   122

5.2   Two definitions of the angle θ are illustrated. Many more are
       possible.   123

5.3   Choosing the Jacobian determines both the (a) ψ and the (b)
       θ coordinate.   124

5.4   The vector $\frac{\partial f}{\partial \psi}$ is proportional to the projection of ∇f in the
       direction orthogonal to the other remaining coordinates, and
       not in the $\nabla \psi$ direction.   126

5.5   In the inverse representation we consider the cylindrical coor-
       dinates R and Z to be functions of ψ and θ.   128

5.6   Points on a constant Ψ contour are found using root-finding
       techniques.   132

5.7   With straight field line angular coordinates, magnetic field
       lines on a surface ψ=const. appear as straight lines when plot-
       ted in (θ, ϕ) space.   134

6.1 $\mathbf{u}_C$ is the velocity of a fluid element with a given ψ,θ value
       relative to a fixed Cartesian frame.   152

7.1   Comparison of the convergence properties of the Backward
       Euler method (BTCS) and the non-linear implicit Newton it-
       erative method.   179

7.2   Cylindrical symmetry restricts the allowable terms in a Taylor
       series expansion in x and y about the origin.   180

7.3   Temperatures are defined at equal intervals in $\delta\Phi$, offset $\frac{1}{2}\delta\Phi$
       from the origin so that the condition of zero flux can be im-
       posed there.   181

7.4   Hybrid DuFort–Frankel/implicit method.   191

8.1   In ideal MHD, plasma is surrounded by either a vacuum region
       or a pressureless plasma, which is in turn surrounded by a
       conducting wall.   196

8.2 $\boldsymbol{\xi}(\mathbf{x}, t)$ is the displacement field.   197

8.3   The physical boundary conditions are applied at the perturbed boundary, which is related to the unperturbed boundary through the displacement field $\boldsymbol{\xi}(\mathbf{x},t)$. 198

8.4   A typical ideal MHD spectrum. 202

8.5   Straight circular cylindrical geometry with periodicity length $2\pi R$. 214

8.6   All physical quantities must satisfy periodicity requirements $\Phi(\psi,\theta,\beta) = \Phi(\psi,\theta,\beta+2\pi) = \Phi(\psi,\theta+2\pi,\beta-2\pi q)$. 229

8.7   A solution with the correct periodicity properties is constructed by taking a linear superposition of an infinite number of offset aperiodic solutions. 230

9.1   In the combined DuFort–Frankel/leapfrog method, the space-time points at odd and even values of $n+j$ are completely decoupled. 240

9.2   Staggering variables in space and time can remove the decoupled grid problem from leapfrog. 241

9.3   The upwind method corresponds to linear interpolation for the point where the characteristic curve intersects time level n as long as $s \equiv u\delta t/\delta x \le 1$ 243

9.4   Characteristics for the Alfvén and fast magnetoacoustic waves for fully explicit and partially implicit methods. 250

10.1 Example of a $k=2$ and $k=2+N$ mode aliasing on a grid with $N=8$. The two modes take on the same values at the grid points. 273

11.1 Linear finite elements ϕ_j equals 1 at node j, 0 at all others, and are linear in between. A half element is associated with the boundary nodes at $j=0$ and $j=N$. 291

11.2 Expressions needed for overlap integrals for functions (a) and derivatives (b) for linear elements. 292

11.3 C^1 Hermite cubic functions enforce continuity of $v(x)$ and $v'(x)$. A linear combination of these two functions is associated with node j. 296

11.4 The basic cubic B-spline is a piecewise cubic with continuous value, first, and second derivatives. 299

11.5 Overlapping elements for cubic B-spline. A given element overlaps with itself and six neighbors. 300

11.6 Boundary conditions at $x=0$ are set by specifying ϕ_{-1} (Dirichlet) or constraining a linear combination of ϕ_{-1} and ϕ_0 (Neumann). 301

11.7 Band structure of the matrix g_{kj} that arises when the Galerkin method for a second-order differential equation is implemented using cubic B-spline elements. 304

11.8 Rectangular elements cannot be locally refined without introducing hanging nodes. Triangular elements can be locally refined, but require unstructured mesh data structures in programming. 305

11.9 Each arbitrary quadrilateral element in the Cartesian space (x, y) is mapped into the unit square in the logical space for that element, (ξ, η). 306

11.10 Relation between global numbering and local numbering in one dimension. 307

11.11 Reduced quintic finite element. 311

List of Tables

5.1   The poloidal angle θ is determined by the function $h(\psi, \theta)$ in the Jacobian as defined in Eq. (5.40). 133

5.2   Once the θ coordinate is determined, the ψ coordinate can be determined by the function $f(\psi)$ in the Jacobian definition, Eq. (5.40). 133

11.1   Exponents of ξ and η for the reduced quintic expansion $\phi(\xi, \eta) = \sum_{i=1}^{20} a_i \xi^{m_i} \eta^{n_i}$. 312

Preface

What is computational physics? Here, we take it to mean techniques for simulating continuous physical systems on computers. Since mathematical physics expresses these systems as partial differential equations, an equivalent statement is that computational physics involves solving systems of partial differential equations on a computer.

This book is meant to provide an introduction to computational physics to students in plasma physics and related disciplines. We present most of the basic concepts needed for numerical solution of partial differential equations. Besides numerical stability and accuracy, we go into many of the algorithms used today in enough depth to be able to analyze their stability, efficiency, and scaling properties. We attempt to thereby provide an introduction to and *working knowledge* of most of the algorithms presently in use by the plasma physics community, and hope that this and the references can point the way to more advanced study for those interested in pursuing such endeavors.

The title of the book starts with *Computational Methods...*, not *All Computational Methods....* Perhaps it should be *Some Computational Methods...* because it admittedly does not cover all computational methods being used in the field. The material emphasizes mathematical models where the plasma is treated as a conducting fluid and not as a kinetic gas. This is the most mature plasma model and also arguably the one most applicable to experiments. Many of the basic numerical techniques covered here are also appropriate for the equations one encounters when working in a higher-dimensional phase space. The book also emphasizes toroidal confinement geometries, particularly the tokamak, as this is the most mature and most successful configuration for confining a high-temperature plasma.

There is not a clear dividing line between computational and theoretical plasma physics. It is not possible to perform meaningful numerical simulations if one does not start from the right form of the equations for the questions being asked, and it is not possible to develop new advanced algorithms unless one has some understanding of the underlying mathematical and physical properties of the equation systems being solved. Therefore, we include in this book many topics that are not always considered "computational physics," but which are essential for a computational plasma physicist to understand. This more theoretical material, such as occurs in Chapters 1, 6, and 8 as well as parts of Chapters 4 and 5, can be skipped if students are exposed to it in other courses, but they may still find it useful for reference and context.

The author has taught a semester class with the title of this book to graduate students at Princeton University for over 20 years. The students are mostly from the Plasma Physics, Physics, Astrophysics, and Mechanical and Aerospace Engineering departments. There are no prerequisites, and most students have very little prior exposure to numerical methods, especially for partial differential equations. The material in the book has grown considerably during the 20-plus years, and there is now too much material to cover in a one-semester class. Chapters 2, 3, 7, 9, and perhaps 10 and 11 form the core of the material, and an instructor can choose which material from the other chapters would be suitable for his students and their needs and interests. A two-semester course covering most of the material is also a possibility.

Computers today are incredibly powerful, and the types of equations we need to solve to model the dynamics of a fusion plasma are quite complex. The challenge is to be able to develop suitable algorithms that lead to stable and accurate solutions that can span the relevant time and space scales. This is one of the most challenging research topics in modern-day science, and the payoffs are enormous. It is hoped that this book will help students and young researchers embark on productive careers in this area.

The author is indebted to his many colleagues and associates for innumerable suggestions and other contributions. He acknowledges in particular M. Adams, R. Andre, J. Breslau, J. Callen, M. Chance, J. Chen, C. Cheng, C. Chu, J. DeLucia, N. Ferraro, G. Fu, A. Glasser, J. Greene, R. Grimm, G. Hammett, T. Harley, S. Hirshman, R. Hofmann, J. Johnson, C. Kessel, D. Keyes, K. Ling, S. Lukin, D. McCune, D. Monticello, W. Park, N. Pomphrey, J. Ramos, J. Richter, K. Sakaran, R. Samtaney, D. Schnack, M. Sekora, S. Smith, C. Sovinec, H. Strauss, L. Sugiyama, D. Ward, and R. White.

Stephen C. Jardin
Princeton, NJ

List of Symbols

$\mathbf{x}$	position coordinate	ρ	mass density
$\mathbf{v}$	velocity coordinate	n	number density when quasi-neutrality is assumed
t	time		
f_j	probability distribution function for species j	e	electron charge
		$\mathbf{u}$	fluid (mass) velocity
q_j	electrical charge of species j	p	sum of electron and ion pressures
m_j	mass of species j		
$\mathbf{E}$	electric field	$\mathbf{J}$	electrical current density
$\mathbf{B}$	magnetic field	$\mathbf{q}$	sum of electron and ion random heat fluxes
μ_0	permeability of free space		
ϵ_0	permittivity of free space	$\mathbf{W}$	rate of strain tensor
c	speed of light in vacuum	π_i^{gyr}	ion gyroviscous stress tensor
ρ_q	electrical charge density	$\mathbf{I}$	identity matrix
C_j	net effect of scattering on species j	$\mathbf{u}_*$	ion magnetization velocity
		$\mathbf{q}_{\wedge j}$	collision-independent heat flux for species j
$C_{jj'}$	effect of scattering due to collisions betwen particles of species j and j'		
		μ	isotropic shear viscosity
		μ_c	isotropic volume viscosity
S_j	source of particles, momentum, or energy for species j	$\mu_\parallel$	parallel viscosity
		$\eta_\parallel$	parallel resistivity
n_j	number density of species j	$\eta_\perp$	perpendicular resistivity
$\mathbf{u}_j$	fluid velocity of species j	η	average (isotropic) resistivity
p_j	pressure of species j		
π_j	stress tensor for species j	Z	charge state of ion
Q_j	external source of energy density for species j	$\ln\lambda$	Coulomb logarithm
		T_j	temperature of species j
$\hat{\mathbf{M}}_j$	external source of momentum density for species j	$\kappa_\parallel^j$	parallel thermal conductivity of species j
$Q_{\Delta j}$	collisional source of energy density for species j	$\kappa_\perp^j$	perpendicular thermal conductivity of species j
$\mathbf{q}_j$	random heat flux vector for species j	$\mathbf{R}_j$	plasma friction force acting on species j
k_B	Boltzmann constant	λ_H	hyper-resistivity coefficient
ω_{pe}	electron plasma frequency	τ_R	resistive diffusion time
λ_D	Debye length	τ_A	Alfvén transit time

S	magnetic Lundquist number	B_i	i^{th} Cartesian component of the magnetic field
γ	adiabatic index	c_S	sound speed
s	entropy density	V_A	Alfvén velocity
$\hat{\mathbf{x}}_i$	unit vector associated with Cartesian coordinate x_i		

Chapter 1

Introduction to Magnetohydrodynamic Equations

1.1   Introduction

A high-temperature magnetized plasma such as exists in modern magnetic fusion experiments is one of the most complex media in existence. Although adequately described by classical physics (augmented with atomic and nuclear physics as required to describe the possible interaction with neutral atoms and the energy release due to nuclear reactions), the wide range of time scales and space scales present in the most general mathematical description of plasmas make meaningful numerical simulations which span these scales enormously difficult. This is the motivation for deriving and solving reduced systems of equations that purport to describe plasma phenomena over restricted ranges of time and space scales, but are more amenable to numerical solution.

The most basic set of equations describing the six dimension plus time $(\mathbf{x}, \mathbf{v}, t)$ phase space probability distribution function $f_j(\mathbf{x}, \mathbf{v}, t)$ for species j of indistinguishable charged particles (electrons or a particular species of ions) is a system of Boltzmann equations for each species:

$$\frac{\partial f_j(\mathbf{x}, \mathbf{v}, t)}{\partial t} + \nabla \cdot (\mathbf{v} f_j(\mathbf{x}, \mathbf{v}, t)) + \nabla_v \cdot \left[\frac{q_j}{m_j} (\mathbf{E} + \mathbf{v} \times \mathbf{B}) f_j(\mathbf{x}, \mathbf{v}, t) \right]$$
$$= C_j + S_j. \quad (1.1)$$

Here m_j is the particle mass and q_j is the particle charge for species j. The collision operator $C_j = \sum_{j'} C_{jj'}$ represents the effect of scattering due to collisions between particles of species j and $j\prime$. External sources of particles, momentum, and energy are represented by S_j. The electric and magnetic fields $\mathbf{E}(\mathbf{x}, t)$ and $\mathbf{B}(\mathbf{x}, t)$ are obtained by solving the free space Maxwell's equations

(SI units):

$$\frac{\partial \mathbf{B}}{\partial t} = -\nabla \times \mathbf{E}, \tag{1.2}$$

$$\nabla \times \mathbf{B} = \frac{1}{c^2}\frac{\partial \mathbf{E}}{\partial t} + \mu_0 \mathbf{J}, \tag{1.3}$$

$$\nabla \cdot \mathbf{E} = \frac{1}{\epsilon_0}\rho_q, \tag{1.4}$$

$$\nabla \cdot \mathbf{B} = 0, \tag{1.5}$$

with charge and current density given by the following integrals over velocity space:

$$\rho_q = \sum_j q_j \int d^3 \mathbf{v} f_j(\mathbf{x}, \mathbf{v}, t), \tag{1.6}$$

$$\mathbf{J} = \sum_j q_j \int d^3 \mathbf{v} \mathbf{v} f_j(\mathbf{x}, \mathbf{v}, t). \tag{1.7}$$

Here $\epsilon_0 = 8.8542 \times 10^{-12} Fm^{-1}$ and $\mu_0 = 4\pi \times 10^{-7} Hm^{-1}$ are the permittivity and permeability of free space, and $c = (\epsilon_0\mu_0)^{-1/2} = 2.9979 \times 10^8 ms^{-1}$ is the speed of light.

These equations form the starting point for some studies of fine-scale plasma turbulence and for studies of the interaction of imposed radio-frequency (RF) electromagnetic waves with plasma. However, the focus here will be to take velocity moments of Eq. (1.1), and appropriate sums over species, so as to obtain fluid-like equations that describe the macroscopic dynamics of magnetized high-temperature plasma, and to discuss techniques for their numerical solution.

The first three velocity moments correspond to conservation of particle number, momentum, and energy. Operating on Eq. (1.1) with the velocity space integrals $\int d^3\mathbf{v}$, $\int d^3\mathbf{v}m_j\mathbf{v}$, and $\int d^3\mathbf{v}m_jv^2/2$ yields the following moment equations:

$$\frac{\partial n_j}{\partial t} + \nabla \cdot (n_j\mathbf{u}_j) = \hat{S}_j, \tag{1.8}$$

$$\frac{\partial}{\partial t}(m_jn_j\mathbf{u}_j) + \nabla \cdot (m_jn_j\mathbf{u}_j\mathbf{u}_j) + \nabla p_j + \nabla \cdot \boldsymbol{\pi}_j$$
$$= q_jn_j(\mathbf{E} + \mathbf{u}_j \times \mathbf{B}) + \mathbf{R}_j + \hat{\mathbf{M}}_j, \tag{1.9}$$

$$\frac{\partial}{\partial t}\left(\frac{3}{2}p_j + \frac{1}{2}m_jn_ju_j^2\right) + \nabla \cdot \left(\frac{1}{2}m_jn_ju_j^2\mathbf{u}_j + \frac{5}{2}p_j\mathbf{u}_j + \boldsymbol{\pi}_j \cdot \mathbf{u}_j + \mathbf{q}_j\right)$$
$$= (\mathbf{R}_j + q_jn_j\mathbf{E}) \cdot \mathbf{u}_j + Q_j + Q_{\Delta j}. \tag{1.10}$$

Here, we have introduced for each species the number density, the fluid velocity, the scalar pressure, the stress tensor, the heat flux density, and the external sources of particles, momentum, and energy density which are defined as:

$$n_j = \int d^3v f_j(\mathbf{x}, \mathbf{v}, t), \tag{1.11}$$

$$\mathbf{u}_j = \frac{1}{n_j} \int d^3v \mathbf{v} f_j(\mathbf{x}, \mathbf{v}, t), \tag{1.12}$$

$$p_j = \frac{1}{3} \int d^3v m_j |\mathbf{v} - \mathbf{u}_j|^2 f_j(\mathbf{x}, \mathbf{v}, t), \tag{1.13}$$

$$\boldsymbol{\pi}_j = \left[\int d^3v m_j (\mathbf{v} - \mathbf{u}_j)(\mathbf{v} - \mathbf{u}_j) f_j(\mathbf{x}, \mathbf{v}, t) \right] - p_j \mathbf{I}, \tag{1.14}$$

$$\mathbf{q}_j = \frac{1}{2} \int d^3v m_j |\mathbf{v} - \mathbf{u}_j|^2 (\mathbf{v} - \mathbf{u}_j) f_j(\mathbf{x}, \mathbf{v}, t). \tag{1.15}$$

$$\hat{S}_j = \int d^3v S_j, \tag{1.16}$$

$$\hat{\mathbf{M}}_j = \int d^3v m_j \mathbf{v} S_j, \tag{1.17}$$

$$Q_j = \frac{1}{2} \int d^3v m_j v^2 S_j. \tag{1.18}$$

The species temperature is defined to be $k_B T_j = p_j/n_j$, where the Boltzmann constant is $k_B = 1.602 \times 10^{-19} J[eV]^{-1}$. The collisional sources of momentum density and heat density are defined as

$$\mathbf{R}_j = \sum_{j'} \int d^3v m_j (\mathbf{v} - \mathbf{u}_j) C_{jj'}, \tag{1.19}$$

$$Q_{\Delta j} = \frac{1}{2} \sum_{j'} \int d^3v m_j |\mathbf{v} - \mathbf{u}_j|^2 C_{jj'}. \tag{1.20}$$

We have used the property that particle collisions neither create nor destroy particles, momentum, or energy, but only transfer momentum or energy between species. These conservation properties imply

$$\sum_j \mathbf{R}_j = 0, \tag{1.21}$$

and for a two-component electron-ion plasma,

$$Q_{\Delta e} + Q_{\Delta i} = \mathbf{R}_e \cdot \frac{\mathbf{J}}{ne}. \tag{1.22}$$

1.2   Magnetohydrodynamic (MHD) Equations

The fluid equations given in the last section are very general; however, they are incomplete and still contain too wide a range of time and space scales to be useful for many purposes. To proceed, we make a number of simplifying approximations that allow the omission of the terms in Maxwell's equations that lead to light waves and high-frequency plasma oscillations. These approximations are valid provided the velocities (v), frequencies (ω), and length scales of interest (L) satisfy the inequalities:

$$v \ll c,$$
$$\omega \ll \omega_{pe},$$
$$L \gg \lambda_D.$$

Here ω_{pe} is the electron plasma frequency, and λ_D is the Debye length. These orderings are normally well satisfied for macroscopic phenomena in modern fusion experiments. We also find it convenient to change velocity variables from the electron and ion velocities $\mathbf{u}_e$ and $\mathbf{u}_i$ to the fluid (mass) velocity and current density, defined as:

$$\mathbf{u} = \frac{m_e n_e \mathbf{u}_e + m_i n_i \mathbf{u}_i}{m_e n_e + m_i n_i} \approx \mathbf{u}_i; \quad \mathbf{J} = n_i q_i \mathbf{u}_i + n_e q_e \mathbf{u}_e. \tag{1.23}$$

The approximation on the left is normally valid since $m_i \gg m_e$ and $n_e \sim n_i$.

The resulting mathematical description is often called *extended magneto-hydrodynamics*, or just simply *magnetohydrodynamics* (MHD). For simplicity, we consider here only a single species of ions with unit charge. The field equations in MHD are then (SI units):

$$\frac{\partial \mathbf{B}}{\partial t} = -\nabla \times \mathbf{E}, \tag{1.24}$$

$$\nabla \cdot \mathbf{B} = 0, \tag{1.25}$$

$$\nabla \times \mathbf{B} = \mu_0 \mathbf{J}, \tag{1.26}$$

$$n_i = n_e = n. \tag{1.27}$$

Equations (1.24) and (1.25) for the magnetic field $\mathbf{B}$ are exact and always valid. Note that Eq. (1.25) can be regarded as an initial condition for Eq. (1.24). Equation (1.26) can be taken as a defining equation for the electrical current density $\mathbf{J}$, effectively replacing both Eqs. (1.3) and (1.7). Equation (1.27), stating the equivalence of the electron and ion number density, is referred to as *quasineutrality* and similarly replaces Eq. (1.6). Note that there is no equation for the divergence of the electric field, $\nabla \cdot \mathbf{E}$, in MHD. It is effectively replaced by the quasineutrality condition, Eq. (1.27).

The fluid equations are the continuity equation for the number density

$$\frac{\partial n}{\partial t} + \nabla \cdot (n\mathbf{u}) = S_m, \qquad (1.28)$$

the internal energy equation for the plasma pressure

$$
\begin{aligned}
\frac{3}{2}\frac{\partial p}{\partial t} \;+\; & \nabla \cdot \left(\mathbf{q} + \frac{3}{2}p\mathbf{u}\right) = -p\nabla \cdot \mathbf{u} + \left(\frac{3}{2}\nabla p_e - \frac{5}{2}\frac{p_e}{n}\nabla n\right) \cdot \frac{\mathbf{J}}{ne} \\
\;+\; & \mathbf{R}_e \cdot \frac{\mathbf{J}}{ne} - \boldsymbol{\pi} : \nabla\mathbf{u} + \boldsymbol{\pi}_e : \nabla\frac{\mathbf{J}}{ne} + Q,
\end{aligned}
\qquad (1.29)
$$

the electron internal energy equation for the electron pressure

$$
\begin{aligned}
\frac{3}{2}\frac{\partial p_e}{\partial t} \;+\; & \nabla \cdot \left(\mathbf{q}_e + \frac{3}{2}p_e\mathbf{u}\right) = -p_e\nabla \cdot \mathbf{u} + \left(\frac{3}{2}\nabla p_e - \frac{5}{2}\frac{p_e}{n}\nabla n\right) \cdot \frac{\mathbf{J}}{ne} \\
\;+\; & \mathbf{R}_e \cdot \frac{\mathbf{J}}{ne} - \boldsymbol{\pi}_e : \nabla\left(\mathbf{u} - \frac{\mathbf{J}}{ne}\right) + Q_{\Delta ei} + Q_e,
\end{aligned}
\qquad (1.30)
$$

and the force balance equation for the fluid velocity

$$nm_i\left(\frac{\partial \mathbf{u}}{\partial t} + \mathbf{u}\cdot\nabla\mathbf{u}\right) + \nabla\cdot(\boldsymbol{\pi}_e + \boldsymbol{\pi}_i) = -\nabla p + \mathbf{J}\times\mathbf{B}. \qquad (1.31)$$

In addition, from the momentum equation for the electrons, Eq. (1.9) with $j = e$, we have the *generalized Ohm's law* equation, which in the limit of vanishing electron mass ($m_e = 0$) becomes:

$$\mathbf{E} + \mathbf{u}\times\mathbf{B} = \frac{1}{ne}\left[\mathbf{R}_e + \mathbf{J}\times\mathbf{B} - \nabla p_e - \nabla\cdot\boldsymbol{\pi}_e\right]. \qquad (1.32)$$

The variables $\mathbf{B}$, $\mathbf{u}$, n, p, p_e are the fundamental variables in that they obey time advancement equations. The variables $\mathbf{E}$, $\mathbf{J}$, and the ion pressure $p_i = p - p_e$ are auxiliary variables which we define only for convenience. We have also introduced the total heat flux $\mathbf{q} = \mathbf{q}_e + \mathbf{q}_i$ and the total non-isotropic stress tensor $\boldsymbol{\pi} = \boldsymbol{\pi}_e + \boldsymbol{\pi}_i$.

To proceed with the solution, one needs *closure* relations for the remaining terms; the collisional friction term $\mathbf{R}_e$, the random heat flux vectors $\mathbf{q}_j$, the anisotropic part of the stress tensor $\boldsymbol{\pi}_j$, and the equipartition term $Q_{\Delta ei}$. It is the objective of transport theory to obtain closure expressions for these in terms of the fundamental variables and their derivatives. There is extensive and evolving literature on deriving closures that are valid in different parameter regimes. We will discuss some of the more standard closures in the following sections. We note here that the external sources of particles, total energy, and electron energy S_m, Q, and Q_e must also be supplied, as must initial and boundary conditions.

1.2.1   Two-Fluid MHD

The set of MHD equations as written in Section 1.2 is not complete because of the closure issue. The most general closures that have been proposed are presently too difficult to solve numerically. However, there are some approximations that are nearly always valid and are thus commonly applied. Other approximations are valid over limited time scales and are useful for isolating specific phenomena.

The two-fluid magnetohydrodynamic equations are obtained by taking an asymptotic limit of the extended MHD equations in which first order terms in the ratio of the ion Larmor radius to the system size are retained. This is sometimes called the *finite Larmor radius* or FLR approximation. The principal effects are to include expressions for the parts of the ion stress tensor $\boldsymbol{\pi}_i$ and ion and electron heat fluxes $\mathbf{q}_i$ and $\mathbf{q}_e$ that do not depend on collisions. This contribution to the ion stress tensor is known as the *gyroviscous stress* or *gyroviscosity* . Let $\mathbf{b}$ be a unit vector in the direction of the magnetic field. The general form of gyroviscosity, assuming isotropic pressure and negligible heat flux, can be shown to be [1, 2, 3]:

$$\boldsymbol{\pi}_i^{gyr} = \frac{m_i p_i}{4eB} \left\{ \mathbf{b} \times \mathbf{W} \cdot (\mathbf{I} + 3\mathbf{bb}) + [\mathbf{b} \times \mathbf{W} \cdot (\mathbf{I} + 3\mathbf{bb})]^\dagger \right\}. \qquad (1.33)$$

Here, the rate of strain tensor is

$$\mathbf{W} = \nabla \mathbf{u} + (\nabla \mathbf{u})^\dagger - \frac{2}{3} \mathbf{I} \nabla \cdot \mathbf{u}. \qquad (1.34)$$

The expressions for the gyroviscosity in Eqs. (1.33) and (1.34) are quite complex and difficult to implement. A common approximation is to replace Eq. (1.33) with what has become known as the *gyroviscous cancellation* approximation [4, 5],

$$\nabla \cdot \boldsymbol{\pi}_i^{gyr} \approx -m_i n \mathbf{u}_* \cdot \nabla \mathbf{u}, \qquad (1.35)$$

where the ion magnetization velocity is defined by

$$\mathbf{u}_* = -\frac{1}{ne} \nabla \times \left(\frac{p_i}{B^2} \mathbf{B} \right). \qquad (1.36)$$

At the same order in this expansion, it can be shown that for each species there exists a heat flux $\mathbf{q}$ that is independent of the collision frequency:

$$\mathbf{q}_{\wedge j} = \frac{5}{2} \frac{p_j k_B}{q_j B} \mathbf{b} \times \nabla T_j. \qquad (1.37)$$

The remaining contributions to the ion stress tensor are dependent on the collisionality regime and magnetic geometry in a complex way. These are often approximated by an isotropic part and a parallel part as follows:

$$\boldsymbol{\pi}_i^{iso} = -\mu \left[\nabla \mathbf{u} + \nabla \mathbf{u}^\dagger \right] - 2(\mu_c - \mu)(\nabla \cdot \mathbf{u}) \mathbf{I}, \qquad (1.38)$$

$$\nabla \cdot \boldsymbol{\pi}_i^{iso} = -\mu \nabla^2 \mathbf{u} - (2\mu_c - \mu)\nabla(\nabla \cdot \mathbf{u}), \qquad (1.39)$$

$$\boldsymbol{\pi}_i^{\parallel} = \mu_\parallel (\mathbf{b} \cdot \mathbf{W} \cdot \mathbf{b})(\mathbf{I} - 3\mathbf{bb}). \qquad (1.40)$$

We note the positivity constraints $\mu \geq 0$, $\mu_c \geq \frac{2}{3}\mu$, $\mu_{\parallel} \geq 0$.

The electron gyroviscosity can normally be neglected due to the small electron mass. However, for many applications, an electron parallel viscosity can be important. This is similar in form to the ion one in Eq. (1.40) but with a coefficient $\mu_{\parallel}^e$. In addition, it has been shown by several authors that in certain problems involving two-fluid magnetic reconnection, an electron viscosity term known as *hyper-resistivity* is required to avoid singularities from developing in the solution [6, 7]. It is also useful for modeling the effect of fundamentally three-dimensional reconnection physics in a two-dimensional simulation [8, 9]. Introducing the coefficient λ_H, we can take the hyper-resistivity to be of the form:

$$\boldsymbol{\pi}_e^{hr} = \lambda_H \eta_{\parallel} ne \nabla \mathbf{J}. \tag{1.41}$$

If h is the smallest dimension that can be resolved on a numerical grid, then it has been proposed that $\lambda_H \sim h^2$ is required for the current singularity to be resolvable, while at the same time vanishing in the continuum limit $h \to 0$.

The plasma friction force is generally taken to be of the form:

$$\mathbf{R}_e = ne \left(\eta_{\parallel} \mathbf{J}_{\parallel} + \eta_{\perp} \mathbf{J}_{\perp} \right), \tag{1.42}$$

where $\parallel$ and $\perp$ are relative to the magnetic field direction. The classical values for these resistivity coefficients are [1]:

$$\eta_{\perp} = 1.03 \times 10^{-4} Z \ln \lambda [T(ev)]^{-3/2} \ \Omega \ m, \tag{1.43}$$

$$\eta_{\parallel} = 0.51 \ \eta_{\perp}. \tag{1.44}$$

Here Z is the effective charge of the ion species and $\ln \lambda \sim 20$ is the Coulomb logarithm [10]. The electron-ion temperature equilibration term is related to the perpendicular resistivity by:

$$Q_{\Delta ei} = \frac{3e^2 n}{m_i} \eta_{\perp} nk_B(T_i - T_e). \tag{1.45}$$

The forms used for the viscosity coefficients μ, μ_c, $\mu_{\parallel}$ and the heat conduction coefficients $\kappa_{\parallel}^e$, $\kappa_{\perp}^e$, $\kappa_{\parallel}^i$, $\kappa_{\perp}^i$ can be the classical value for the collisional regime [1], but are normally chosen according to other *anomalous* (or empirical) models or so as to perform parametric studies. However, from basic physical considerations (that the particles essentially free-stream with very long collisional mean-free-paths parallel to the magnetic field but not perpendicular to it) we normally have:

$$\kappa_{\parallel}^{e,i} \gg \kappa_{\perp}^{e,i}. \tag{1.46}$$

The standard two-fluid MHD model can now be summarized as follows.

Using the equations of Section 1.2, we use the following closure model:

$$\mathbf{R}_e = ne\left(\eta_\parallel \mathbf{J}_\parallel + \eta_\perp \mathbf{J}_\perp\right), \tag{1.47}$$

$$\mathbf{q}_e = -\kappa_\parallel^e \nabla_\parallel T_e - \kappa_\perp^e \nabla_\perp T_e + \mathbf{q}_{\wedge e}, \tag{1.48}$$

$$\mathbf{q}_i = -\kappa_\parallel^i \nabla_\parallel T_i - \kappa_\perp^i \nabla_\perp T_i + \mathbf{q}_{\wedge i}, \tag{1.49}$$

$$\boldsymbol{\pi}_i = \boldsymbol{\pi}_i^{gyr} + \boldsymbol{\pi}_i^{iso} + \boldsymbol{\pi}_i^{\parallel}, \tag{1.50}$$

$$\boldsymbol{\pi}_e = \boldsymbol{\pi}_e^{\parallel} + \boldsymbol{\pi}_e^{hr}, \tag{1.51}$$

$$Q_{\Delta ei} = \frac{3e^2 n}{m_i} \eta_\perp n k_B (T_i - T_e). \tag{1.52}$$

To complete this model, one must specify the 11 scalar functions $\eta_\parallel$, $\eta_\perp$, $\kappa_\parallel^e$, $\kappa_\perp^e$, $\kappa_\parallel^i$, $\kappa_\perp^i$, μ, μ_c, $\mu_\parallel$, $\mu_\parallel^e$, and λ_H. These can be either constants or functions of the macroscopic quantities being evolved in time according to the transport model being utilized.

1.2.2 Resistive MHD

The resistive MHD model treats the electrons and ions as a single fluid with pressure $p = p_e + p_i$. This model can formally be derived by taking a limiting case of the two-fluid equations. The first limit is that of *collision dominance*, which allows one to neglect the parallel viscosities compared to ∇p. The second limit is that of *zero Larmor radius*. This implies the neglect of the gyroviscous stress and the collision-independent perpendicular heat fluxes $\mathbf{q}_{\wedge j}$ as well as most terms involving $\mathbf{J}/ne$ compared to those involving $\mathbf{u}$. It follows that the electron hyperviscosity is also not required. With these simplifications, neglecting external sources, and introducing the mass density $\rho \equiv nm_i$, the equations of Section 1.2 become:

$$\frac{\partial \rho}{\partial t} + \nabla \cdot (\rho \mathbf{u}) = 0, \tag{1.53}$$

$$\rho\left(\frac{\partial \mathbf{u}}{\partial t} + \mathbf{u} \cdot \nabla \mathbf{u}\right) = -\nabla p + \mathbf{J} \times \mathbf{B} - \nabla \cdot \boldsymbol{\pi}_i^{iso}, \tag{1.54}$$

$$\frac{3}{2}\frac{\partial p}{\partial t} + \nabla \cdot \left(\mathbf{q} + \frac{3}{2}p\mathbf{u}\right) = -p\nabla \cdot \mathbf{u} - \boldsymbol{\pi}_i^{iso} : \nabla \mathbf{u} + \eta J^2, \tag{1.55}$$

$$\frac{\partial \mathbf{B}}{\partial t} = \nabla \times (\mathbf{u} \times \mathbf{B} - \eta \mathbf{J}), \tag{1.56}$$

$$\mathbf{J} = \frac{1}{\mu_0}\nabla \times \mathbf{B}, \tag{1.57}$$

$$\mathbf{q}_i = -\kappa_\| \nabla_\| T - \kappa_\perp \nabla_\perp T. \tag{1.58}$$

Here the fluid temperature is defined as $T = p/2nk_B$ and the viscosity term in Eq. (1.54) is evaluated using Eq. (1.39). This model requires only the five transport coefficients: η, $\kappa_\|^e$, $\kappa_\perp$, μ, and μ_c.

1.2.3 Ideal MHD

A further approximation has to do with the smallness of the plasma resistivity and the time scales over which resistive effects are important, $\tau_R = \mu_0 a^2/\eta$, where a is the minor radius or other typical global dimension. If we non-dimensionalize the MHD equations, using the Alfvén velocity $V_A = B/\sqrt{\mu_0 n M_i}$, a, and the Alfvén time $\tau_A = a/V_A$, we find that the plasma resistivity becomes multiplied by the inverse *magnetic Lundquist number* S^{-1}, where

$$S \equiv \frac{\tau_R}{\tau_A}. \tag{1.59}$$

In modern fusion experiments, this number is typically in the range $S \sim 10^6 - 10^{12}$.

The other dissipative quantities $\mathbf{q}_j$ and $\boldsymbol{\pi}^{iso}$ are related to the resistivity and thus also become multiplied by S^{-1}. Although these terms are very important for the longer time dynamics of the plasma or to describe *resistive instabilities* that involve internal boundary layers [11], if we are only interested in the fastest time scales present in the equations, we can neglect all these dissipative terms which scale as S^{-1}. Doing so leaves the *ideal MHD equations* [14].

These equations have been extensively studied by both physicists and mathematicians. They have a seemingly simple symmetrical structure with well-defined mathematical properties, but at the same time can be exceedingly rich in the solutions they admit. Adding back the dissipative and dispersive terms will enlarge the class of possible solutions by making the equations higher order, but will not fundamentally change the subset of non-dissipative solutions found here for macroscopic motions. It is therefore important to understand the types of solutions possible for these equations before studying more complex equation sets.

The ideal MHD equations can be written (in SI units), as follows:

$$\frac{\partial \rho}{\partial t} + \nabla \cdot \rho \mathbf{u} = 0, \tag{1.60}$$

$$\frac{\partial \mathbf{B}}{\partial t} = \nabla \times (\mathbf{u} \times \mathbf{B}), \tag{1.61}$$

$$\rho \left(\frac{\partial \mathbf{u}}{\partial t} + \mathbf{u} \cdot \nabla \mathbf{u} \right) + \nabla p = \mathbf{J} \times \mathbf{B}, \tag{1.62}$$

$$\frac{\partial p}{\partial t} + \mathbf{u} \cdot \nabla p + \gamma p \nabla \cdot \mathbf{u} = 0, \qquad (1.63)$$

where the current density is given by $\mathbf{J} \equiv \mu_o^{-1} \nabla \times \mathbf{B}$. Here we have introduced $\gamma = \frac{5}{3}$, which is the ratio of specific heats, sometimes called the adiabatic index. It is sometimes useful to also define another variable

$$s \equiv p/\rho^{\gamma},$$

the entropy per unit mass. It then follows from Eqs. (1.60) and (1.63) that s obeys the equation

$$\frac{\partial s}{\partial t} + \mathbf{u} \cdot \nabla s = 0. \qquad (1.64)$$

Eq. (1.64) can be used to replace Eq. (1.63).

1.2.4 Other Equation Sets for MHD

The equation sets corresponding to the closures listed in the previous sections are by no means exhaustive. Higher-order closures exist which involve integrating the stress tensor and higher-order tensors in time [15]. These become very complex, and generally require a subsidiary kinetic calculation to complete the closure of the highest-order tensor quantities. Additional closures of a more intermediate level of complexity exist in which the pressure stress tensor remains diagonal but is allowed to have a different form parallel and perpendicular to the magnetic field [16]. These allow the description of some dynamical effects that arise from the fact that the underlying kinetic distribution function is not close to a Maxwellian.

There are also widely used sets of starting equations for computational MHD that are in many ways easier to solve than the ones presented here. The *reduced MHD* equations follow from an additional expansion in the inverse aspect ratio of a toroidal confinement device [17, 18]. These are discussed more in Chapters 9 and 11.

The *surface-averaged* MHD equations are discussed in Chapter 6. These are a very powerful system of equations that are valid for modeling the evolution of an MHD stable system with good magnetic flux surfaces over long times characterized by $\tau_R \sim S \tau_A$. The equations are derived by an asymptotic expansion in which the plasma inertia approaches zero. This leads to considerable simplification, effectively removing the ideal MHD wave transit time scales from the problem. At the same time, the flux surface averaging allows the implementation of low collisionality closures that would not otherwise be possible.

1.2.5 Conservation Form

A system of partial differential equations is said to be in *conservation form* if all of the terms can be written as divergences of products of the dependent variable, i.e., in the form

$$\frac{\partial}{\partial t}(\cdots) + \nabla \cdot (\cdots) = 0.$$

This form is useful for several purposes. It allows one to use Gauss's theorem to obtain global conservation laws and the appropriate boundary conditions and jump conditions across shocks or material interfaces. It can also be used as a starting point to formulate non-linear numerical solutions which maintain the exact global conservation properties of the original equations.

By using elementary vector manipulations, we obtain the conservation form of the MHD equations of Section 1.1, also incorporating the common approximations discussed in Section 1.2. Assuming the external source terms S_m, S_e, S_{ee}, etc. are also zero, we have

$$\frac{\partial}{\partial t}\rho + \nabla \cdot (\rho \mathbf{u}) = 0, \tag{1.65}$$

$$\frac{\partial}{\partial t}B_i + \nabla \cdot [\mathbf{E} \times \hat{\mathbf{x}}_i] = 0, \tag{1.66}$$

$$\frac{\partial}{\partial t}(\rho \mathbf{u}) + \nabla \cdot \left[\rho \mathbf{u}\mathbf{u} + \left(p + \frac{B^2}{2\mu_0}\right)\mathbf{I} + \boldsymbol{\pi} - \frac{1}{\mu_0}\mathbf{B}\mathbf{B} \right] = 0, \tag{1.67}$$

$$\frac{\partial}{\partial t}\left(\frac{1}{2}\rho u^2 + \frac{3}{2}p + \frac{B^2}{2\mu_0}\right) + \nabla \cdot \left[\mathbf{u}\left(\frac{1}{2}\rho u^2 + \frac{5}{2}p\right)\right.$$
$$\left. + \mathbf{u} \cdot \boldsymbol{\pi} + \frac{1}{\mu_0}\mathbf{E} \times \mathbf{B} + \mathbf{q} \right] = 0. \tag{1.68}$$

These express conservation of mass, magnetic flux, momentum, and energy. Here $\hat{\mathbf{x}}_i$ is the Cartesian unit vector for $i = 1, 2, 3$, and B_i is the corresponding Cartesian component of the magnetic field; $B_i = \hat{\mathbf{x}}_i \cdot \mathbf{B}$. The global conservation laws are obtained by simply integrating Eq. (1.65)–(1.68) over a volume and using Gauss's theorem to convert divergences to surface integrals over the boundary. We obtain

$$\frac{\partial}{\partial t}\int \rho d^3\mathbf{x} = -\int dS\hat{\mathbf{n}} \cdot \rho \mathbf{u}, \tag{1.69}$$

$$\frac{\partial}{\partial t}\int B_i d^3\mathbf{x} = \int dS(\hat{\mathbf{n}} \times \hat{\mathbf{x}}_i) \cdot \mathbf{E}, \tag{1.70}$$

$$\frac{\partial}{\partial t}\int \rho \mathbf{u} d^3\mathbf{x} = -\int dS\hat{\mathbf{n}} \cdot \left[\rho \mathbf{u}\mathbf{u} + (p + \frac{B^2}{2\mu_o})\mathbf{I} + \boldsymbol{\pi} - \frac{1}{\mu_o}\mathbf{B}\mathbf{B}\right]. \tag{1.71}$$

$$\frac{\partial}{\partial t} \int \left(\frac{1}{2}\rho u^2 + \frac{3}{2}p + \frac{B^2}{2\mu_0} \right) d^3\mathbf{x} \;=$$

$$-\int dS\hat{n} \cdot \left[\mathbf{u} \left(\frac{1}{2}\rho u^2 + \frac{5}{2}p \right) \;+\; \mathbf{u}\cdot\boldsymbol{\pi} + \frac{1}{\mu_o}\mathbf{E}\times\mathbf{B} + \mathbf{q} \right]. \quad (1.72)$$

We note in passing that Stokes' theorem can also be applied directly to any surface integral of Eq. (1.24) to obtain the flux conservation relation in a more familiar form. Also, if the resistivity is zero so that $\mathbf{E}+\mathbf{u}\times\mathbf{B}=0$, then this implies that the electric field in the frame of the fluid is zero, and a conservation law can be used to show that the magnetic flux in any fluid element is conserved. But the global relation, Eq. (1.70), will of course still hold.

As discussed above and further in Section 2.6, a numerical solution based on the conservation form of the equations offers some advantages over other formulations in that globally conserved quantities can be exactly maintained in a computational solution. However, in computational magnetohydrodynamics of highly magnetized plasmas, the conservative formulation is not normally the preferred starting point for the following reasons: While the mass and magnetic flux conservation equations, Eqs. (1.65) and (1.66) (in some form) offer clear advantages, the conservation form for the momentum and energy equations, Eqs. (1.67) and (1.68), can lead to some substantial inaccuracies. The primary difficulty is due to the difference in magnitude of the kinetic energy, pressure, and magnetic energy terms in a strongly magnetized plasma. It can be seen that there is no explicit equation to advance the pressure in the set of Eqs. (1.65)–(1.68). The pressure evolution equation is effectively replaced by Eq. (1.68) to advance the total energy density:

$$\epsilon = \left(\frac{1}{2}\rho u^2 + \frac{3}{2}p + \frac{B^2}{2\mu_0} \right). \quad (1.73)$$

In order to obtain the pressure, one must use Eq. (1.73) so that:

$$p = \frac{2}{3}\left(\epsilon - \frac{1}{2}\rho u^2 - \frac{B^2}{2\mu_0} \right). \quad (1.74)$$

While this is mathematically correct, it can happen that if $B^2 \gg \mu_0 p$, the relatively small computational errors acquired in computing $\mathbf{B}$ will be amplified in the computation of p. Similar considerations may apply to the momentum equation, Eq. (1.67), where high-order cancellations need to occur. This is discussed more in Chapter 9.

1.2.6 Boundary Conditions

The global conservation laws Eqs. (1.65)–(1.68) apply to any volume and can be used to obtain boundary conditions and jump conditions across interfaces by applying them appropriately. For example, if we have two regions,

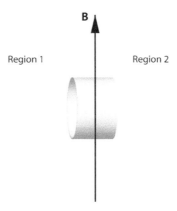

FIGURE 1.1: Gaussian pill box is used to derive jump conditions between two regions.

within both of which the MHD equations are satisfied, separated by a boundary parallel to $\mathbf{B}$, we can construct a Gaussian pill box in a frame that is stationary with respect to the interface as shown in Figure 1.1. Taking the limit as the width and volume approach zero but the surface area remains finite gives the jump conditions. Using the notation that $[[a]]$ is the jump in the value of the scalar or vector component a when crossing an interface, we have:

$$\left[\left[\rho\hat{\mathbf{n}} \cdot \mathbf{u}\right]\right] = 0, \tag{1.75}$$

$$\left[\left[\hat{\mathbf{n}} \times \mathbf{E}\right]\right] = 0, \tag{1.76}$$

$$\left[\left[\left(p + \frac{1}{2\mu_0}B^2\right)\hat{\mathbf{n}} + \hat{\mathbf{n}} \cdot \boldsymbol{\pi} + \rho\hat{n} \cdot \mathbf{uu}\right]\right] = 0, \tag{1.77}$$

$$\left[\left[\hat{\mathbf{n}} \cdot \mathbf{u}\left(\frac{1}{2}\rho u^2 + \frac{5}{2}p\right) + \hat{\mathbf{n}} \cdot \mathbf{u} \cdot \boldsymbol{\pi} + \frac{1}{\mu_0}\hat{\mathbf{n}} \cdot \mathbf{E} \times \mathbf{B} + \hat{\mathbf{n}} \cdot \mathbf{q}\right]\right] = 0. \tag{1.78}$$

Subsets of these equations apply across more general interfaces as appropriate. For example, Eq. (1.76), which followed only from Eq. (1.24), which is universally valid, can be used to show that the tangential electric field is continuous across a fluid-wall interface as well. Also, Eq. (1.25) can be used to show that the normal component of the magnetic field must always be continuous.

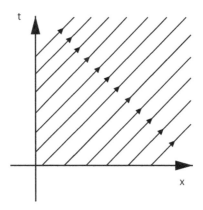

FIGURE 1.2: Characteristic curves $dx/dt = u$. All information is propagated along these lines.

1.3   Characteristics

The ideal MHD equations are a quasilinear symmetric hyperbolic system of equations and thus have real characteristics. Because they are real, the characteristic directions or characteristic manifolds have important physical meaning since all information is propagated along them. In Section 1.3.1 we review the nature of the ideal MHD characteristics.

To establish the concepts, let us first consider the simplest possible hyperbolic equation in one-dimension plus time,

$$\frac{\partial s}{\partial t} + u\frac{\partial s}{\partial x} = 0, \tag{1.79}$$

with u a constant. The characteristics of this equation are just the straight lines $dx/dt = u$ as shown in Figure 1.2. All information is propagated along these lines so that if the boundary data $s(x, 0) = f(x)$ are given for some range of x, then $s(x, t) = f(x - ut)$ is the solution at a later time. The characteristic curves, as shown in Figure 1.2, form a one-parameter family that fills all space. To find the solution at some space-time point (x, t) one need only trace back along the characteristic curve until a boundary data point is intersected.

We note that the boundary data that define the solution may be given many different ways. The data may be given entirely along the t axis, in which case they are commonly referred to as *boundary conditions*, or they may be given entirely along the x axis in which case they are referred to as *initial conditions*. However, more general cases are also allowable, the only restriction is that all characteristic curves intersect the boundary data curves once and only once. In particular, the boundary data curve is not allowed

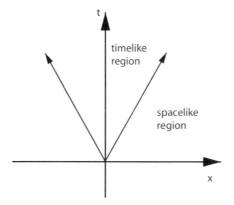

FIGURE 1.3: Space is divided into two regions by characteristic curves.

to be *tangent* to the characteristic curves anywhere. We will see in the next section that this restriction allows us to determine the characteristics for a more complicated system.

Finally, let us review another simple concept that is of importance for hyperbolic equations. It is that information always travels at a finite velocity. Consider the simple wave equation

$$\frac{\partial^2 \psi}{\partial t^2} - c^2 \frac{\partial^2 \psi}{\partial x^2} = 0, \tag{1.80}$$

where c is a constant. This could be written as two coupled first-order equations,

$$\frac{\partial \psi}{\partial t} = c \frac{\partial \omega}{\partial x},$$
$$\frac{\partial \omega}{\partial t} = c \frac{\partial \psi}{\partial x}.$$

Or, by defining $f \equiv \psi + \omega$ and $g \equiv \psi - \omega$, as two uncoupled first-order equations,

$$\frac{\partial f}{\partial t} = c \frac{\partial f}{\partial x},$$
$$\frac{\partial g}{\partial t} = -c \frac{\partial g}{\partial x}.$$

Comparing these with Eq. (1.79), it is clear that Eq. (1.80) has two characteristics, corresponding to $dx/dt = \pm c$. Information emanating from a point will in general travel along the two characteristics. These will divide spacetime into two regions relative to that point, a *timelike* region and a *spacelike* region as shown in Figure 1.3. A given point can only be influenced by a finite

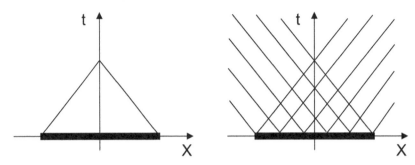

FIGURE 1.4: Domain of dependence (l) and domain of influence (r).

region of the boundary data for which that point is timelike. This region is the *domain of dependence* of the point. Similarly, a given region of boundary data can only affect points that are timelike with respect to at least part of that region. This is the *domain of influence* of that region; see Figure 1.4.

1.3.1   Characteristics in Ideal MHD

We saw that in one spatial dimension the characteristics were lines, as in Figure 1.2. In two spatial dimensions, the characteristics will be two-dimensional surfaces in three-dimensional space time, as shown in Figure 1.5. In full three-dimensional space, the characteristics will be three-dimensional manifolds, in four-dimensional space time $(\mathbf{r}, t)$. They are generated by the motion of surfaces in ordinary three-dimensional space $\mathbf{r}$. Let us determine these characteristics for the ideal MHD equations [19, 20].

We start with the ideal MHD equations, in the following form:

$$\rho \left(\frac{\partial \mathbf{v}}{\partial t} + \mathbf{v} \cdot \nabla \mathbf{v} \right) = -\nabla p - \frac{1}{\mu_o} \nabla \mathbf{B} \cdot \mathbf{B} + \frac{1}{\mu_0} (\mathbf{B} \cdot \nabla) \mathbf{B}, \tag{1.81}$$

$$\frac{\partial \mathbf{B}}{\partial t} + \mathbf{v} \cdot \nabla \mathbf{B} = \mathbf{B} \cdot \nabla \mathbf{v} - \mathbf{B} \nabla \cdot \mathbf{v}, \tag{1.82}$$

$$\frac{\partial p}{\partial t} + \mathbf{v} \cdot \nabla p = -\gamma p \nabla \cdot \mathbf{v}, \tag{1.83}$$

$$\frac{\partial s}{\partial t} + \mathbf{v} \cdot \nabla s = 0. \tag{1.84}$$

We use the fact that if the boundary data are given only along the characteristic curves, then the solution cannot be determined away from those curves. Let us assume that the boundary data for variables $\mathbf{v}$, $\mathbf{B}$, p, s are given on

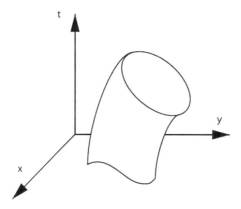

FIGURE 1.5: Characteristics in two spatial dimensions.

the three-dimensional surface $\phi(\mathbf{r}, t) = \phi_0$, and ask under what conditions is this insufficient to determine the solution away from this surface.

$$\phi(\mathbf{r}, t) = \phi_0, \tag{1.85}$$

We perform a coordinate transformation to align the boundary data surface with one of the coordinates. We consider ϕ as a coordinate and introduce additional coordinates χ, σ, τ within the three-dimensional boundary data manifold. Thus we transform

$$(\mathbf{r}, t) \rightarrow (\phi, \chi, \sigma, \tau). \tag{1.86}$$

On the boundary data manifold $\phi = \phi_0$, we specify $\mathbf{v}(\mathbf{r}, t) = \mathbf{v}_0(\chi, \sigma, \tau)$, $\mathbf{B}(\mathbf{r}, t) = \mathbf{B}_0(\chi, \sigma, \tau)$, $p(\mathbf{r}, t) = p_0(\chi, \sigma, \tau)$, and $s(\mathbf{r}, t) = s_0(\chi, \sigma, \tau)$. Since $\mathbf{v}_0$, $\mathbf{B}_0$, p_0, and s_0 are known functions, the derivatives with respect to χ, σ, and τ are also known. We ask under what conditions can the solutions $\mathbf{v}(\phi, \chi, \sigma, \tau)$, $\mathbf{B}(\phi, \chi, \sigma, \tau)$, $p(\phi, \chi, \sigma, \tau)$, and $s(\phi, \chi, \sigma)$ be obtained away from the boundary $\phi = \phi_0$?

We look for a power series solution of the form

$$\mathbf{v}(\phi, \chi, \sigma, \tau) = \mathbf{v}_0(\chi, \sigma, \tau) + (\phi - \phi_0)\frac{\partial \mathbf{v}}{\partial \phi}\bigg|_{\phi_0}$$
$$+ (\chi - \chi_0)\frac{\partial \mathbf{v}}{\partial \chi}\bigg|_{\phi_0} + (\sigma - \sigma_0)\frac{\partial \mathbf{v}}{\partial \sigma}\bigg|_{\phi_0} + (\tau - \tau_0)\frac{\partial \mathbf{v}}{\partial \tau}\bigg|_{\phi_0} + \cdots,$$

and similarly for $\mathbf{B}$, p, and s. The problem is solvable if the normal derivatives $\partial \mathbf{v}/\partial \phi|_{\phi_0}$, $\partial \mathbf{B}/\partial \phi|_{\phi_0}$, $\partial p/\partial \phi|_{\phi_0}$, and $\partial s/\partial \phi|_{\phi_0}$ can be constructed since all surface derivatives are known and higher-order derivatives can be constructed by differentiating the original PDEs.

Using the chain rule and using subscripts to denote partial derivatives with respect to the time t, $\phi_t = \partial\phi/\partial t$, etc., we can calculate

$$\frac{\partial \mathbf{v}}{\partial t} = \frac{\partial \mathbf{v}}{\partial \phi}\phi_t + \frac{\partial \mathbf{v}}{\partial \chi}\chi_t \cdots ,$$

$$\mathbf{v}\cdot\nabla\mathbf{v} = \mathbf{v}\cdot\nabla\phi\frac{\partial \mathbf{v}}{\partial \phi} + \mathbf{v}\cdot\nabla\chi\frac{\partial \mathbf{v}}{\partial \chi} + \cdots ,$$

$$\nabla\cdot\mathbf{v} = \nabla\phi\cdot\frac{\partial \mathbf{v}}{\partial \phi} + \nabla\chi\cdot\frac{\partial \mathbf{v}}{\partial \chi} + \cdots ,$$

etc. It is convenient to define some new notation. Define the *spatial normal*,

$$\hat{\mathbf{n}} = \nabla\phi/\mid\nabla\phi\mid,$$

the *characteristic speed*,

$$u \equiv -\left(\phi_t + \mathbf{v}\cdot\nabla\phi\right)/|\nabla\phi|,$$

which is the normal velocity of the characteristic measured with respect to the fluid moving with velocity $\mathbf{v}$, and let a prime denote the normal derivative,

$$(\ )' \equiv \frac{\partial}{\partial \phi}(\ ).$$

Using this notation, the ideal MHD equations take the form

$$-\rho u \mathbf{v}' + \hat{\mathbf{n}}p' + \frac{1}{\mu_0}\hat{\mathbf{n}}\mathbf{B}\cdot\mathbf{B}' - \frac{1}{\mu_0}\hat{\mathbf{n}}\cdot\mathbf{B}\mathbf{B}' = \cdots , \qquad (1.87)$$

$$-u\mathbf{B}' - \hat{\mathbf{n}}\cdot\mathbf{B}\mathbf{v}' + \mathbf{B}\hat{\mathbf{n}}\cdot\mathbf{v}' = \cdots , \qquad (1.88)$$

$$-up' + \gamma p\hat{\mathbf{n}}\cdot\mathbf{v}' = \cdots , \qquad (1.89)$$

$$-us' = \cdots , \qquad (1.90)$$

where the right side contains only known derivative terms. For definiteness, now choose $\mathbf{B}$ along the z axis and $\hat{\mathbf{n}}$ in the $(\hat{\mathbf{x}}, \hat{\mathbf{z}})$ plane so that in Cartesian coordinates,

$$\mathbf{B} = (0,0,B),$$
$$\hat{\mathbf{n}} = (n_x, 0, n_z).$$

We define the quantities $V_A \equiv B/\sqrt{\mu_0\rho}$ and $c_S \equiv \sqrt{\gamma p/\rho}$, and premultiply Eq. (1.88) by $\sqrt{\rho/\mu_0}$ and Eq. (1.89) by c_S^{-1}. This allows the system of equations Eqs. (1.87)–(1.90) to be written as

$$\mathbf{A}\cdot\mathbf{X} = \cdots , \qquad (1.91)$$

where

$$\mathbf{A} = \begin{bmatrix} -u & 0 & 0 & -n_z V_A & 0 & n_x V_A & n_x c_S & 0 \\ 0 & -u & 0 & 0 & -n_z V_A & 0 & 0 & 0 \\ 0 & 0 & -u & 0 & 0 & 0 & n_z c_S & 0 \\ -n_z V_A & 0 & 0 & -u & 0 & 0 & 0 & 0 \\ 0 & -n_z V_A & 0 & 0 & -u & 0 & 0 & 0 \\ n_x V_A & 0 & 0 & 0 & 0 & -u & 0 & 0 \\ n_x c_S & 0 & n_z c_S & 0 & 0 & 0 & -u & 0 \\ 0 & 0 & 0 & 0 & 0 & 0 & 0 & -u \end{bmatrix}$$

and

$$\mathbf{X} = \begin{bmatrix} \rho v'_x \\ \rho v'_y \\ \rho v'_z \\ \sqrt{\rho/\mu_0} B'_x \\ \sqrt{\rho/\mu_0} B'_y \\ \sqrt{\rho/\mu_0} B'_z \\ \frac{1}{c_S} p' \\ s' \end{bmatrix}.$$

Note that the matrix in Eq. (1.91) is symmetric, which guarantees that the eigenvalues will be real and that the system is hyperbolic. The characteristics are obtained when the determinant vanishes so that solutions cannot be propagated away from the boundary data manifold $\phi = \phi_0$. The determinant is given by

$$D = u^2 \left(u^2 - V_{An}^2 \right) \left[u^4 - \left(V_A^2 + c_S^2 \right) u^2 + V_{An}^2 c_S^2 \right] = 0, \qquad (1.92)$$

where we have used the relation $n_x^2 + n_z^2 = 1$ and let $V_{An}^2 = n_z^2 V_A^2$. The eight roots are given by:

$$u = u_0 = \pm 0 \quad \text{entropy disturbances,}$$

$$u = u_A = \pm V_{An} \quad \text{Alfvén waves,}$$

$$u = u_s = \pm \left\{ \frac{1}{2} \left(V_A^2 + c_S^2 \right) - \frac{1}{2} \left[\left(V_A^2 + c_S^2 \right)^2 - 4 V_{An}^2 c_S^2 \right]^{1/2} \right\}^{1/2} \quad \text{slow wave,}$$

$$u = u_f = \pm \left\{ \frac{1}{2} \left(V_A^2 + c_S^2 \right) + \frac{1}{2} \left[\left(V_A^2 + c_S^2 \right)^2 - 4 V_{An}^2 c_S^2 \right]^{1/2} \right\}^{1/2} \quad \text{fast wave.}$$

The latter two roots are also known as the slow magnetoacoustic and fast magnetoacoustic waves.

In normal magnetically confined fusion plasmas, we can take the *low-β limit*, $c_S^2 \ll V_A^2$, which implies

$$u_s^2 \cong n_z^2 c_S^2,$$
$$u_f^2 \cong V_A^2 + n_x^2 c_S^2.$$

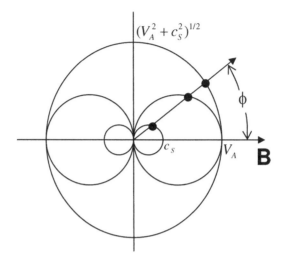

FIGURE 1.6: Reciprocal normal surface diagram in low-β limit.

The solutions following Eq. (1.92) are represented pictorially by the *reciprocal normal surface diagram* of Figure 1.6. The intersection points u_j give the speed of a plane wavefront whose normal $\hat{n}$ is at an angle ψ with the magnetic field **B**.

We list here some properties of the characteristic speeds of Eq. (1.92) and Figure 1.6:

1. $|u_0| \le |u_s| \le |u_A| \le |u_f| < \infty$.
2.   For propagation along **B**, $V_{An} = V_A$, and

$$\begin{aligned} |u_s| &= \min(V_A, c_S), \\ |u_A| &= V_A, \\ |u_f| &= \max(V_A, c_S). \end{aligned}$$

3.   For propagation perpendicular to **B**, $V_{An} = 0$, and

$$\begin{aligned} |u_s| &= |u_A| = 0, \\ |u_f| &= \left(V_A^2 + c_S^2\right)^{1/2} . \end{aligned}$$

4.   If $V_A = 0$ (no magnetic field) the gas dynamics equations are obtained:

$$|u_s|, |u_A| \to 0,$$
$$|u_f| \to c_S \quad \text{(ordinary sound wave } c_S\text{)}.$$

5.   Incompressible plasma limit is $\gamma \to \infty, \text{or } c_S \to \infty$:

$$|u_s| \to |u_A| \cdots \quad \text{these coincide,}$$
$$|u_f| \to \infty \cdots \quad \text{instantaneous propagation.}$$

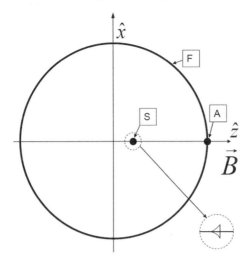

FIGURE 1.7: Ray surface diagram in low-β limit.

It is also instructive to compute a *ray surface diagram*. Suppose the initial disturbance is on a circle of radius R_0. After a time t, the equation for the wave surface is

$$z \cos \phi + x \sin \phi = R_0 \pm u_j(\phi)t. \tag{1.93}$$

We can take the ϕ derivative of Eq. (1.93) to obtain

$$-z \sin \phi + x \cos \phi = \pm \frac{d}{d\phi} u_j(\phi)t. \tag{1.94}$$

Now, invert Eqs. (1.93) and (1.94) for (x, z) to obtain

$$
\begin{aligned}
x &= R_0 \sin \phi + \left[\sin \phi u_j + \cos \phi \frac{d}{d\phi} u_j(\phi) \right] t, \\
z &= R_0 \cos \phi + \left[\cos \phi u_j - \sin \phi \frac{d}{d\phi} u_j(\phi) \right] t.
\end{aligned}
$$

Next, we let $R_0 \to 0$ represent a point disturbance and plot $[x(\phi), z(\phi)]$ for $0 < \phi < 2\pi$ to obtain the ray surface diagram in Figure 1.7. This shows clearly the extreme anisotropy of the Alfvén wave and also the slow waves in the low-β limit $c_S^2 \ll V_A^2$. Point disturbances just travel along the magnetic field.

Finally, let us consider the eigenvectors of the matrix in Eq. (1.91). The eigenvectors corresponding to each of the 4-pair of eigenvalues have physical significance. On a characteristic manifold, relations exist between $\mathbf{v}'$, $\mathbf{B}'$, p', s'.

Discontinuities satisfying these relations are propagated along with the characteristics. Substitution of the roots of Eq. (1.92) into the matrix in Eq. (1.91) yields the following eigenvectors:

$$
\begin{bmatrix}
\rho v_x' \\
\rho v_y' \\
\rho v_z' \\
\sqrt{\rho/\mu_0}B_x' \\
\sqrt{\rho/\mu_0}B_y' \\
\sqrt{\rho/\mu_0}B_z' \\
\frac{1}{c_S}p' \\
s'
\end{bmatrix}
=
\overset{\text{Entropy}}{\begin{bmatrix} 0 \\ 0 \\ 0 \\ 0 \\ 0 \\ 0 \\ 0 \\ \pm 1 \end{bmatrix}},
\overset{\text{Alfvén}}{\begin{bmatrix} 0 \\ 1 \\ 0 \\ 0 \\ \pm 1 \\ 0 \\ 0 \\ 0 \end{bmatrix}},
\overset{\text{Magnetoacoustic}}{\begin{bmatrix} n_x u^2/(u^2 - V_A^2) \\ 0 \\ n_z \\ -n_x u V_A n_z/(u^2 - V_A^2) \\ 0 \\ n_x^2 u V_A/(u^2 - V_A^2) \\ u/c_S \\ 0 \end{bmatrix}}.
$$

$$(1.95)$$

Note that the Alfvén wave is purely transverse, only perturbing **v** and **B** perpendicular both to the propagation direction and to the equilibrium magnetic field, while the magnetoacoustic waves involve perturbations in the other two directions. Note also that the different eigenvectors are orthogonal (see Problem 1.5).

It is instructive to again take the low-β limit $c_S^2 \ll V_A^2$, and also to examine propagation parallel to ($n_z = 1$, $n_x = 0$), and perpendicular to ($n_z = 0$, $n_x = 1$) the equilibrium magnetic field. We find

$$
\begin{bmatrix}
\rho v_x' \\
\rho v_y' \\
\rho v_z' \\
\sqrt{\rho/\mu_0}B_x' \\
\sqrt{\rho/\mu_0}B_y' \\
\sqrt{\rho/\mu_0}B_z' \\
\frac{1}{c}p' \\
s'
\end{bmatrix}
=
\overset{\substack{Fast \\ n_z = 0 \\ n_x = 1}}{\begin{bmatrix} 1 \\ 0 \\ 0 \\ 0 \\ 0 \\ 1 \\ c_S/V_A \\ 0 \end{bmatrix}},
\overset{\substack{Fast \\ n_z = 1 \\ n_x = 0}}{\begin{bmatrix} 1 \\ 0 \\ 0 \\ \pm 1 \\ 0 \\ 0 \\ 0 \\ 0 \end{bmatrix}},
\overset{\substack{Slow \\ n_z = 1 \\ n_x = 0}}{\begin{bmatrix} 0 \\ 0 \\ 1 \\ 0 \\ 0 \\ 0 \\ \pm 1 \\ 0 \end{bmatrix}}.
$$

$$(1.96)$$

It is seen that in a low-β ($\mu_0 p \ll B^2$) magnetized plasma, there exists a dramatic difference in the nature of the three non-trivial families of characteristics. All three waves can propagate parallel to the magnetic field, with the fast wave and the Alfvén wave having approximately the same velocity, but different polarities with respect to the perturbed velocities and fields, and the slow wave being much slower. The slow wave is the only one that involves a velocity component parallel to the background magnetic field.

The fast wave alone can propagate perpendicular to the background magnetic field and does so by compressing and expanding the field. Note that since the energy in the magnetic field is $B^2/2\mu_0$, perturbing the background field by an amount δB_z will require an energy $B_z\delta B_z/\mu_0 \gg \delta B_z^2/2\mu_0$. Thus, from energetic considerations we expect these perpendicularly propagating disturbances to be of very small amplitude compared to other disturbances that do not compress the background field. However, because the length scales perpendicular to the field are normally much smaller than those parallel to the field (see Problem 1.6), the time scales associated with the fast wave can be much shorter than those associated with either the Alfvén or slow waves. This is the fundamental reason for the stiffness of the ideal MHD equations when applied to a low-β magnetic confinement device.

1.3.2 Wave Dispersion Relation in Two-Fluid MHD

The additional terms present in the two-fluid MHD model of Section 1.2.1 introduce many new effects that are not present in the ideal MHD equations. The resistivities and viscosities generally damp the wave motion found in the last section, but resistivity can also allow plasma instabilities by relaxing the flux constrains of ideal MHD [11].

Here we discuss the effect of the non-dissipative terms in Ohm's law, Eq. (1.32), that are present in the two-fluid model but not in the ideal MHD description,

$$\mathbf{E} + \mathbf{u} \times \mathbf{B} = \frac{1}{ne}\left[\mathbf{J} \times \mathbf{B} - \nabla p_e + \cdots\right]. \tag{1.97}$$

When we non-dimensionalize the equations, these new terms on the right bring in a new dimensionless parameter,

$$d_i \equiv \frac{1}{\Omega_{ci}\tau_A}, \tag{1.98}$$

where $\Omega_{ci} = eB_0/M_i$ is the ion cyclotron frequency. The parameter d_i is called the *ion skin depth*. With these terms included, commonly called the *Hall terms*, the equations are no longer purely hyperbolic and become dispersive, i.e., different wavelength disturbances will propagate at different velocities.

If we linearize the equations about an equilibrium state with no flow and assume a periodic time and space dependence $\sim \exp i(\omega t - \mathbf{k}\cdot\mathbf{x})$, and now take the velocity u to be the phase velocity, $u \equiv \omega/k$, the analogue of the dispersion relation, Eq. (1.92) (after removing the entropy roots), becomes [21]

$$\begin{aligned} D &= \left(u^2 - V_{An}^2\right)\left[u^4 - \left(V_A^2 + c_S^2\right)u^2 + V_{An}^2 c_S^2\right] \\ &\quad - V_{An}^2 u^2 d_i^2 k^2\left(u^2 - c_S^2\right) = 0. \end{aligned} \tag{1.99}$$

Equation (1.99) is a cubic equation in u^2. In Figure 1.8 we show the roots as a function of the parameter $d_i k$ corresponding to the low-β parameters of

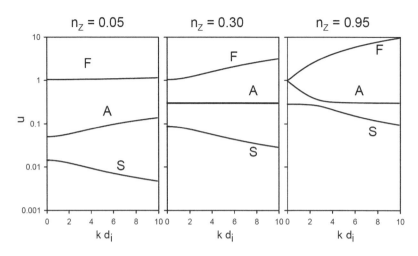

FIGURE 1.8: Typical dispersion relation for low-β two-fluid MHD for different angles of propagation relative to the background magnetic field.

Figure 1.6 with $c_S = 0.3V_A$ for three angles of propagation relative to the magnetic field, $n_z = V_{An}/V_A$. The curves are labeled according to which ideal MHD characteristic they match onto in the limit $k \to 0$. It is seen that the Hall terms increase the propagation speed of the fast wave, either increase or decrease the propagation velocity of the Alfvén wave, and decrease the velocity of the slow wave. The modification is most pronounced for propagation parallel to the background magnetic field. Only wavelengths such that $kd_i > 1$ are substantially affected.

1.4   Summary

The most fundamental mathematical description of plasma involves solving for the six dimension plus time phase-space distribution function $f_j(\mathbf{x}, \mathbf{v}, t)$ for each particle species j together with Maxwell's equation. The magnetohydrodynamic equations reduce this complexity by taking velocity moments to obtain a small system of fluid-like equations. In order for these equations to be complete, one needs to invoke closures that are obtained by subsidiary kinetic transport calculations. Three standard closures for two-fluid MHD, resistive MHD, and ideal MHD have been presented. The MHD equations can be written in conservation form reflecting global conservation of mass, magnetic flux, momentum, and energy. The conservation form of the equations is not always the best starting point for numerical solution, but they do provide an invaluable check on the global accuracy of the solutions. The ideal MHD equations are a subset of the more complete system of MHD equations. Their simplified mathematic structure allows rigorous analysis that reveals their solutions in terms of characteristic curves. Each characteristic curve emanating from a point separates spacetime into spacelike and timelike regions for that characteristic. These characteristics have implications for the type of numerical method used to solve the equations. Comparison of the fast wave to the Alfvén and slow waves clearly shows the underlying cause of the stiffness of the ideal MHD equations when applied to a low-β plasma. The Hall terms that appear in Ohm's law in the two-fluid description can make the wave speeds dependent on the wavenumber, k. However, if the wavelengths are long compared to the ion skin depth d_i, there is little change in the propagation velocities.

Problems

1.1: Derive the energy conservation equation, Eq. (1.68), starting from Eqs. (1.10) and (1.24).

1.2: Consider a circular cross section periodic cylinder with minor radius a and periodicity length $2\pi R$. (In cylindrical coordinates, the plasma occupies the region $0 < r < a$, $0 < Z < 2\pi R$, $0 < \theta < 2\pi$ with θ and Z being periodic.) The normal component of the magnetic field vanishes at the boundary $r = a$. Show that it is possible to have an arbitrarily large magnetic field **B** and self consistent current density **J** that satisfy the *force-free* condition $\mathbf{J} \times \mathbf{B} = 0$.

1.3: Consider a *low-β* plasma configuration where the magnetic field and current density are nearly force free (as in Problem 1.2) but where there exists a small pressure $\mu_0 p \ll B^2$. If the *relative error* in computing **B** is δB and if the conservative form of the energy equation is used, Eq. (1.63), what would be the resulting relative error in calculating p?

1.4: Verify that the following free-space Maxwell equations form a symmetric hyperbolic system and find their characteristics and eigenvectors.

$$\frac{\partial \mathbf{B}}{\partial t} = -\nabla \times \mathbf{E},$$

$$\frac{\partial \mathbf{E}}{\partial t} = c^2 \nabla \times \mathbf{B}.$$

1.5: Show explicitly that the eigenvectors corresponding to the fast magnetoacoustic and slow magnetoacoustic roots, as given in Eq. (1.95), are orthogonal to one another.

1.6: In the geometry of Problem 1.2 above, suppose we have a strong nearly uniform magnetic field in the $\hat{\mathbf{z}}$ direction, B_z, and a weaker radially dependent magnetic field in the θ direction, $B_\theta(r)$, with $B_z \gg B_\theta$. Define the *safety factor* $q(r)$ as $q = rB_z/RB_\theta$. For a given radial position r, estimate the ratio of times it takes for the fast wave and for the Alfvén wave to propagate from $(\theta = 0, Z = 0)$ to $(\theta = \pi, Z = 0)$.

Chapter 2

Introduction to Finite Difference Equations

2.1 Introduction

There are several distinctly different methods for obtaining numerical solutions to systems of partial differential equations. The three primary categories we consider here are *finite difference* methods, *finite element* methods, and *spectral* methods. Of course, within each of these three categories are many variations, not all of which we will be able to develop. *Finite volume* methods are basically finite difference methods applied to the conservation form of the equations, and are discussed somewhat in Section 2.6. We begin this chapter by discussing finite difference equations. There are several good references for this introductory material [22, 23, 24, 25].

Let us consider an initial value problem given by the model partial differential equation

$$\frac{\partial \phi}{\partial t} + u \frac{\partial \phi}{\partial x} = \alpha \frac{\partial^2 \phi}{\partial x^2}. \tag{2.1}$$

Here, u and α are non-negative real constants, and $\phi(x,t)$ is the unknown. Eq. (2.1) is typical of equations describing realistic systems in that it is of mixed type. If α were zero it would be *hyperbolic*, while if u were zero, it would be *parabolic* [26].

Equations like this occur in many applications. For example, this could apply to heat conduction in a moving medium in one dimension. In this case, ϕ would be the temperature, u would be the convection velocity, and α would be the thermal diffusivity. Or, it could also apply to the evolution of the magnetic vector potential in a plasma. In this case, there is only x variation, ϕ is the $\hat{z}$-component of the vector potential, A_z, u is the x-component of the plasma velocity, and α is the plasma resistivity. Besides Eq. (2.1), we also need boundary conditions, i.e., $\phi(0,t)$ and $\phi(L,t)$, and initial conditions $\phi(x,0)$ for $0 < x < L$.

What do we mean by a numerical solution? Let δx and δt be fixed spatial and time intervals. Let j and n be integer indices. We want to determine the solution at all discrete space-time points $x_j = j\delta x$, $t^n = n\delta t$ that lie

27

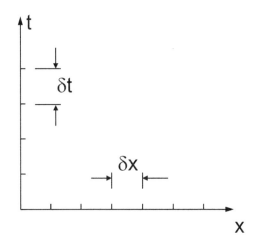

FIGURE 2.1: Space and time are discretized into a finite number of points.

within our domain; see Figure 2.1. We denote by ϕ_j^n the discrete solution that approximates the continuous solution $\phi(x_j, t^n)$.

To start, we replace the differential equation by a difference equation. In this case, we will use centered spatial differences and forward time differences. Using a Taylor's series expansion, we approximate

$$\left.\frac{\partial\phi}{\partial x}\right|_j^n \simeq \frac{\phi(x_j + \delta x, t^n) - \phi(x_j - \delta x, t^n)}{2\delta x} = \frac{\phi_{j+1}^n - \phi_{j-1}^n}{2\delta x},$$

$$\left.\frac{\partial\phi}{\partial t}\right|_j^n \simeq \frac{\phi(x_j, t^n + \delta t) - \phi(x_j, t^n)}{\delta t} = \frac{\phi_j^{n+1} - \phi_j^n}{\delta t},$$

$$\left.\frac{\partial^2\phi}{\partial x^2}\right|_j^n = \frac{\partial}{\partial x}\left\{\frac{\partial\phi}{\partial x}\right\} \simeq \frac{\left.\frac{\partial\phi}{\partial x}\right|_{j+1/2}^n - \left.\frac{\partial\phi}{\partial x}\right|_{j-1/2}^n}{\delta x} = \frac{\phi_{j+1}^n - 2\phi_j^n + \phi_{j-1}^n}{\delta x^2}.$$

The entire difference equation is then

$$\frac{\phi_j^{n+1} - \phi_j^n}{\delta t} + u\frac{\phi_{j+1}^n - \phi_{j-1}^n}{2\delta x} - \alpha\frac{\phi_{j+1}^n - 2\phi_j^n + \phi_{j-1}^n}{\delta x^2} = 0,$$

which we can recast in a form which is set to advance forward in time from initial conditions

$$\phi_j^{n+1} = \phi_j^n - \frac{u\delta t}{2\delta x}\left(\phi_{j+1}^n - \phi_{j-1}^n\right) + \frac{\alpha\delta t}{\delta x^2}\left(\phi_{j+1}^n - 2\phi_j^n + \phi_{j-1}^n\right). \qquad (2.2)$$

This now allows us to fill in the space-time net of Figure 2.1 by starting with the initial conditions $\phi_j^0; j = 0, N$ and using the difference equation to calculate the interior points $\phi_j^1; j = 1, N - 1$, with the points ϕ_0^1 and ϕ_N^1 being supplied by boundary conditions. This process is then repeated to obtain the time points $n = 2$ and so forth.

We note that Eq. (2.2) has two dimensionless terms,

$$S_1 = \frac{u \delta t}{\delta x}, \qquad S_2 = \frac{\alpha \delta t}{\delta x^2},$$

that are associated with convection and with dissipation. These are purely numerical parameters in that they depend on the size of the discrete space and time intervals. If we attempted to solve this equation on a computer, we would find that for some values of S_1 and S_2, *good* solutions occur, but not so for others. For different values of these parameters one would find either (i) bounded oscillations, (ii) unbounded growth, or (iii) unbounded oscillations. The latter two categories of behavior we call *numerical instabilities*, and they must be avoided to obtain meaningful solutions.

2.2   Implicit and Explicit Methods

There are two distinct classes of finite difference methods, explicit and implicit. In an *explicit* method, all spatial derivatives are evaluated at the old time points. An example of this is Eq. (2.2). In an *implicit* method, some spatial derivatives are evaluated at new time points as, for example, in this equation

$$\phi_j^{n+1} = \phi_j^n - \frac{u \delta t}{2 \delta x} \left(\phi_{j+1}^{n+1} - \phi_{j-1}^{n+1} \right) + \frac{\alpha \delta t}{\delta x^2} \left(\phi_{j+1}^{n+1} - 2\phi_j^{n+1} + \phi_{j-1}^{n+1} \right). \qquad (2.3)$$

The unknowns at the advanced time, $\phi_j^{n+1}; j = 0, \cdots, N$, are seen to be coupled together by the spatial derivative operators in Eq. (2.3), so that solving for these unknowns requires inversion of a matrix. This is true of implicit methods in general, while it is not the case for explicit methods.

We will see that generally implicit methods are more *stable* and allow larger time steps δt, but explicit methods require less computational effort for each time step. Note that explicit and implicit methods have *domains of dependence* that are quite different from one another as shown in Figures 2.2a and 2.2b. These are both generally different from that of the differential equation.

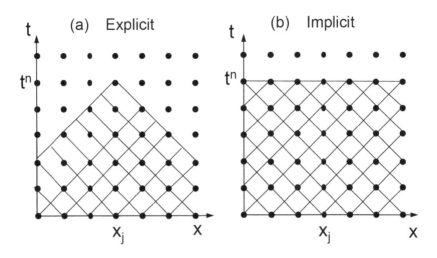

FIGURE 2.2: Domain of dependence of the point (x_j, t^n) for an (a) explicit and (b) implicit finite difference method.

2.3 Errors

The partial differential equation we are considering, Eq. (2.1), can be written in operator form as

$$L\{\phi\} = 0,$$

where in our case,

$$L = \frac{\partial}{\partial t} + u\frac{\partial}{\partial x} - \alpha\frac{\partial^2}{\partial x^2}. \tag{2.4}$$

In the finite difference approximation, the exact solution, ϕ, is approximated by a *discrete* state vector, ϕ_Δ, which is defined only on integer space-time points, thus

$$\phi \Rightarrow \phi_\Delta,$$

and the exact operator L gets replaced by a finite difference operator

$$L \Rightarrow L_\Delta,$$

giving the finite difference approximation to the partial differential equation

$$L_\Delta\{\phi_\Delta\} = 0.$$

There are two sources of error in obtaining numerical solutions, and these behave quite differently. These are (*i*) truncation error

$$T_\Delta \equiv L_\Delta\{\phi\} - L\{\phi\},$$

and (*ii*) machine error, or round-off error. Say ϕ_N is the actual solution which is obtained on the computer. Then the total error in the solution is

$$E = \phi - \phi_N = (\phi - \phi_\Delta) + (\phi_\Delta - \phi_N).$$

The first term on the right is the error due to the truncation error in the operator, and the second term is due to the finite arithmetic precision in the machine. As $T_\Delta \to 0$, we will show that $\phi_\Delta \to \phi$ under some circumstances, but $\phi_N \neq \phi_\Delta$ unless the arithmetic is of infinite precision.

2.4 Consistency, Convergence, and Stability

Here we define some fundamental concepts in the study of numerical analysis [22].

1. A finite difference operator L_Δ is said to be *consistent* with the differential operator L if $T_\Delta \to 0$ as $\delta t, \delta x \to 0$. In our example, Eq. (2.2), we have

$$L\{\phi\} = \frac{\partial \phi}{\partial t} + u\frac{\partial \phi}{\partial x} - \alpha\frac{\partial^2 \phi}{\partial x^2}$$

$$L_\Delta\{\phi\} = \frac{\phi_j^{n+1} - \phi_j^n}{\delta t} + \cdots .$$

By expanding $L_\Delta\{\phi\}$ in a Taylor's series about ϕ_j^n, we can compute the truncation error

$$T_\Delta = L_\Delta\{\phi\} - L\{\phi\} = \frac{1}{2}\delta t\frac{\partial^2 \phi}{\partial t^2} + \frac{1}{6}u\delta x^2\frac{\partial^3 \phi}{\partial x^3} - \frac{\alpha}{12}\delta x^2\frac{\partial^4 \phi}{\partial x^4} + \cdots . \quad (2.5)$$

If all the derivatives of the true solution are *bounded*, then this scheme is consistent since $T_\Delta \to 0$ as $\delta t, \delta x \to 0$.

2. If ϕ is a solution to $L\{\phi\} = 0$, and ϕ_Δ is a solution to $L_\Delta\{\phi_\Delta\} = 0$, then ϕ_Δ *converges* to ϕ if $\phi_\Delta \to \phi$ as $\delta t, \delta x \to 0$. This implies performing a series of calculations with differing values of δt and δx and comparing the results. Note that consistency applies to the operator and convergence applies to the solution.

3. *Stability* is the requirement that the solution remain bounded, even in the limit $\delta t \to 0$, $n \to \infty$ with $n\delta t$ finite. In other words, say the finite difference equation is of the form

$$L_\Delta\{\phi_\Delta\} = B_1\phi_\Delta^{n+1} - B_0\phi_\Delta^n = 0,$$

where B_1 and B_0 are finite difference operators. We can define the new operator

$$C(\delta t, \delta x) \equiv B_1^{-1} B_0,$$

so that the finite difference equation becomes

$$\phi_\Delta^{n+1} = C(\delta t, \delta x)\phi_\Delta^n.$$

The scheme is *stable* if for some $\tau > 0$, the infinite set of operators

$$[C(\delta t, \delta x)]^n$$

is uniformly bounded for $0 < \delta t < \tau$, $0 < n\delta t < T$ as we take the limit $n \to \infty$. Note that the concept of stability makes no reference to the differential equations one wishes to solve, but is a property solely of the sequence of difference equations.

4. These three concepts are related by the *Lax equivalence theorem*, the most celebrated theorem in numerical analysis. It says: "Given a properly posed initial value problem and a finite difference approximation to it that satisfies the *consistency* condition, then *stability* is the necessary and sufficient condition for *convergence*" [22].

The proof of this involves two steps as follows:

(i) First note that the error $E = \phi - \phi_\Delta$ satisfies the same difference equation as ϕ_Δ, but with an inhomogeneous term due to the local truncation error T_Δ,

$$L_\Delta\{E\} = L_\Delta\{\phi\} - L_\Delta\{\phi_\Delta\} = T_\Delta.$$

Therefore, it follows from the property of stability that the error must depend continuously on its data. This implies that there is some constant c for which

$$\|E\| \leq c(\|T_\Delta\| + \|E(0)\|).$$

(ii) By the property of consistency of the operator L_Δ, the right side approaches zero as $\delta t, \delta x \to 0$. Thus, convergence is proved.

2.5   Von Neumann Stability Analysis

In the last section we discussed the Lax equivalence theorem, which says that it is enough to show that a finite difference approximation is *consistent* and *stable* in order to guarantee convergence. It is normally quite straightforward to show consistency. The leading order truncation error is readily calculated by a Taylor series expansion, and we need only show that it is proportional to some powers of δx and δt. This whole procedure can often be done by inspection.

In contrast, it is *not* normally obvious whether or not a given finite difference approximation to an initial value problem is stable. One of the simplest and most powerful techniques to determine stability is that due to von Neumann. The method is to perform a *finite Fourier transform* [27] to convert the spatial differences into a *multiplicative factor*. The time advancement equation for each Fourier harmonic is then put into a form so that the new time level is expressed as an *amplification factor* times the amplitude of the harmonic at the old time level. This amplification factor describes the growth in going from one time step to the next. Then, the criterion for stability is that the absolute value of this factor be less than or equal to unity for all allowable Fourier components.

Suppose that the solution vector at time level n is given by ϕ_j^n; $j = 0, 1, \cdots, N - 1$. We define, for integer k values, the finite Fourier transform vector

$$\tilde{\phi}_k^n = \frac{1}{N} \sum_{j=0}^{N-1} \phi_j^n \exp(2\pi i k j / N); \qquad k = 0, \cdots, N - 1, \qquad (2.6)$$

where $i = \sqrt{-1}$. The corresponding inverse transform is

$$\phi_j^n = \sum_{k=0}^{N-1} \tilde{\phi}_k^n \exp(-2\pi i k j / N); \qquad j = 0, \cdots, N - 1. \qquad (2.7)$$

This is shown to be an exact transform pair by substitution of Eq. (2.6) into Eq. (2.7),

$$
\begin{aligned}
\phi_j^n &= \sum_{k=0}^{N-1} \frac{1}{N} \sum_{j'=0}^{N-1} \phi_{j'}^n \exp\left(2\pi i k j'/N\right) \exp\left[-\left(2\pi i k j/N\right)\right] \\
&= \frac{1}{N} \sum_{j'=0}^{N-1} \phi_{j'}^n \sum_{k=0}^{N-1} \exp\left[-2\pi i k(j - j')/N\right] \\
&= \frac{1}{N} \sum_{j'=0}^{N-1} \phi_{j'}^n N \delta_{jj'} \\
&= \phi_j^n.
\end{aligned}
$$

Note that Eqs. (2.6) and (2.7) are just matrix multiplications,

$$
\begin{bmatrix} \tilde{\phi}_0 \\ \vdots \\ \tilde{\phi}_{N-1} \end{bmatrix} = \begin{bmatrix} & \vdots & \\ \cdots & \frac{1}{N}\exp(2\pi i k j/N) & \cdots \\ & \vdots & \end{bmatrix} \cdot \begin{bmatrix} \phi_0 \\ \vdots \\ \phi_{N-1} \end{bmatrix}.
$$

The matrices corresponding to Eqs. (2.6) and (2.7) are exact inverses of one another.

Let us apply this technique to the difference equation introduced in Section 2.1,

$$\phi_j^{n+1} = \phi_j^n - \frac{1}{2}S_1(\phi_{j+1}^n - \phi_{j-1}^n) + S_2(\phi_{j+1}^n - 2\phi_j^n + \phi_{j-1}^n), \qquad (2.8)$$

where, we recall, $S_1 = u\delta t/\delta x$ and $S_2 = \alpha\delta t/\delta x^2$. Assuming that the coefficients S_1 and S_2 are constants, we transform each ϕ_j^n in Eq. (2.8) according to the definition (2.7) and select out only a single Fourier component k. After dividing by the common factor $\exp[-(2\pi ikj/N)]$, we obtain the algebraic equation

$$
\begin{aligned}
\tilde{\phi}_k^{n+1} &= \tilde{\phi}_k^n \Big\{ 1 - \frac{1}{2}S_1 \left[\exp(-2\pi ik/N) - \exp(2\pi ik/N)\right] \\
&\quad + S_2 \left[\exp(-2\pi ik/N) - 2 + \exp(2\pi ik/N)\right] \Big\},
\end{aligned}
$$

or

$$\tilde{\phi}_k^{n+1} = \tilde{\phi}_k^n \left[1 - iS_1 \sin\theta_k + S_2(2\cos\theta_k - 2)\right], \qquad (2.9)$$

where we have used the identity $\exp(i\alpha) = \cos\alpha + i\sin\alpha$ and have introduced the k dependent angle

$$\theta_k = -2\pi k/N.$$

Equation (2.9) can be written in the form

$$\tilde{\phi}_k^{n+1} = r\tilde{\phi}_k^n, \qquad (2.10)$$

where

$$r = 1 + 2S_2(\cos\theta_k - 1) - iS_1 \sin\theta_k \qquad (2.11)$$

is the amplification factor. For stability, this must satisfy $|r| \leq 1$ (or $|r| \leq 1 + O(\delta t)$ when the differential equation allows exponential growth). As θ_k varies from 0 to 2π, Eq. (2.11) traces out an ellipse in the complex plane as shown in Figure 2.3. The stability criterion that $|r| \leq 1$ translates into the condition that the ellipse in Figure 2.3 lies inside the unit circle.

Define the x coordinate of the center of the ellipse as $x_0 = 1 - 2S_2$. The ellipse is then defined by

$$\frac{(x - x_0)^2}{4S_2^2} + \frac{y^2}{S_1^2} = 1.$$

We look for the point(s) where this curve intersects the unit circle

$$x^2 + y^2 = 1.$$

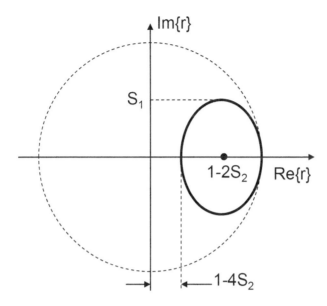

FIGURE 2.3: Amplification factor, r (solid), must lie within the unit circle (dashed) in the complex plane for stability.

Eliminating y^2 and solving the quadratic for x^2 gives

$$x = \begin{cases} 1, \\ \dfrac{S_1^2(1 - 4S_2) + 4S_2^2}{S_1^2 - 4S_2^2}. \end{cases}$$

For stability, we require $|x| \geq 1$ so that there is no y intersection. This leads to the two conditions

$$S_1^2 \leq 2S_2 \qquad \rightarrow \qquad \delta t \leq \frac{2\alpha}{u^2}, \tag{2.12}$$

$$S_2 \leq \frac{1}{2} \qquad \rightarrow \qquad \delta t \leq \frac{\delta x^2}{2\alpha}. \tag{2.13}$$

The finite difference method described by Eq. (2.8) is therefore conditionally stable, with the requirement that the time step δt satisfy the inequalities in Eqs. (2.12) and (2.13). We will return to a more systematic study of the numerical stability of different methods applied to different types of equations in later chapters.

The von Neumann stability analysis presented here strictly applies to linear problems with constant coefficients. However, since the most unstable modes are normally those with the smallest wavelengths on the grid, it is an excellent indicator of *local stability* if the local values of the coefficients are used in evaluating the stability criteria. For non-constant coefficients, it should be regarded as as a *necessary* condition for stability but not a *sufficient* one. For non-linear problems, one normally would reduce the time step by some small factor below that predicted by the von Neumann condition to ensure stability.

2.5.1 Relation to Truncation Error

There are several interesting limits of the analysis just presented. First, consider $u = 0$, $\alpha \neq 0$, so that there is no convection. The stability criterion becomes

$$\delta t \leq \frac{\delta x^2}{2\alpha},$$

which we will see in Chapter 7 is the criterion expected for a forward-time centered-space (FTCS) method applied to a parabolic equation. Secondly, consider the limit where $\alpha = 0$, $u \neq 0$, which corresponds to no dissipation. The criterion becomes

$$\delta t \leq 0,$$

so that FTCS is always *unstable* when applied to a pure convection problem.

Here we provide some additional insight as to why this is. Recall, from Eq. (2.5), that the leading order truncation error for $\alpha = 0$ is

$$T_\Delta = \frac{1}{2}\delta t \frac{\partial^2 \phi}{\partial t^2} + \frac{1}{6}u\delta x^2 \frac{\partial^3 \phi}{\partial x^3} + \cdots .$$

Hence the equation actually being approximated by the difference equations is

$$\frac{\partial \phi}{\partial t} + u\frac{\partial \phi}{\partial x} = -\frac{1}{2}\delta t \frac{\partial^2 \phi}{\partial t^2} - \frac{1}{6}u\delta x^2 \frac{\partial^3 \phi}{\partial x^3} + \cdots . \qquad (2.14)$$

The differential equation itself can be used to eliminate time derivatives in favor of space derivatives on the right side of Eq. (2.14), i.e.,

$$\frac{\partial^2}{\partial t^2} \quad \rightarrow \quad u^2\frac{\partial^2}{\partial x^2} + O\left(\delta t, \delta x^2\right).$$

To first order in δt and δx, the finite difference equation we are considering is therefore actually approximating the differential equation

$$\frac{\partial \phi}{\partial t} + u\frac{\partial \phi}{\partial x} = -\left(\frac{u^2\delta t}{2}\right)\frac{\partial^2 \phi}{\partial x^2}, \qquad (2.15)$$

where the term in parentheses is the effective diffusion coefficient. The fact that it is negative makes the problem mathematically ill posed, which is the source of the "numerical" instability [28].

2.5.2 Higher-Order Equations

As an example of a higher-order system, let us consider the inviscid fluid equations (with $\gamma = \frac{5}{3}$), which are a subset of the ideal MHD equations, but with $\mathbf{B} = 0$. In one dimension these equations, in conservation form, are

$$\frac{\partial \rho}{\partial t} + \frac{\partial}{\partial x} \rho u = 0,$$

$$\frac{\partial}{\partial t} \rho u + \frac{\partial}{\partial x} \left(\rho u^2 + p \right) = 0,$$

$$\frac{\partial}{\partial t} \left(\frac{1}{2} \rho u^2 + \frac{3}{2} p \right) + \frac{\partial}{\partial x} \left[u \left(\frac{1}{2} \rho u^2 + \frac{5}{2} p \right) \right] = 0. \tag{2.16}$$

It is convenient to define a *fluid state vector* $\mathbf{U}$ and a *flux vector* $\mathbf{F}$ by

$$\mathbf{U} = \begin{bmatrix} \rho \\ m \\ e \end{bmatrix}, \qquad \mathbf{F} = \begin{bmatrix} m \\ \frac{2}{3} \left(e + \frac{m^2}{\rho} \right) \\ \frac{m}{\rho} \left(\frac{5}{3} e - \frac{m^2}{3\rho} \right) \end{bmatrix},$$

where ρ, $m = \rho u$, $e = \frac{1}{2} \rho u^2 + \frac{3}{2} p$ are the mass density, momentum density, and total energy density, respectively. Rewriting u and p in terms of these variables, we can rewrite Eq. (2.16) in the conservation form as

$$\frac{\partial \mathbf{U}}{\partial t} + \frac{\partial \mathbf{F}}{\partial x} = 0,$$

or, in the linearized form, as

$$\frac{\partial \mathbf{U}}{\partial t} + \mathbf{A} \cdot \frac{\partial \mathbf{U}}{\partial x} = 0. \tag{2.17}$$

Here, the *Jacobian matrix* $\mathbf{A} = \frac{\partial \mathbf{F}}{\partial \mathbf{U}}$ is given by

$$\mathbf{A} = \begin{bmatrix} 0 & 1 & 0 \\ -\frac{2}{3} \frac{m^2}{\rho^2} & \frac{4}{3} \frac{m}{\rho} & \frac{2}{3} \\ \frac{m}{\rho} \left(-\frac{5}{3} \frac{e}{\rho} + \frac{2}{3} \frac{m^2}{\rho^2} \right) & \left(\frac{5}{3} \frac{e}{\rho} - \frac{m^2}{\rho^2} \right) & \frac{5}{3} \frac{m}{\rho} \end{bmatrix}.$$

The eigenvalues of $\mathbf{A}$, which satisfy the characteristic equation,

$$|\mathbf{A} - \lambda_A \mathbf{I}| = 0,$$

are given by

$$\lambda_A = u + c_S, u - c_S, u,$$

where

$$c_S^2 = \frac{5}{3} p/\rho$$

is the sound speed. For illustration, let us consider the *Lax–Friedrichs* difference method (see Section 9.2.1), which is defined by

$$\mathbf{U}_j^{n+1} = \frac{1}{2}\left(\mathbf{U}_{j+1}^n + \mathbf{U}_{j-1}^n\right) - \frac{\delta t}{2\delta x}\mathbf{A}\cdot\left(\mathbf{U}_{j+1}^n - \mathbf{U}_{j-1}^n\right). \tag{2.18}$$

As in the scalar case, we transform Eq. (2.18) and consider a single harmonic. We make the transformation

$$\mathbf{U}_j^n \to \tilde{\mathbf{U}}_\mathbf{k} r^n \exp(-2\pi ikj/N),$$

where r will be the amplification factor. After dividing by the common exponential factor, Eq. (2.18) becomes

$$r\tilde{\mathbf{U}}_\mathbf{k} = \cos\theta_k\tilde{\mathbf{U}}_\mathbf{k} - \frac{i\delta t}{\delta x}\sin\theta_k\mathbf{A}\cdot\tilde{\mathbf{U}}_\mathbf{k}, \tag{2.19}$$

where $\theta_k = -2\pi k/N$. We multiply on the left by the matrix $\mathbf{T}$, chosen so that the matrix

$$\mathbf{D} = \mathbf{T}\cdot\mathbf{A}\cdot\mathbf{T}^{-1}$$

is a diagonal matrix. This is a *similarity transformation*, which leaves the eigenvalues unchanged. Defining the vector $\tilde{\mathbf{V}}_\mathbf{k} = \mathbf{T}\cdot\tilde{\mathbf{U}}_\mathbf{k}$, Eq. (2.19) becomes

$$\left\{(r - \cos\theta_k)\mathbf{I} + \frac{i\delta t}{\delta x}\sin\theta_k\mathbf{D}\right\}\cdot\tilde{\mathbf{V}}_\mathbf{k} = 0. \tag{2.20}$$

Since the matrix in brackets in Eq. (2.20) is diagonal, there is no mixing of the components of the vector $\tilde{\mathbf{V}}_\mathbf{k}$ and we can consider each component separately. The diagonal elements of $\mathbf{D}$ are just the eigenvalues of $\mathbf{A}$; $\lambda_A = u, u \pm c_S$. The condition that the determinant of the matrix in brackets vanish gives the equation for the amplification factor

$$r = \cos\theta_k - i\frac{\delta t}{\delta x}\lambda_A\sin\theta_k. \tag{2.21}$$

The condition that $|r| \leq 1$ for all values of θ_k gives the stability criterion

$$\delta t \leq \frac{\delta x}{|u| + c_S}. \tag{2.22}$$

Equation (2.22) is known as the Courant–Friedrichs–Lewy or CFL condition. We will return to it in Chapter 9. Pictorially, it can be seen to be the condition that the characteristic curves passing backward in time from a point lie within the numerical domain of dependence of that point as depicted in Figure 2.2.

2.5.3 Multiple Space Dimensions

Here, we present an example of the von Neumann stability analysis being used on an equation involving more than one spatial dimension. Consider the two-dimension convection equation:

$$\frac{\partial \phi}{\partial t} + u\frac{\partial \phi}{\partial x} + v\frac{\partial \phi}{\partial y} = 0. \tag{2.23}$$

Let $x_j = j\delta x$ and $y_l = l\delta y$ and denote $\phi(x_j, y_l, t^n)$ by $\phi_{j,l}^n$. Assume for simplicity that there are N zones in both the x and y directions. The *leapfrog* method (see Section 9.2.4) applied to this equation is

$$\phi_{j,l}^{n+1} - \phi_{j,l}^{n-1} \quad + \quad \frac{u\delta t}{\delta x}\left(\phi_{j+1,l}^n - \phi_{j-1,l}^n\right)$$

$$+ \quad \frac{v\delta t}{\delta y}(\phi_{j,l+1}^n - \phi_{j,l-1}^n) = 0. \tag{2.24}$$

To examine the numerical stability, we apply a finite Fourier transform to both the j and the l index. Thus, making the substitution

$$\phi_{j,l}^n \to \tilde{\phi}_{k,m} r^n \exp(-2\pi i k j/N - 2\pi i m l/N)$$

in Eq. (2.24) and dividing through by the common exponential factor, we have

$$r - \frac{1}{r} + 2iS_x \sin\theta_k + 2iS_y \sin\theta_m = 0,$$

where $S_x = u\delta t/\delta x$, $S_y = v\delta t/\delta y$, $\theta_k = -2\pi k/N$, $\theta_m = -2\pi m/N$. This is a quadratic equation in r that can be written

$$r^2 + 2ibr - 1 = 0,$$

where $b = S_x \sin\theta_k + S_y \sin\theta_m$ is purely real. The criterion that the amplification factor $|r| \le 1$ is equivalent to the condition $|b| \le 1$ for all values of θ_k and θ_m, which leads to the stability criterion

$$\delta t \le \frac{1}{|u|/\delta x + |v|/\delta y}. \tag{2.25}$$

2.6 Accuracy and Conservative Differencing

Stability is not the only property that makes one finite difference approximation to a partial differential equation superior to another. Other considerations have to do with *efficiency* and *accuracy*. These concepts are often closely intertwined and difficult to quantify. For example, an *implicit* method

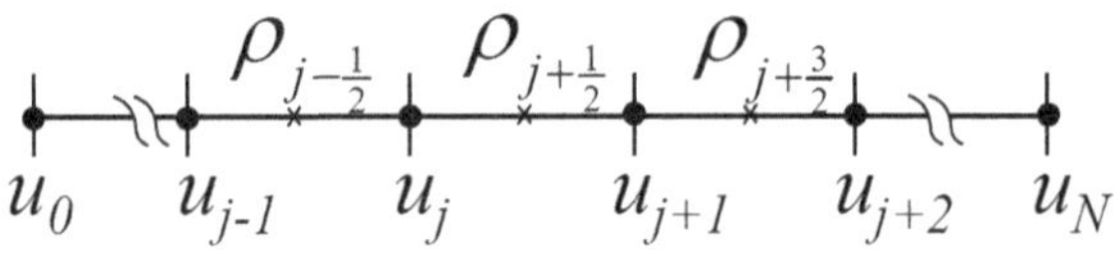

FIGURE 2.4: Density and velocity variables are defined at staggered locations.

may appear to be preferable to an *explicit* method if it allows a much less severe restriction on the time step, or perhaps no restriction at all. But the implicit method normally takes much more computational effort per time step, and its accuracy may be inferior. Also, there are different measures of the accuracy. One scheme that is classified as higher accuracy than another may actually be less accurate in certain global measures of the accuracy.

We normally classify the accuracy of a difference scheme by the powers of δt and δx that appear as multipliers in the leading order truncation error. For example, the method defined in Eq. (2.2) that has the truncation error as defined in Eq. (2.5) is *first order* accurate in δt and *second order* accurate in δx.

Let us now discuss the benefits of *conservative differencing*, which can greatly increase one global measure of the accuracy. As an example, consider the mass density equation in one spatial dimension. This can be written in two forms: conservative

$$\frac{\partial \rho}{\partial t} + \frac{\partial}{\partial x}(\rho u) = 0, \tag{2.26}$$

or non-conservative

$$\frac{\partial \rho}{\partial t} + u\frac{\partial \rho}{\partial x} + \rho\frac{\partial u}{\partial x} = 0. \tag{2.27}$$

Mathematically, these two forms are equivalent, but the finite difference equations that arise from these two are not. The conservative form, Eq. (2.26) is superior in several respects which we now discuss.

In converting Eq. (2.26) to a finite difference form, it is natural to stagger the locations where ρ and u are defined on the spatial mesh. If u is defined at cell boundaries or integer values of j, then ρ is defined at cell centers, or half integer values, $j + \frac{1}{2}$, as shown in Figure 2.4. If the value of ρ is needed at an integer j value, it is averaged over the two adjacent values: $\rho_j \equiv (1/2) \times (\rho_{j-\frac{1}{2}} + \rho_{j+\frac{1}{2}})$. With this convention, the centered space, centered time (using leapfrog) finite difference approximation to Eq. (2.26) is given by

$$\frac{\rho_{j+\frac{1}{2}}^{n+1} - \rho_{j+\frac{1}{2}}^{n-1}}{2\delta t} + \frac{(\rho u)_{j+1}^n - (\rho u)_j^n}{\delta x} = 0. \tag{2.28}$$

The main advantage of Eq. (2.28) stems from the fact that the spatial

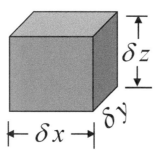

FIGURE 2.5: Finite volume method defines the densities at the centers of cells and the fluxes at the centers of their corresponding faces.

derivative term is the difference between two fluxes, each of which appears in exactly the same form in the difference equation for a neighboring j value, but with the opposite sign. Global conservation laws are exactly obeyed, as can be seen by summing Eq. (2.28) over all j values and noting that all the fluxes evaluated at interior cell boundaries cancel exactly, leaving only the difference in fluxes evaluated at the two boundaries

$$\delta x \sum_{j=0}^{N-1} \rho_{j+\frac{1}{2}}^{n+1} = \delta x \sum_{j=0}^{N-1} \rho_{j+\frac{1}{2}}^{n-1} - 2\delta t \left[(\rho u)_N^n - (\rho u)_0^n \right]. \tag{2.29}$$

If the fluxes evaluated at the two boundaries were prescribed to be zero for all time, then the sum of the mass in all the cells would be exactly constant between successive time steps. This global conservation property would not be maintained by the finite difference form of Eq. (2.27).

In two or more dimensions, a scalar equation in conservation form takes the form

$$\frac{\partial \rho}{\partial t} + \nabla \cdot \mathbf{F} = 0. \tag{2.30}$$

In the case of the mass density equation, $\mathbf{F} = \rho \mathbf{u}$, where $\mathbf{u}$ is the velocity vector. Let us consider Cartesian coordinates in three dimensions (x, y, z). Letting $x_i = i\delta x$, $y_j = j\delta y$, and $z_k = k\delta z$, if we integrate Eq. (2.30) over the volume of a single cell bounded by $[x_i, x_{i+1}]$, $[y_j, y_{j+1}]$, $[z_k, z_{k+1}]$, as shown in Figure 2.5, and apply Gauss's theorem, we obtain

$$\frac{\partial}{\partial t} \int dV \rho + \sum_{i=1}^{6} \int dS_i \hat{\mathbf{n}}_i \cdot \mathbf{F} = 0, \tag{2.31}$$

where the integral is over the volume $dV = dxdydz$ centered at $(x_{i+\frac{1}{2}}, y_{j+\frac{1}{2}}, z_{k+\frac{1}{2}})$. Replacing the volume integral by the value of the density at the center of the cell times the cell volume, and the surface integrals by the value of the flux at the center of the face times the surface area, after dividing by the cell volume we obtain the finite difference equation

$$
\frac{\partial}{\partial t}\rho_{i+\frac{1}{2},j+\frac{1}{2},k+\frac{1}{2}} = \; - \; \frac{1}{\delta x}\left[\hat{\mathbf{x}}\cdot\mathbf{F}_{i+1,j+\frac{1}{2},k+\frac{1}{2}} - \hat{\mathbf{x}}\cdot\mathbf{F}_{i,j+\frac{1}{2},k+\frac{1}{2}}\right]
$$
$$
- \; \frac{1}{\delta y}\left[\hat{\mathbf{y}}\cdot\mathbf{F}_{i+\frac{1}{2},j+1,k+\frac{1}{2}} - \hat{\mathbf{y}}\cdot\mathbf{F}_{i+\frac{1}{2},j,k+\frac{1}{2}}\right]
$$
$$
- \; \frac{1}{\delta z}\left[\hat{\mathbf{z}}\cdot\mathbf{F}_{i+\frac{1}{2},j+\frac{1}{2},k+1} - \hat{\mathbf{z}}\cdot\mathbf{F}_{i+\frac{1}{2},j+\frac{1}{2},k}\right].
$$

We see that the natural location for the density to be defined is at the cell centers, $(x_{i+\frac{1}{2}}, y_{j+\frac{1}{2}}, z_{k+\frac{1}{2}})$, and that the natural locations for each of the fluxes is in the center of its corresponding face. These conventions are the basis for the *finite volume method*, which is the natural finite difference method when a system of equations is expressible in conservative form. This is most readily extended to non-Cartesian coordinates by starting with the coordinate independent equation, Eq. (2.31).

The conservation form also has some additional advantages over the non-conservation form. One is that only the product ρu has to be specified on the computational boundary, not ρ and u separately. This helps in avoiding over-specifying boundary conditions. Also, by staggering the variables, we have been able to approximate centered derivatives over only a single grid spacing, which leads to a more efficient difference scheme. Finally, the conservation form is preferable when discontinuities are present, for example, due to material interfaces. Differences are only taken of quantities that are continuous across these interfaces.

2.7 Summary

Finite difference equations are based on a truncated Taylor series expansion of the solution about discrete space-time points, with the missing terms being the truncation error. The finite difference equation is consistent with the original partial differential equation if the truncation error approaches zero as the time step and zone spacing approach zero. The finite difference solution will approach the true solution to the PDE if the finite difference equation is consistent and numerically stable. The von Neumann stability analysis is a powerful technique to determine numerical stability, although it is rigorously applicable only to linear equations with constant coefficients. It basically amounts to computing a dispersion relation for the finite difference equation. Other things being equal, one should try and take advantage of a conservative formulation of a problem by using conservative differencing. This allows global conservation relations to be exactly preserved in the finite difference solution. This is the basis of the finite volume method of discretization.

Problems

2.1: Verify that the three eigenvalues of the matrix $\mathbf{A}$ in Eq. (2.5.2) have the values u, $u + c_S$, $u - c_S$, where $c_S^2 = \frac{5}{3}p/\rho$. Compute the corresponding eigenvectors.

2.2: Consider a finite difference method that can be written as

$$\phi_j^{n+1} = A\phi_{j+1}^n + B\phi_{j-1}^n, \tag{2.32}$$

for some constants A and B. What are the conditions on A and B for the method of Eq. (2.32) to be stable?

2.3: Consider the one-dimensional system of conservation laws:

$$\frac{\partial \mathbf{U}}{\partial t} + \frac{\partial \mathbf{F}(\mathbf{U})}{\partial x} = 0, \tag{2.33}$$

or

$$\frac{\partial \mathbf{U}}{\partial t} + \mathbf{A} \cdot \frac{\partial \mathbf{U}}{\partial x} = 0. \tag{2.34}$$

Here, $\mathbf{U}(x, t)$ is the vector of unknowns, and $\mathbf{A} = \partial \mathbf{F}/\partial \mathbf{U}$ is a real, positive definite Jacobian matrix with real eigenvalues λ_A. Let $\mathbf{U}_j^n$ be the finite difference approximation to $\mathbf{U}(x_j, t^n)$, where $x_j = j\delta x$ and $t^n = n\delta t$, with δt and δx being the time step and zone size, respectively.

Now, consider the following finite difference method to advance the solution at an interior point from time level t^n to time level t^{n+1}:

$$\mathbf{U}_j^* = \mathbf{U}_j^n - \frac{\delta t}{\delta x}\mathbf{A} \cdot \left[\mathbf{U}_{j+1}^n - \mathbf{U}_j^n\right]$$

$$\mathbf{U}_j^{**} = \mathbf{U}_j^* - \frac{\delta t}{\delta x}\mathbf{A} \cdot \left[\mathbf{U}_j^* - \mathbf{U}_{j-1}^*\right]$$

$$\mathbf{U}_j^{n+1} = \frac{1}{2}\left(\mathbf{U}_j^n + \mathbf{U}_j^{**}\right). \tag{2.35}$$

(a) Show that the method is second order accurate in both δt and δx.
(b) Use the von Neumann stability analysis technique to derive a stability condition on the maximum allowable value of δt in terms of δx and the maximum eigenvalue max λ_A.
HINT: First convert the finite difference method, Eq. (2.35), into a single-step method by substitution. Treat the matrix $\mathbf{A}$ as a constant. In evaluating the truncation error, use the equation (2.34) to eliminate second-order time derivatives in favor of spatial derivatives.

2.4: Consider the heat conduction equation, Eq. (2.1) with $u = 0$. Apply the central difference in time (leapfrog) and centered difference in space method:

$$\phi_j^{n+1} = \phi_j^{n-1} + \frac{2\delta t \alpha}{\delta x^2}\left[\phi_{j+1}^n - 2\phi_j^n + \phi_{j-1}^n\right]. \tag{2.36}$$

Show that the scheme is unstable for any value of δt.

2.5: Consider the anisotropic diffusion equation:

$$\frac{\partial T}{\partial t} = \kappa_x \frac{\partial^2 T}{\partial x^2} + \kappa_y \frac{\partial^2 T}{\partial y^2}$$

where $\kappa_x \neq \kappa_y$. Write down a forward-time centered-space finite difference approximation to this and calculate the maximum stable time step.

Chapter 3

Finite Difference Methods for Elliptic Equations

3.1 Introduction

In this chapter we discuss numerical methods for solving elliptic equations, such as Poisson's equation or the Grad–Shafranov equation. A general mathematical classification of equation types as elliptic, parabolic, or hyperbolic can be found elsewhere [26, 29]. Elliptic equations need boundary conditions of either the Dirichlet or Neumann type specified on a closed surface to possess unique solutions.

Elliptic equations often arise as the steady-state, time-independent limit of hyperbolic or parabolic equations. They also arise in incompressible flow problems or in a general formulation of the time-dependent MHD equations when the fluid velocity and magnetic vector potential are expressed in terms of potentials and stream function variables [30]. In this case they need to be solved each time step as is shown in Chapter 9 and efficiency is of paramount importance.

The numerical formulation of an elliptic equation always reduces to a discrete matrix equation. There are many different algorithms for solving these matrix equations that have been developed from several very different approaches. These same types of matrix equations arise in implicit formulations of parabolic and hyperbolic equations (or systems of equations) as shown in Chapters 7 and 9. Methods for their efficient solution are therefore of central importance.

No one method is superior for all applications. Some methods are very simple to program but are inefficient compared to other more complex methods. These may be useful in exploratory programming, but not in production applications. Or, as in the multigrid method described in Section 3.6, relatively simple methods may form the basis for more complex methods when combined with other techniques. Some matrices have symmetry or other properties such as strong diagonal dominance that can be exploited using methods that are not generally applicable. In the remainder of this chapter we give an overview of the most common methods that are in use today.

3.2   One-Dimensional Poisson's Equation

Consider first Poisson's equation in 1D:

$$\frac{\partial^2 u}{\partial x^2} = R(x), \tag{3.1}$$

for $0 < x < L$, where $R(x)$ is given over the interval. Equation (3.1) is an ordinary differential equation (ODE) rather than a partial differential equation (PDE) since there is only one independent variable, x. Since it is a second-order ODE, it requires two boundary values. If these boundary values are specified at the same boundary point, such as $u(0)$ and the derivative $u'(0)$, then it is referred to as an *initial value problem*. There is an extensive literature on ODE methods for integrating the solution of initial value problems from 0 to L [31], such as the Runge–Kutta method. These will not be discussed further here.

3.2.1   Boundary Value Problems in One Dimension

However, if the problem is such that one boundary value is specified at each boundary, then it becomes a *boundary value problem* and as such is similar to boundary value problems in two or more dimensions. We proceed to discuss this case as an introduction to boundary value problems in multiple dimensions. Consider the case where we specify one value at each boundary: either $u(0)$ or $u'(0)$, and either $u(L)$ or $u'(L)$. However, note that the two derivatives $u'(0)$ and $u'(L)$ cannot both be prescribed independently, since there is an integral constraint relating these that is obtained by integrating Eq. (3.1),

$$\int_0^L \frac{\partial^2 u}{\partial x^2} dx = \left. \frac{\partial u}{\partial x} \right|_0^L = u'(L) - u'(0) = \int_0^L R(x) dx. \tag{3.2}$$

This is known as a *solvability constraint*. Note also that even if Eq. (3.2) was satisfied exactly, the solution would only be determined up to a constant if only the derivative boundary conditions were specified.

To solve Eq. (3.1) by finite differences, we use the notation

$$u_j = u(x_j), \qquad R_j = R(x_j); \qquad j = 0, \cdots, N,$$

where $x_j = jh$, and $h = L/N$. A finite difference approximation to Eq. (3.1) is obtained by Taylor series analysis. Expanding about u_j, we have

$$u_{j+1} = u_j + h \left. \frac{\partial u}{\partial x} \right|_j + \frac{h^2}{2} \left. \frac{\partial^2 u}{\partial x^2} \right|_j + \frac{h^3}{3!} \left. \frac{\partial^3 u}{\partial x^3} \right|_j + \frac{h^4}{4!} \left. \frac{\partial^4 u}{\partial x^4} \right|_j + \cdots,$$

$$u_{j-1} = u_j - h \frac{\partial u}{\partial x}\bigg|_j + \frac{h^2}{2} \frac{\partial^2 u}{\partial x^2}\bigg|_j - \frac{h^3}{3!} \frac{\partial^3 u}{\partial x^3}\bigg|_j + \frac{h^4}{4!} \frac{\partial^4 u}{\partial x^4}\bigg|_j + \cdots .$$

Adding these and dividing by h^2, we obtain

$$\frac{\partial^2 u}{\partial x^2}\bigg|_j = \frac{u_{j+1} - 2u_j + u_{j-1}}{h^2} - \frac{h^2}{12} \frac{\partial^4 u}{\partial x^4}\bigg|_j \cdots .$$

For $h^2 \ll 1$, the second term on the right, which is the leading order in h truncation error, is negligible. The finite difference approximation to Eq. (3.1), accurate to second order in h, is thus given by

$$u_{j+1} - 2u_j + u_{j-1} = h^2 R_j. \tag{3.3}$$

Equation (3.3) is called a tridiagonal matrix equation. It occurs in many applications. When written in matrix form, the coefficient matrix has only three diagonal bands of possible non-zero values as illustrated below:

$$
\begin{bmatrix}
-B_0 & A_0 & & & \\
C_1 & -B_1 & A_1 & & \\
& \cdots & & & \\
& & C_{N-1} & -B_{N-1} & A_{N-1} \\
& & & C_N & -B_N
\end{bmatrix}
\cdot
\begin{bmatrix}
u_0 \\
u_1 \\
\vdots \\
u_{N-1} \\
u_N
\end{bmatrix}
=
\begin{bmatrix}
D_0 \\
D_1 \\
\vdots \\
D_{N-1} \\
D_N
\end{bmatrix}. \tag{3.4}
$$

For our case, we have $A_j = 1$, $B_j = 2$, $C_j = 1$, $D_j = h^2 R_j$ for $j = 1, \cdots, N-1$. The values of the coefficients for $j = 0$ and $j = N$ are determined by the boundary conditions as follows: Suppose the values of $u(0)$ and $u(L)$ are given (Dirichlet conditions). This is incorporated into the above matrix by setting $B_0 = B_N = -1$, $A_0 = C_N = 0$, $D_0 = u(0)$, $D_N = u(L)$. Incorporation of derivative (Neumann) boundary conditions is discussed in Section 3.3.1 and in Problem 3.1. We call this matrix *sparse* because it has mostly zero entries, and the non-zero elements occur in particular places. These locations form the *sparseness pattern*.

3.2.2 Tridiagonal Algorithm

There is a simple yet powerful technique for solving general finite difference equations that are of the tridiagonal form. Let us rewrite the matrix equation, Eq. (3.4), in the form

$$A_j u_{j+1} - B_j u_j + C_j u_{j-1} = D_j; \qquad j = 0, \ldots, N. \tag{3.5}$$

Equation (3.5) is a second-order difference equation since it involves three values of u_j. To solve, we look for vectors E_j and F_j that satisfy the first-order difference equation:

$$u_j = E_j u_{j+1} + F_j, \tag{3.6}$$

or, letting $j \rightarrow j - 1$

$$u_{j-1} = E_{j-1}u_j + F_{j-1}. \tag{3.7}$$

Next, using Eq. (3.7) to eliminate u_{j-1} in Eq. (3.5), and comparing with Eq. (3.6), we see that E_j and F_j, must obey the recurrence relations.

$$E_j = \left[\frac{A_j}{B_j - C_j E_{j-1}} \right], \tag{3.8}$$

$$F_j = \left[\frac{C_j F_{j-1} - D_j}{B_j - C_j E_{j-1}} \right]. \tag{3.9}$$

Equations (3.6), (3.8), and (3.9) now provide a system of first-order difference equations which are equivalent to the second-order difference equation, Eq. (3.5), but which are inherently stable, and which incorporate the boundary conditions in a natural way.

Since $C_0 = 0$, E_0 and F_0 are well defined as follows:

$$\begin{aligned} E_0 &= A_0/B_0, \\ F_0 &= -D_0/B_0. \end{aligned} \tag{3.10}$$

This is sufficient initialization to advance the difference equation, Eq. (3.9), for E_j and F_j from $j = 1$ to $j = N - 1$. The right boundary condition is then utilized by noting that since $A_N = 0$, Eq. (3.5) can be used to eliminate u_N in Eq. (3.6) and we have

$$u_{N-1} = \left[\frac{B_N F_{N-1} - D_N E_{N-1}}{B_N - E_{N-1} C_N} \right]. \tag{3.11}$$

Since the E_j and F_j are all known, the recursion relation, Eq. (3.6), can then be stepped backwards from $j = N - 2$ to $j = 1$ to determine u_j at every location. The entire process requires only $5 \times N$ arithmetic operations to obtain an exact solution to the difference equation.

3.3   Two-Dimensional Poisson's Equation

Let us now discuss the 2D Poisson's equation for the unknown $u(x, y)$ in a rectangular domain Ω with Dirichlet boundary conditions applied on the boundary curve $\delta\Omega$. We consider the equation:

$$\nabla^2 u = \frac{\partial^2 u}{\partial x^2} + \frac{\partial^2 u}{\partial y^2} = R(x, y), \tag{3.12}$$

with boundary values

$$\begin{aligned} u(0, y), u(L, y) &\quad ; \quad 0 < y < L, \\ u(x, 0), u(x, L) &\quad ; \quad 0 < x < L, \end{aligned} \tag{3.13}$$

and a given inhomogeneous term;

$$R(x, y); \qquad 0 < x, y < L.$$

We note here that the problem could also be posed such that the boundary conditions given by Eq. (3.13) were replaced by Neumann boundary conditions in which the values of the normal derivative are specified on the boundary rather than the function. This is normally denoted by:

$$\frac{\partial u}{\partial n} = b_1 \ \text{ on } \ \delta\Omega. \tag{3.14}$$

This is discussed in Section 3.3.1

The problem we consider is to solve for $u(x, y)$ in the interior. We divide space into N intervals of length $h = L/N$ in each direction and adopt the conventional notation:

$$\begin{aligned}
x_j &= jh; & j &= 0, \cdots, N, \\
y_k &= kh; & k &= 0, \cdots, N, \\
u_{jk} &= u(x_j, y_k); & R_{jk} &= R(x_j, y_k).
\end{aligned}$$

The second-order accurate finite difference equation, again derived by a Taylor series expansion, is given by

$$u_{j+1,k} + u_{j-1,k} + u_{j,k+1} + u_{j,k-1} - 4u_{j,k} = h^2 R_{j,k}. \tag{3.15}$$

Let the $(N + 1)^2$ vector of unknowns be denoted by $\mathbf{x}$. When Eq. (3.15) is written in matrix form,

$$\mathbf{A} \cdot \mathbf{x} = \mathbf{b}, \tag{3.16}$$

the matrix $\mathbf{A}$ is no longer tridiagonal but is sparse with a definite sparseness pattern as shown below:

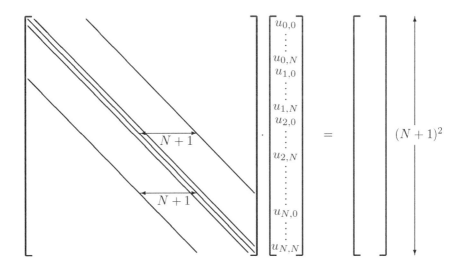

This matrix has non-zero elements on the diagonal and on either side of it, but also has non-zero diagonal bands located $(N + 1)$ elements away from the main diagonal. As in the 1D case, this general matrix will be modified to incorporate the boundary conditions. In particular, for Dirichlet boundary conditions, the rows that correspond to boundary values will be replaced by rows with all zeros, except for 1s on the diagonal, and the boundary value will be inserted in the right-side vector.

This matrix equation must be inverted to solve for the unknowns (although the actual inverse is not normally calculated). We consider several methods of solving this system: (1) direct inversion as if the matrix were full using *Gauss elimination* or *LU decomposition*, (2) direct inversion using *block tridiagonal* methods, (3) several approaches to *iterative methods*, and (4) direct methods based on *transform techniques*. But first we include a brief discussion of the inclusion of Neumann boundary conditions in a 2D problem.

3.3.1   Neumann Boundary Conditions

As discussed in the 1D case in Section 3.2, if Neumann boundary conditions are specified, the inhomogeneous term must satisfy a solvability condition that follows from the application of Gauss's theorem, in this case to Eq. (3.12) using Eq. (3.14):

$$\int_\Omega R(x,y)dxdy = \int_\Omega \nabla \cdot \nabla u \, dxdy = \int_{\delta\Omega} \frac{\partial u}{\partial n}d\ell = \int_{\delta\Omega} b_1 d\ell. \qquad (3.17)$$

As for the 1D case, even with the solvability constraint satisfied, the solution is determined only up to a constant. We can see this mathematically because both the equation being solved and the boundary conditions now depend only on the derivative of the function and not on the function itself. For example, in the rectangular geometry being considered, the boundary conditions could take the form of:

$$\frac{\partial u}{\partial x}(0, y_k) \approx \frac{u_{1,k} - u_{0,k}}{h} = b_1(0, y_k), \qquad (3.18)$$

or,

$$\frac{\partial u}{\partial x}(0, y_k) \approx \frac{-u_{2,k} + 4u_{1,k} - 3u_{0,k}}{2h} = b_1(0, y_k). \qquad (3.19)$$

One of these would replace the row of the matrix corresponding to the boundary value $u_{0,k}$ for $k = 1, \cdots, N$. Equation (3.18) is first order accurate and Eq. (3.19) is second order accurate. We see that if $\mathbf{x} = u_{j,k}$ is any solution vector that satisfies the difference equation, Eq. (3.15), and either Eq. (3.18) or Eq. (3.19), then $\mathbf{x}+c\mathbf{x}_0$ would also be a solution, where c is any real number and $\mathbf{x}_0$ is a vector consisting of all 1s. If this is the case, we say that $\mathbf{x}_0$ is a *null vector* or belongs to the *null space* of $\mathbf{A}$, i.e.,

$$\mathbf{A} \cdot \mathbf{x}_0 = 0. \qquad (3.20)$$

The nonuniqueness of the solution of the Neumann problem implies that the matrix $\mathbf{A}$ in Eq. (3.16) is singular.

If the problem is properly formulated, this non-uniqueness may not be a mathematical difficulty. For example, if the unknown $u(x, y)$ corresponds to a potential or stream function, then only the gradient of u, ∇u has physical meaning. However, the non-uniqueness could lead to difficulties in some of the numerical methods for solving the discrete equations that are given in the following sections. One possible remedy is to replace one of the rows of the matrix $\mathbf{A}$ (usually corresponding to a boundary point) with a row that arbitrarily fixes one component of $\mathbf{x}$. This renders the resulting system nonsingular, but is not totally satisfactory because the accuracy of the solution will depend on what equation is deleted. Another remedy [32] is to note that we can regularize the solution by adding a small, positive definite term proportional to ϵu^2 to the variational statement of the problem, where ϵ is some small positive number. The regularized problem is stated as finding the function u that minimizes the functional

$$I(u) = \int \left[|\nabla u|^2 + \epsilon u^2 + 2uR(x, y) \right] dx dy. \tag{3.21}$$

This results in Eq. (3.12) being replaced by the Euler equation for Eq. (3.21):

$$\nabla^2 u - \epsilon u = R(x, y). \tag{3.22}$$

For ϵ small enough, the solution to the finite difference form of Eq. (3.22) will be close to the particular solution of Eq. (3.12) that minimizes ϵu^2 as well as satisfying the differential equation.

We now describe a convenient way to impose second-order accurate Neumann boundary conditions, to ensure that the solvability constraint is satisfied to machine accuracy, and to derive an exact discrete form of Gauss's theorem. We use conservative differencing to derive the finite difference equations, and define the finite difference grid so that the physical boundary lies midway between two grid points as illustrated in Figure 3.1. Suppose that the elliptic equation we wish to solve, in 2D Cartesian geometry, is given by:

$$\frac{\partial}{\partial x} \left(f(x, y) \frac{\partial u}{\partial x} \right) + \frac{\partial}{\partial y} \left(g(x, y) \frac{\partial u}{\partial y} \right) = R(x, y). \tag{3.23}$$

Consider a domain $0 < x < L_x$, $0 < y < L_y$ with $f(x, y)$, $g(x, y)$, and $R(x, y)$ all given, and with Neumann boundary conditions specified. The grid spacings $\delta x = L_x/N$ and $\delta y = L_y/N$ are defined in Figure 3.1. Second-order, conservative finite differencing of Eq. (3.23) gives:

$$\frac{1}{\delta x} \left[f(x, y) \frac{\partial u}{\partial x} \bigg|_{j+\frac{1}{2}, k} - f(x, y) \frac{\partial u}{\partial x} \bigg|_{j-\frac{1}{2}, k} \right]$$

$$+ \frac{1}{\delta y} \left[g(x, y) \frac{\partial u}{\partial y} \bigg|_{j, k+\frac{1}{2}} - g(x, y) \frac{\partial u}{\partial y} \bigg|_{j, k-\frac{1}{2}} \right] = R_{i, j}, \tag{3.24}$$

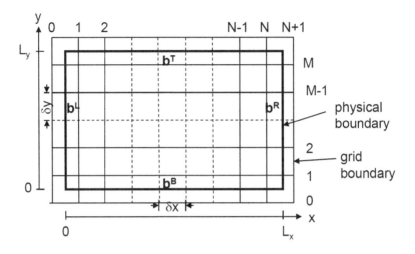

FIGURE 3.1: The computational grid can be shifted from the physical boundary by a half grid spacing to naturally incorporate Neumann boundary conditions.

or,

$$\frac{1}{\delta x^2}\left[f_{j+\frac{1}{2},k}\left(u_{j+1,k}-u_{j,k}\right)-f_{j-\frac{1}{2},k}\left(u_{j,k}-u_{j-1,k}\right)\right]$$
$$+\frac{1}{\delta y^2}\left[g_{j,k+\frac{1}{2}}\left(u_{j,k+1}-u_{j,k}\right)-g_{j,k-\frac{1}{2}}\left(u_{j,k}-u_{j,k-1}\right)\right] \quad = \quad R_{i,j}.$$

$$(3.25)$$

Let the Neumann boundary conditions for the geometry of Figure 3.1 be specified in terms of the four boundary arrays for the right, left, top, and bottom boundaries, b_k^R, b_k^L, b_j^T, and b_j^B, as follows:

$$b_k^R = \frac{(u_{N+1,k}-u_{N,k})}{\delta x}; \qquad k=1,\cdots,M,$$

$$b_k^L = \frac{(u_{0,k}-u_{1,k})}{\delta x}; \qquad k=1,\cdots,M, \qquad (3.26)$$

$$b_j^T = \frac{(u_{j,M+1}-u_{j,M})}{\delta y}; \qquad j=1,\cdots,N,$$

$$b_j^B = \frac{(u_{j,0}-u_{j,1})}{\delta y}; \qquad j=1,\cdots,N.$$

The finite difference matrix consistent with the boundary conditions is obtained by just replacing one of the four terms on the left of Eq. (3.25) with the appropriate boundary value from Eq. (3.26) when that term lies on the boundary. In this form, we can sum both sides of Eq. (3.25), using Eq. (3.26) and all the internal cancellations that come from summing a difference equation which uses conservative differencing. This yields an exact finite difference form of Gauss's theorem that must be satisfied by the boundary conditions and the inhomogeneous term:

$$\sum_{k=1}^{N} \delta y \left[f_{N+\frac{1}{2},k} b_k^R + f_{\frac{1}{2},k} b_k^L \right] + \sum_{j=1}^{M} \delta x \left[g_{j,M+\frac{1}{2}} b_j^T + g_{j,\frac{1}{2}} b_j^B \right]$$
$$= \sum_{k=1}^{N} \sum_{j=1}^{M} \delta x \delta y R_{i,j}. \tag{3.27}$$

Equation (3.27) is the discrete form of the solvability constraint given by Eq. (3.17).

3.3.2 Gauss Elimination

Let us consider the matrix equation, Eq. (3.16), where $\mathbf{A}$ is an $M \times M$ square matrix, and $\mathbf{x}$ and $\mathbf{b}$ are vectors of length M. Equation (3.15) is of this form if we let $M = (N+1)^2$. In this section we proceed as if $\mathbf{A}$ is a full matrix and consider a general algorithm which is a specific implementation of Gauss elimination for solving for the unknown vector $\mathbf{x}$. This method will also work for a matrix having the sparseness pattern of Eq. (3.15), but will not be nearly as efficient as other methods which take advantage of this sparseness.

We illustrate the method when A is a 3×3 matrix, and will generalize from there. Note that we assume here that the diagonal elements of the matrices are not zero, as is normally the case for matrices arising from applying the finite difference method to elliptic equations. For the more general case of Gauss elimination where no such assumptions are made and *pivoting* is required, the reader is referred to one of numerous books on the subject [31].

Consider then the matrix equation

$$\begin{bmatrix} a_{11} & a_{12} & a_{13} \\ a_{21} & a_{22} & a_{23} \\ a_{31} & a_{32} & a_{33} \end{bmatrix} \cdot \begin{bmatrix} x_1 \\ x_2 \\ x_3 \end{bmatrix} = \begin{bmatrix} b_1 \\ b_2 \\ b_3 \end{bmatrix}. \tag{3.28}$$

The first step is to multiply row 1 by $-m_{21} \equiv -a_{21}/a_{11}$ and to add this to row 2:

$$\begin{bmatrix} a_{11} & a_{12} & a_{13} \\ 0 & a_{22}' & a_{23}' \\ a_{31} & a_{32} & a_{33} \end{bmatrix} \cdot \begin{bmatrix} x_1 \\ x_2 \\ x_3 \end{bmatrix} = \begin{bmatrix} b_1 \\ b_2' \\ b_3 \end{bmatrix},$$

where

$$a_{22}' = a_{22} - m_{21}a_{12},$$
$$a_{23}' = a_{23} - m_{21}a_{13},$$
$$b_2' = b_2 - m_{21}b_1.$$

Next, multiply row 1 by $-m_{31} \equiv -a_{31}/a_{11}$ and add to row 3 to obtain

$$
\begin{bmatrix} a_{11} & a_{12} & a_{13} \\ 0 & a'_{22} & a'_{23} \\ 0 & a'_{32} & a'_{33} \end{bmatrix} \cdot \begin{bmatrix} x_1 \\ x_2 \\ x_3 \end{bmatrix} = \begin{bmatrix} b_1 \\ b'_2 \\ b'_3 \end{bmatrix},
$$

where

$$
\begin{aligned}
a'_{32} &= a_{32} - m_{31}a_{12}, \\
a'_{33} &= a_{33} - m_{31}a_{13}, \\
b'_3 &= b_3 - m_{31}b_1.
\end{aligned}
$$

Finally, multiply row 2 by $-m_{32} = -a'_{32}/a'_{22}$ and add to row 3 giving

$$
\begin{bmatrix} a_{11} & a_{12} & a_{13} \\ 0 & a'_{22} & a'_{23} \\ 0 & 0 & a''_{33} \end{bmatrix} \cdot \begin{bmatrix} x_1 \\ x_2 \\ x_3 \end{bmatrix} = \begin{bmatrix} b_1 \\ b'_2 \\ b''_3 \end{bmatrix}, \tag{3.29}
$$

where

$$
\begin{aligned}
a''_{33} &= a'_{33} - m_{32}a'_{23}, \\
b''_3 &= b'_3 - m_{32}b'_2.
\end{aligned}
$$

This final matrix in Eq. (3.29) we call $\mathbf{U}$. It is of *upper triangular* type in that all the non-zero entries are on or above the diagonal.

Note that by saving the elements m_{ij} we could form the *lower triangular matrix* $\mathbf{L}$, by adding 1s on the diagonal:

$$
\mathbf{L} = \begin{bmatrix} 1 & 0 & 0 \\ m_{21} & 1 & 0 \\ m_{31} & m_{32} & 1 \end{bmatrix}. \tag{3.30}
$$

The matrices $\mathbf{L}$ and $\mathbf{U}$ have the property that they are factors of the original matrix $\mathbf{A}$:

$$
\mathbf{L} \cdot \mathbf{U} = \mathbf{A}.
$$

This is called the *LU decomposition* of the matrix $\mathbf{A}$.

Once the matrix, Eq. (3.28), has been transformed to upper triangular form, Eq. (3.29), we can perform the *back substitution* step to solve for the unknowns, i.e.,

$$
\begin{aligned}
x_3 &= \frac{b''_3}{a''_{33}}, \\
x_2 &= \frac{b'_2 - a'_{23}x_3}{a'_{22}}, \\
x_1 &= \frac{b_1 - a_{12}x_2 - a_{13}x_3}{a_{11}}.
\end{aligned}
$$

We can easily generalize this example from a 3×3 to an $M \times M$ matrix where we recall that $M = (N + 1)^2$ for the matrix Eq. (3.16). The number of multiplications needed to reduce the system to *triangular form* is then, for M large

$$\frac{1}{3}M^3 \cong \frac{1}{3}N^6, \tag{3.31}$$

whereas the number of multiplications needed to solve the final back-substitution step is, for M large,

$$\frac{1}{2}M^2 \cong \frac{1}{2}N^4. \tag{3.32}$$

It is illustrative to evaluate this in terms of how long it would take on a modern computer. Commodity computers can perform approximately 10^8 arithmetic operations per second. For $N = 100$, which is a typical value, this would require approximately 3300 sec, or 1 hour of dedicated computer time to reduce to upper triangular form, although the back-substitution step would require only about 0.5 sec. This is far too much computational effort to spend on such a simple equation, especially since we will see that there are much more efficient methods for solving the same equation.

We note here that if one were solving many equations of the form of Eq. (3.16) that involve the same matrix $\mathbf{A}$, but have differing right-hand-side vectors $\mathbf{b}$, it would normally be worthwhile to form the LU decomposition of the matrix $\mathbf{A}$. Once this is formed, we could solve the system for any right-hand-side vector $\mathbf{b}$ by performing only two back-substitution steps, each requiring $\sim \frac{1}{2} M^2$ operations. To solve the system

$$\mathbf{L} \cdot \mathbf{U} \cdot \mathbf{x} = \mathbf{b},$$

we could first solve

$$\mathbf{L} \cdot \mathbf{y} = \mathbf{b}$$

for the vector $\mathbf{y}$ by back substitution, starting from the top, and would then solve

$$\mathbf{U} \cdot \mathbf{x} = \mathbf{y}$$

for the vector $\mathbf{x}$ by normal back substitution, starting from the bottom.

We remark here that the estimates made in this section, Eqs. (3.31) and (3.32), assumed that the matrix $\mathbf{A}$ was a full matrix. In reality, the matrix in Eq. (3.16) that comes from a 2D elliptic equation finite differenced on a structured rectangular grid will have a band structure as illustrated after Eq. (3.16). If the band width is of order N, then this will reduce the operation count, Eq. (3.31), from $\sim N^6$ to $\sim N^4$.

3.3.3 Block-Tridiagonal Method

There is a simple direct method of solving the finite difference equation, Eq. (3.15), that is a generalization of the tridiagonal technique discussed in Section 3.2.2. For each k value, we define a vector of length $N+1$ that contains the unknowns at all the j values with that k, i.e., we define the length $N+1$ vector for each k as follows:

$$\mathbf{U}_k = \begin{bmatrix} u_{0,k} \\ u_{1,k} \\ \vdots \\ u_{N-1,k} \\ u_{N,k} \end{bmatrix}.$$

The finite difference equation (3.15) can then be written in the block form

$$\mathbf{A}_k \cdot \mathbf{U}_{k+1} - \mathbf{B}_k \cdot \mathbf{U}_k + \mathbf{C}_k \cdot \mathbf{U}_{k-1} = \mathbf{D}_k.$$

Here, $\mathbf{B}_k$ is a tridiagonal $(N+1) \times (N+1)$ matrix and $\mathbf{A}_k$ and $\mathbf{C}_k$ are both diagonal matrices. If we assume that all boundary conditions are of the Dirichlet type, then these matrices are of the form:

$$\mathbf{B}_k = \begin{bmatrix} -1 & & & & & \\ -1 & 4 & -1 & & & \\ & -1 & 4 & -1 & & \\ & & \cdots & & & \\ & & & \cdots & & \\ & & & -1 & 4 & -1 \\ & & & & -1 & \end{bmatrix} \quad N+1 \tag{3.33}$$

$$\mathbf{A}_k = \mathbf{C}_k = \begin{bmatrix} 0 & & & & \\ & 1 & & & \\ & & 1 & & \\ & & & \cdots & \\ & & & & \cdots \\ & & & 1 & \\ & & & & 0 \end{bmatrix} \quad N+1 \qquad k = 1, \cdots, N-1.$$

For $k = 0$ and $k = N$, $\mathbf{B}_0 = \mathbf{B}_N = -\mathbf{I}$ (the identity matrix) and $\mathbf{C}_0 = \mathbf{A}_0 = \mathbf{C}_N = \mathbf{A}_N = 0$. The matrix $\mathbf{D}$ is also suitably modified to include the boundary values.

In a manner closely analogous to that discussed in Section 3.2.2, we seek to find the rank $(N+1)$ square matrices $\mathbf{E}_k$ and the length $(N+1)$ vectors $\mathbf{F}_k$ such that

$$\mathbf{U}_k = \mathbf{E}_k \cdot \mathbf{U}_{k+1} + \mathbf{F}_k. \tag{3.34}$$

In a matrix generalization of the procedure that led to the scalar equations in Eq. (3.9), we find that $\mathbf{E}_k$ and $\mathbf{F}_k$ obey the recurrence relations

$$
\begin{aligned}
\mathbf{E}_k &= [\mathbf{B}_k - \mathbf{C}_k \cdot \mathbf{E}_{k-1}]^{-1} \cdot \mathbf{A}_k, \\
\mathbf{F}_k &= [\mathbf{B}_k - \mathbf{C}_k \cdot \mathbf{E}_{k-1}]^{-1} \cdot (\mathbf{C}_k \cdot \mathbf{F}_{k-1} - \mathbf{D}_k).
\end{aligned}
\tag{3.35}
$$

To initialize, we note that $\mathbf{E}_0$ and $\mathbf{F}_0$ are well defined since $\mathbf{C}_0 = 0$, and the analogue of Eq. (3.11) is:

$$
\mathbf{U}_{N-1} = \left[\mathbf{I} - \mathbf{E}_{N-1}\mathbf{B}_N^{-1}\mathbf{C}_N\right]^{-1} \cdot \left[\mathbf{F}_{N-1} - \mathbf{E}_{N-1}\mathbf{B}_N^{-1}\mathbf{D}_N\right].
\tag{3.36}
$$

The majority of the work for this method involves inverting the matrices in brackets in Eq. (3.35) for each of the $(N+1)\,k$ values. This requires approximately N inversions with $\sim 1/3\,N^3$ operations each, for a total of $\sim 1/3\,N^4$ operations to solve the complete system, a substantial improvement over Gauss elimination on a full matrix, but the same scaling as for Gauss elimination taking into account the banded structure.

3.3.4 General Direct Solvers for Sparse Matrices

Recently there has been considerable progress in developing efficient general direct solvers for sparse matrices. The general algorithms have complex logic as required for efficient handling of the *fill-in* of the factors L and U. A typical solver consists of four distinct steps: (i) an ordering step that reorders the rows and columns such that the factors suffer little fill, or such that the matrix has special structure such as block triangular form; (ii) an analysis step or symbolic factorization that determines the non-zero structures of the factors and creates suitable data structures for the factors; (iii) numerical factorization that computes the L and U factors; and (iv) a solve step that performs forward and back substitution using the factors.

There are many different algorithmic possibilities for each step. Excellent review papers are available [33, 34, 35]. Modern "black box" packages such as SUPER_LU [36] or MUMPS [37] that exploit new architectural features, such as memory hierarchy and parallelism, are available and many are able to obtain outstanding speeds, close to those theoretically possible on given platforms. Physicists or other application scientists who require such a solver are advised to seek out the documentation and try one of these remarkable packages.

3.4 Matrix Iterative Approach

In the next several sections we discuss iterative methods for solving the difference equations that arise from the approximation of elliptic equations.

Iterative methods have an advantage over direct methods in that they are quite general, and don't necessarily require a particular band structure or that the coefficient matrices be independent of one coordinate as do the fast direct methods that utilize transform techniques. Also they benefit from a good initial guess to the solution so that, for example, if an elliptic equation has to be solved at each time step as part of a time advancement solution, the values at the previous time step can be used effectively to initialize the iteration. We consider several different formalisms for discussing and analyzing these methods. In this section we discuss the matrix iterative approach, which uses matrix techniques. In the next section we present the physical approach, which exploits an analogy between iterative techniques for solving elliptic equations and solving a corresponding time-dependent equation to the steady state. In the following sections we discuss the multigrid and Krylov space methods.

We have seen that the finite difference equations that approximate an elliptic equation can be written in matrix form

$$\mathbf{A} \cdot \mathbf{x} = \mathbf{b}, \tag{3.37}$$

where $\mathbf{b}$ is a known vector which arises from the inhomogeneous term, $\mathbf{x}$ is the solution vector which is to be determined, and $\mathbf{A}$ is a sparse, diagonally dominant matrix

$$|a_{ii}| \geq \sum_{j \neq i} |a_{ij}|. \tag{3.38}$$

We want first to explore iterative methods of the form [38, 39]

$$\mathbf{x}^{(n+1)} = \mathbf{P} \cdot \mathbf{x}^{(n)} + \mathbf{c}, \tag{3.39}$$

where the iteration is initialized with an initial guess $\mathbf{x}^{(0)}$. Now, Eq. (3.39) must reduce to Eq. (3.37) in the steady state when $\mathbf{x}^{(n+1)} = \mathbf{x}^{(n)}$. Let us suppose that the vectors $\mathbf{c}$ and $\mathbf{b}$ are related by the linear relation

$$\mathbf{c} = \mathbf{T} \cdot \mathbf{b}. \tag{3.40}$$

In the steady state, from Eqs. (3.39) and (3.40), we have

$$\mathbf{x} = \mathbf{P} \cdot \mathbf{x} + \mathbf{T} \cdot \mathbf{b},$$

or

$$\mathbf{T}^{-1} \cdot (\mathbf{I} - \mathbf{P}) \cdot \mathbf{x} = \mathbf{b}.$$

By comparing with Eq. (3.37), we find

$$\mathbf{A} = \mathbf{T}^{-1} \cdot (\mathbf{I} - \mathbf{P}),$$

or

$$\mathbf{P} = \mathbf{I} - \mathbf{T} \cdot \mathbf{A}. \tag{3.41}$$

Choosing an iterative method for solving Eq. (3.37) is therefore equivalent to choosing a matrix $\mathbf{T}$ or a matrix $\mathbf{P}$, which are related by Eq. (3.41). We note that if $\mathbf{T} = \mathbf{A}^{-1}$, the iteration will converge in a single iteration. This suggests that we should make $\mathbf{T}$ "close to" $\mathbf{A}^{-1}$ for rapid convergence.

3.4.1 Convergence

Let $\epsilon^{(n)} = \mathbf{x}^{(n)} - \mathbf{x}$ be the error in the solution at iteration n. Subtracting the exact relation

$$\mathbf{x} = \mathbf{P} \cdot \mathbf{x} + \mathbf{c}$$

from the iterative method in Eq. (3.39) yields an equation for the error:

$$\epsilon^{(n+1)} = \mathbf{P} \cdot \epsilon^{(n)}. \tag{3.42}$$

For convergence, we demand

$$\left|\epsilon^{(n+1)}\right| < \left|\epsilon^{(n)}\right|.$$

We analyze further in terms of the eigenvalues and eigenvectors of $\mathbf{P}$. Say that $\mathbf{P}$ has M distinct eigenvalues λ_i. We can then expand the initial error in terms of the eigenvectors of $\mathbf{P}, \mathbf{S}^{(i)}$

$$\epsilon^{(0)} = \sum_{i=1}^{M} \alpha_i \mathbf{S}^{(i)}.$$

Now, from Eq. (3.42)

$$\begin{aligned} \epsilon^{(n)} &= \mathbf{P} \cdot \epsilon^{(n-1)} = \mathbf{P}^n \cdot \epsilon^{(0)} \\ &= \sum_{i=1}^{M} \alpha_i \lambda_i^n \mathbf{S}^{(i)}. \end{aligned} \tag{3.43}$$

This shows clearly that for convergence we need all the eigenvalues of $\mathbf{P}$ to be less than unity. For rapid convergence, we want the maximum of the absolute values of the eigenvalues, or the *spectral radius*, to be as small as possible.

Let us now consider some specific methods. It is conventional to decompose a given matrix $\mathbf{A}$ into submatrices. For the matrix $\mathbf{A}$ in Eq. (3.37), let

$$\mathbf{A} = \mathbf{D} \cdot (\mathbf{I} + \mathbf{L} + \mathbf{U}), \tag{3.44}$$

where $\mathbf{D}$ is a diagonal matrix, $\mathbf{I}$ is the identity matrix, $\mathbf{L}$ is a lower triangular matrix with elements only below the diagonal, and $\mathbf{U}$ is an upper triangular matrix with elements only above the main diagonal. In terms of these submatrices, we discuss three common iterative methods.

3.4.2 Jacobi's Method

Jacobi's method, in matrix form, corresponds to the following:

$$\mathbf{x}^{(n+1)} = -(\mathbf{L} + \mathbf{U}) \cdot \mathbf{x}^{(n)} + \mathbf{D}^{-1} \cdot \mathbf{b}. \tag{3.45}$$

This is of the form of Eq. (3.39) if we identify

$$\mathbf{P} = \mathbf{I} - \mathbf{D}^{-1} \cdot \mathbf{A} = -(\mathbf{L} + \mathbf{U}), \qquad \mathbf{T} = \mathbf{D}^{-1}.$$

Applied to the finite difference equation in Eq. (3.15), Jacobi's method, Eq. (3.45), becomes:

$$u_{j,k}^{n+1} = \frac{1}{4} \left(u_{j+1,k}^{n} + u_{j-1,k}^{n} + u_{j,k+1}^{n} + u_{j,k-1}^{n} - h^2 R_{j,k} \right). \tag{3.46}$$

Jacobi's method, Eq. (3.45), is a particular example of a more general class of iterative methods called *relaxation*. We return to this and gain additional insight in Section 3.5.1 The method is very easy to program and requires few computations per iteration, but we will see that it requires many iterations to converge and it is thus not competitive with faster methods, at least for production applications.

However, a slight generalization of the Jacobi method is often used as a smoother in the multigrid method (Section 3.5). The *weighted Jacobi method* is defined by

$$\mathbf{x}^{(n+1)} = \mathbf{x}^{n} + \omega \left[-(\mathbf{I} + \mathbf{L} + \mathbf{U}) \cdot \mathbf{x}^{(n)} + \mathbf{D}^{-1} \cdot \mathbf{b} \right]. \tag{3.47}$$

This reduces to the Jacobi method for $\omega = 1$. (It is also sometimes referred to as Richardson relaxation.) We will see in Section 3.5.1 that it is unstable for $\omega > 1$ and generally will converge more slowly than the Jacobi method for $\omega < 1$. However, a value of $\omega = 2/3$ is especially effective in damping grid-scale oscillations and is often used in conjunction with the multigrid method as discussed in Section 3.6.

3.4.3 Gauss–Seidel Method

The *Gauss–Seidel* method is defined by:

$$\mathbf{x}^{(n+1)} = -\mathbf{L} \cdot \mathbf{x}^{(n+1)} - \mathbf{U} \cdot \mathbf{x}^{(n)} + \mathbf{D}^{-1} \cdot \mathbf{b}. \tag{3.48}$$

This has $\mathbf{P} = -(\mathbf{I} + \mathbf{L})^{-1} \cdot \mathbf{U}$, $\mathbf{T} = (\mathbf{I} + \mathbf{L})^{-1} \cdot \mathbf{D}^{-1}$. Applied to the finite difference equation in Eq. (3.15), the Gauss–Seidel method, Eq. (3.48), gives:

$$u_{j,k}^{n+1} = \frac{1}{4} \left(u_{j+1,k}^{n} + u_{j-1,k}^{n+1} + u_{j,k+1}^{n} + u_{j,k-1}^{n+1} - h^2 R_{j,k} \right).$$

The Gauss–Seidel method is very similar to Jacobi's method, but uses the most recent values of the variable being iterated. It thus requires less storage and converges faster than Jacobi's method, normally about twice as fast. Note that it implies a certain order of computation so that non-zero elements of **L** only multiply components of **x** that have already been updated.

This implied ordering of computation presents a difficulty when implementing this algorithm on parallel computers where different parts of the grid would normally be assigned to different processors. For parallelism to be effective, each part of the grid, residing on different processors, would need be updated simultaneously. A variant of the Gauss–Seidel method called *red-black Gauss–Seidel* addresses this by dividing all the points into two sets: a *red* set with $j + k$ even and a *black* set with $j + k$ odd. First, the entire red set of points can all be updated in any order, and then the black set can likewise be updated in any order.

3.4.4 Successive Over-Relaxation Method (SOR)

The *Successive Over-Relaxation* (SOR) method is given by:

$$\mathbf{x}^{(n+1)} = \mathbf{x}^{(n)} + \omega \left[-\mathbf{L} \cdot \mathbf{x}^{(n+1)} - (\mathbf{U} + \mathbf{I}) \cdot \mathbf{x}^{(n)} + \mathbf{D}^{-1} \cdot \mathbf{b} \right]. \qquad (3.49)$$

In this case we have

$$\mathbf{P}(\omega) = (\mathbf{I} + \omega \mathbf{L})^{-1} \cdot \left[(1 - \omega)\mathbf{I} - \omega \mathbf{U} \right], \; \mathbf{T} = \omega(\mathbf{I} + \omega \mathbf{L})^{-1} \cdot \mathbf{D}^{-1}. \qquad (3.50)$$

It can be shown that this method will not converge for the *over-relaxation factor*, ω, outside the range $0 < \omega < 2$.

Applied to the finite difference equation in Eq. (3.15), the SOR method, Eq. (3.49), gives:

$$u_{j,k}^{n+1} = u_{j,k}^n + \frac{\omega}{4} \left(u_{j+1,k}^n + u_{j-1,k}^{n+1} + u_{j,k+1}^n + u_{j,k-1}^{n+1} - 4u_{j,k}^n - h^2 R_{j,k} \right).$$

The SOR method appears to be very similar to the Gauss–Seidel method, but in fact it is much more powerful. For a given sparse matrix, there is an *optimal* value of ω, normally between 1 and 2, for which the SOR method can converge much faster than the Gauss–Seidel or Jacobi methods, making it competitive with more modern methods in certain situations. This will be explored more in Problem 3.6.

3.4.5 Convergence Rate of Jacobi's Method

It is generally not possible to compute the eigenvalues of the matrix **P** in Eq. (3.43) and thus obtain the convergence rate. However, for several of the most simple geometries, operators, and methods, this is possible. We illustrate a particular technique here.

Suppose we are solving the finite difference form of Laplace's equation in a square domain of size $L \times L$. Define a uniform mesh size $h = L/N$ in each direction so that the x and y mesh locations are at $x_j = jh$ and $y_k = kh$. Homogeneous boundary conditions are applied at $j, k = 0$ and N. Let the solution vector at iteration level n be given by: $u^n_{j,k}$; $j, k = 0, 1, \cdots, N$. We define, for integer l and m values, the 2D discrete sine transform [40] vector

$$\tilde{u}^n_{l,m} = \sum_{j=1}^{N-1} \sum_{k=1}^{N-1} u^n_{j,k} \sin \frac{\pi jl}{N} \sin \frac{\pi km}{N}; \qquad l, m = 1, \cdots, N-1. \qquad (3.51)$$

The corresponding exact inverse transform is given by

$$u^n_{j,k} = \frac{4}{N^2} \sum_{l=1}^{N-1} \sum_{m=1}^{N-1} \tilde{u}^n_{l,m} \sin \frac{\pi jl}{N} \sin \frac{\pi km}{N}; \qquad j, k = 1, \cdots, N-1. \qquad (3.52)$$

This representation has the property that $u^n_{j,k}$ can be extended beyond where it is defined and is odd about $j, k = 0$ and N. It is thus compatible with the boundary conditions continuing to be satisfied during the iteration, Eq. (3.46), even when the interior equations are applied at the boundary.

From Eqs. (3.43) and (3.46), we see that the eigenvalues for the iteration matrix $\mathbf{P}$ for Jacobi's method satisfies the equation:

$$\frac{1}{4} \left(u^n_{j+1,k} + u^n_{j-1,k} + u^n_{j,k+1} + u^n_{j,k-1} \right) = \lambda u^n_{j,k}. \qquad (3.53)$$

Substituting from Eq. (3.52) into Eq. (3.53) and using the angle addition formula, we see that each l, m harmonic decouples from the others, and thus the eigenvalues are given by:

$$\lambda_{l,m} = \frac{1}{2} \left[\cos \frac{\pi l}{N} + \cos \frac{\pi m}{N} \right]; \qquad l, m = 1, \cdots, N-1. \qquad (3.54)$$

Since the eigenvalues are all less than 1, the method will converge. However, the largest eigenvalue, corresponding to $l, m = 1, 1$ is very close to 1, given by

$$\lambda_{1,1} = \cos \frac{\pi}{N} \sim 1 - \frac{\pi^2}{2N^2}.$$

This implies that the convergence will be very slow, particularly for N large. The number of iterations required for the error to decay by $1/e$ is

$$N_{it} = \frac{-1}{\ln \lambda_{1,1}} \sim \frac{2N^2}{\pi^2}. \qquad (3.55)$$

We return to this in Section 3.5.1.

3.5   Physical Approach to Deriving Iterative Methods

There is a much more physical approach to developing iterative methods that involves adding time derivative terms to the original elliptic equation and advancing forward in time until a steady state is obtained. We continue to discuss methods of solving Eq. (3.12), which we write here as

$$\frac{\partial^2 u}{\partial x^2} + \frac{\partial^2 u}{\partial y^2} - R(x, y) = 0. \tag{3.56}$$

Assume that Dirichlet boundary conditions are applied on a square $L \times L$ domain. The corresponding second-order centered finite difference equation, assuming $\delta x = \delta y = h = L/N$ is given by Eq. (3.15), or:

$$u_{j+1,k} + u_{j-1,k} + u_{j,k+1} + u_{j,k-1} - 4u_{j,k} - h^2 R_{j,k} = 0. \tag{3.57}$$

In the next two sections we will examine the iterative methods that correspond to adding a first-order and a second-order time derivative term to Eq. (3.56).

3.5.1 First-Order Methods

As a first attempt, let us add a single time derivative term to convert Eq. (3.56) to a parabolic equation, i.e.,

$$\frac{\partial u}{\partial t} = \frac{\partial^2 u}{\partial x^2} + \frac{\partial^2 u}{\partial y^2} - R(x, y). \tag{3.58}$$

We finite difference using forward time and centered space differences to obtain the discrete equation

$$u_{j,k}^{n+1} = u_{j,k}^n + s \left[u_{j+1,k}^n + u_{j-1,k}^n + u_{j,k+1}^n + u_{j,k-1}^n - 4u_{j,k}^n - h^2 R_{j,k} \right]. \tag{3.59}$$

Here, the superscript n is now the time index and we have introduced the parameter $s = \delta t / h^2$. If the iteration defined in Eq. (3.59) converges so that $u_{j,k}^{n+1} = u_{j,k}^n$ everywhere, then the difference equation in Eq. (3.57) will be satisfied.

Two questions must be answered concerning the iteration so defined: (i) How large of a value of s can be used and still obtain a stable iteration and (ii) How many iterations are necessary to converge? To answer (i), we examine the stability of the difference scheme using the 2D von Neumann stability analysis method introduced in Section 2.5.3. Making the substitution

$$u_{j,k}^n = \tilde{u} r^n e^{i(j\theta_l + k\theta_m)} \tag{3.60}$$

in Eq. (3.59) and applying the criterion that $|r| \leq 1$ for all values of θ_l and θ_m we find that for stability

$$|r| = |1 - 8s| \leq 1,$$

or

$$s \leq \frac{1}{4}. \tag{3.61}$$

This implies $\delta t \leq \frac{1}{4} h^2$. Using the equality, the iteration scheme becomes

$$u_{j,k}^{n+1} = \frac{1}{4} \left[u_{j+1,k}^n + u_{j-1,k}^n + u_{j,k+1}^n + u_{j,k-1}^n - h^2 R_{i,j} \right]. \tag{3.62}$$

The iterative method defined by Eq. (3.62) is just Jacobi relaxation, identical to Eq. (3.46). The more general method defined by Eq. (3.59) for s arbitrary is the modified Jacobi or Richardson relaxation. Jacobi relaxation is seen to be a special case of Richardson relaxation optimized by choosing the relaxation parameter s to be the largest value allowed by stability.

To estimate how many iterations are necessary for the method to converge, we look at transient solutions of the corresponding differential equation (3.58). To solve, let

$$u = u^{ss}(x, y) + \tilde{u}(x, y, t),$$

where $u^{ss}(x, y)$ is the steady-state solution which satisfies the boundary condition. Then $\tilde{u}(x, y, t)$ is zero on the boundary, and satisfies the homogeneous equation

$$\frac{\partial \tilde{u}}{\partial t} = \frac{\partial^2 \tilde{u}}{\partial x^2} + \frac{\partial^2 \tilde{u}}{\partial y^2}. \tag{3.63}$$

We look for separable solutions of the form

$$\tilde{u} = e^{-\lambda_{mn} t} \sin \left(\frac{m\pi x}{L} \right) \sin \left(\frac{n\pi y}{L} \right), \tag{3.64}$$

with corresponding decay rates

$$\lambda_{m,n} = \left(n^2 + m^2 \right) \frac{\pi^2}{L^2}. \tag{3.65}$$

The slowest decaying mode has $m = n = 1$. It will decay in a time

$$t = 1/\lambda_{11} = L^2/2\pi^2.$$

The number of iterations required for the error to damp one e-folding time is therefore

$$\frac{t}{\delta t} = \frac{L^2}{2\pi^2} \cdot \frac{4}{h^2} \simeq \frac{2}{\pi^2} N^2. \tag{3.66}$$

This is the same result as obtained in Eq. (3.55) from analysis of the eigenvalues of the matrix $\mathbf{P}$. Since each iteration requires $\sim N^2$ multiplications, the total number of multiplications required to obtain a solution with the error reduced by e^{-1} will scale like N^4.

However, we see from Eq. (3.65) that the short wavelength modes (with $n \sim m \sim N$) decay much faster that these long wavelength modes that limit the overall convergence rate. This is one of the key observations underlying the multigrid method that is discussed in Section 3.6.

3.5.2 Accelerated Approach: Dynamic Relaxation

In order to accelerate convergence, we consider the effect of adding a higher time derivative term to Eq. (3.56), converting it now to a hyperbolic equation:

$$\frac{\partial^2 u}{\partial t^2} + \frac{2}{\tau}\frac{\partial u}{\partial t} = \frac{\partial^2 u}{\partial x^2} + \frac{\partial^2 u}{\partial y^2} - R(x,y). \tag{3.67}$$

We have also included a first-order time derivative term in Eq. (3.67) with a multiplier of $2/\tau$ to provide the possibility of damping. The multiplier will be chosen to optimize the convergence rate. In finite difference form, now using second-order centered differences in both space and time, Eq. (3.67) becomes

$$\frac{u_{j,k}^{n+1} - 2u_{j,k}^{n} + u_{j,k}^{n-1}}{\delta t^2} + \frac{u_{j,k}^{n+1} - u_{j,k}^{n-1}}{\tau \delta t}$$
$$- \left[\frac{u_{j-1,k}^{n} + u_{j+1,k}^{n} + u_{j,k+1}^{n} + u_{j,k-1}^{n} - 4u_{j,k}^{n}}{h^2}\right] + R_{j,k} = 0. \tag{3.68}$$

Applying the von Neumann stability analysis, we find that for stability we require the condition $\delta t \leq h/\sqrt{2}$. Again choosing the equality, the iteration scheme becomes

$$u_{j,k}^{n+1} = -\frac{H}{D}u_{j,k}^{n-1} + \frac{1}{2D}\left[u_{j-1,k}^{n} + u_{j+1,k}^{n} + u_{j,k+1}^{n} + u_{j,k-1}^{n}\right] - \frac{h^2}{2D}R_{j,k},$$

with $H = 1 - h/\sqrt{2}\tau, D = 1 + h/\sqrt{2}\tau$. This is known as *dynamic relaxation* with the parameter τ to be chosen to optimize convergence as described below. To estimate convergence, we again look at the transient decay time. From the homogeneous equation

$$\frac{\partial^2 \tilde{u}}{\partial t^2} + \frac{2}{\tau}\frac{\partial \tilde{u}}{\partial t} = \frac{\partial^2 \tilde{u}}{\partial x^2} + \frac{\partial^2 \tilde{u}}{\partial y^2}$$

we obtain the eigenmode solution

$$\tilde{u} = e^{-\lambda t}\sin\left(\frac{m\pi x}{L}\right)\sin\left(\frac{n\pi y}{L}\right),$$

where λ satisfies

$$\lambda^2 - \frac{2}{\tau}\lambda + \lambda_{mn} = 0,$$

with λ_{mn} being defined in Eq. (3.65). The two roots of this quadratic equation are:

$$\lambda = \frac{1}{\tau} \mp \left[\frac{1}{\tau^2} - \lambda_{mn}\right]^{\frac{1}{2}}.$$

For optimal damping, we choose $\tau^2 = 1/\lambda_{11}$, which gives critical damping of the slowest decaying mode. We thus choose

$$\tau = \frac{L}{\sqrt{2}\pi},$$

and the slowest decaying mode will decay by a factor of e in

$$\frac{\tau}{\delta t} \sim \frac{L}{\pi h} \sim \frac{1}{\pi} N \quad \text{iterations.} \tag{3.69}$$

Since each iteration requires N^2 operations, the total effort to reach a solution will scale like N^3. Comparing with the method of Section 3.5.1 (Jacobi iteration), we see that the accelerated method is faster by a factor of N.

We can now better understand why the Jacobi method converges so slowly, and why the accelerated method offers a dramatic improvement. The Jacobi method corresponds to solving a diffusion equation to a steady state. Diffusion is inherently a slow process, and we will see in Chapter 7 that it requires N^2 time steps for a disturbance to propagate across a grid in solving a diffusion problem with an explicit finite difference method. In contrast, we will see in Chapter 9 that a disturbance can propagate across a grid in N time steps when solving a hyperbolic (or wave propagation) equation. By adding some small damping to the wave equation, we arrive at a system in which the different parts of the mesh communicate with each other much more efficiently and this leads to considerably more rapid decay of the error component of the solution.

3.6 Multigrid Methods

Standard iterative methods for solving sparse matrix equations arising from the finite difference approximation to elliptic partial differential equations such as described in Sections 3.4 and 3.5 are very general and relatively easy to program, and thus have many advantages, especially for exploratory programming and in rapid prototyping. However, they are generally not suitable for production applications since when compared to the most efficient methods, they are not nearly as efficient. *Multigrid methods* [41, 42] are a technique for greatly increasing the speed and efficiency of standard iterative methods, making them competitive with the fastest methods available. There are *three key ideas* on which multigrid methods are based.

The *first* observation is that standard iterative methods benefit greatly from a good initial guess. We exploit this by first solving the equation on a coarse grid, then use the coarse grid solution to initialize a finer grid iteration as shown in Figure 3.2. This process is recursive, going from the coarsest to the finest grid allowable in the problem.

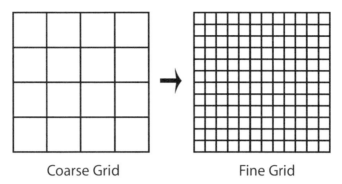

Coarse Grid   Fine Grid

FIGURE 3.2: Multigrid methods use a sequence of grids of different coarseness.

The *second* observation is that on a given mesh, or grid, the short wavelength components of the error decay much faster than the long wavelength components. This is discussed at the end of Section 3.5.1 in conjunction with Eq. (3.65). The coarse mesh is therefore much more efficient in handling the long wavelength components of the error, which it can accurately represent, while the fine mesh is needed for the short wavelength components.

The *third* observation is that the error satisfies the same matrix equation as the unknown. Recall that we are discussing solving Eq. (3.37),

$$\mathbf{A} \cdot \mathbf{x} = \mathbf{b}. \qquad (3.70)$$

Let $\mathbf{x}$ be the exact solution, and $\mathbf{v}$ be the approximate solution after some number of iterations. We define the residual corresponding to the approximate solution as

$$\mathbf{r} = \mathbf{b} - \mathbf{A} \cdot \mathbf{v}.$$

This is a measure of how well the approximate solution satisfies the original equation. Also, define the error as:

$$\mathbf{e} = \mathbf{x} - \mathbf{v}.$$

From the definition of the residual and the error, we can subtract

$$\mathbf{A} \cdot \mathbf{v} = \mathbf{b} - \mathbf{r} \qquad (3.71)$$

from Eq. (3.70) to obtain

$$\mathbf{A} \cdot \mathbf{e} = \mathbf{r}. \qquad (3.72)$$

Since the equation we are solving can be cast in terms of the error and the

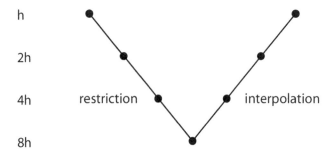

FIGURE 3.3: Basic coarse grid correction method successively restricts solution from finest (h) to coarsest (8h) grid, and then successively interpolates back to finest grid.

residual, we can restrict (fine to coarse) and interpolate (coarse to fine) approximations to the solution between a sequence of grids of different coarseness using appropriate restriction and interpolation operations.

The basic steps of the multigrid method are given here. There are two cycles that are periodically combined to deliver an optimal strategy. These are called *nested iteration* and *coarse grid correction*.

The nested iteration is just a strategy to obtain a good initial guess. We begin by relaxing Eq. (3.70), $\mathbf{A} \cdot \mathbf{x} = \mathbf{b}$, on a very coarse grid to get an approximate solution. We then interpolate this solution onto a finer grid as a good "initial guess" for that grid. Then, Eq. (3.70), $\mathbf{A} \cdot \mathbf{x} = \mathbf{b}$, is relaxed on that grid, and the solution is in turn used as a good initial guess for an even finer grid. This process is repeated until we initialize the unknown on the finest grid.

The second basic cycle is the coarse grid correction. It is illustrated in Figure 3.3. The steps in the coarse grid correction are as follows:

1. Start relaxing Eq. (3.70), $\mathbf{A} \cdot \mathbf{x} = \mathbf{b}$, on the fine grid (h).

2. After a few iterations, stop and compute the residual: $\mathbf{r} = \mathbf{b} - \mathbf{A} \cdot \mathbf{v}$, where $\mathbf{v}$ is the current approximation to the solution $\mathbf{x}$.

3. Map the residual $\mathbf{r}$ onto the coarse grid using a *restriction operator*.

4. Start relaxing the equation for the error, $\mathbf{A}^{2h} \cdot \mathbf{e} = \mathbf{r}$, on the coarse grid (2h) using the *coarse grid operator*.

5. After a few iterations, use the *interpolation operator* to interpolate the error $\mathbf{e}$ back to the fine grid.

6. Form the new approximation to the solution, $\mathbf{x} = \mathbf{e} + \mathbf{v}$.

These basic cycles can be used recursively between a sequence of grids of

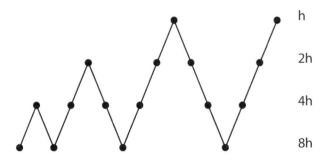

FIGURE 3.4: Full Multigrid V-cycle (FMV).

different coarseness to form a multigrid tree. Figure 3.3 is a basic block called a *V-cycle*. A more advanced cycle that is often used is called the *Full Multigrid V-cycle* (FMV) and is illustrated in Figure 3.4.

Important details in the theory of multigrid methods include the forms of the restriction operator, the interpolation operator, the coarse grid operator, and the relaxation method (or smoother) used. If the solution on the fine grid is denoted by $\mathbf{v}^h$ and that on the coarse grid by $\mathbf{v}^{2h}$, then the standard operator that interpolates from the coarse to the fine grid can be defined as

$$\mathbf{v}^h = I_{2h}^h \mathbf{v}^{2h}. \tag{3.73}$$

The components of $\mathbf{v}^h$ are given by, for $0 \leq j, k \leq N/2 - 1$,

$$
\begin{aligned}
v_{2j,2k}^h &= v_{j,k}^{2h} \\
v_{2j+1,2k}^h &= \frac{1}{2}\left(v_{j,k}^{2h} + v_{j+1,k}^{2h}\right) \\
v_{2j,2k+1}^h &= \frac{1}{2}\left(v_{j,k}^{2h} + v_{j,k+1}^{2h}\right) \\
v_{2j+1,2k+1}^h &= \frac{1}{4}\left(v_{j,k}^{2h} + v_{j+1,k}^{2h} + v_{j,k+1}^{2h} + v_{j+1,k+1}^{2h}\right)
\end{aligned}
$$

The corresponding restriction operator that transfers the solution from the fine grid to the coarse grid is defined as

$$\mathbf{v}^{2h} = I_h^{2h} \mathbf{v}^h. \tag{3.74}$$

Here, the components are given by

$$
\begin{aligned}
v_{j,k}^{2h} &= \frac{1}{16}\left[v_{2j-1,2k-1}^h + v_{2j-1,2k+1}^h + v_{2j+1,2k-1}^h + v_{2j+1,2k+1}^h\right. \\
&+ \left. 2\left(v_{2j,2k-1}^h + v_{2j,2k+1}^h + v_{2j-1,2k}^h + v_{2j+1,2k}^h\right) + 4v_{2j,2k}^h\right].
\end{aligned}
$$

These two operators have the property that they are the transposes of each other, up to a real constant:

$$I_{2h}^h = c\left(I_h^{2h}\right)^T.$$

With the definition of these two operators, the coarse grid operator $\mathbf{A}^{2h}$ can be defined in terms of the fine grid operator $\mathbf{A}^h$ as

$$\mathbf{A}^{2h} = I_h^{2h} \mathbf{A}^h I_{2h}^h. \tag{3.75}$$

With the operators defined this way, Eq. (3.75) gives exactly the same result as one would get if the original problem were discretized on the coarse grid.

There are a number of parameters in the multigrid method that are left undefined or are problem dependent. These include the number of iterations that should be performed on each grid level each cycle, which iteration method (or smoother) to use (Jacobi, some other form of Richardson relaxation, Gauss–Seidel, etc.) and whether to use the V-cycle or the FMV-cycle. As discussed in Section 3.4.2, the weighted Jacobi method with $\omega = 2/3$ gives rapid damping of grid-scale oscillations and is often used as the relaxation operator.

General theoretical considerations show that the majority of work is performed in iterating on the finest grid, where $\sim N^d$ operations are required for a d-dimensional problem with N^d unknowns. If only a few sweeps of the finest grid are needed, the number of V-cycles or FMV-cycles required should scale like $\log N$ for N large. The total amount of work required for a multigrid solution should therefore scale like $N^d \log N$, which is competitive with the best available solvers.

In a variant of multigrid called *algebraic multigrid* [43, 44], black box programs have been developed that work solely with the matrix $\mathbf{A}$ and right-hand-side vector $\mathbf{b}$ without any explicit knowledge of the underlying grid being used or differential equation being solved. These construct their hierarchy of operators directly from the system matrix, where the system is represented as a graph of n nodes where an edge (i, j) is represented by a non-zero coefficient A_{ij}.

3.7 Krylov Space Methods

There is a class of very effective modern iterative methods that use a combination of *preconditioning* and some definition of optimal iteration in a *Krylov Subspace* to obtain rapid convergence [34, 45]. Let us suppose that $\mathbf{A}$ is a matrix close to the identity matrix and that the equation we want to solve is

$$\mathbf{A} \cdot \mathbf{x} = \mathbf{b}. \tag{3.76}$$

We can use the identity matrix to rewrite this as

$$[\mathbf{I} - (\mathbf{I} - \mathbf{A})] \cdot \mathbf{x} = \mathbf{b}. \tag{3.77}$$

Consider a particular form of Richardson iteration defined by

$$\begin{aligned} \mathbf{x}^{(n+1)} &= \mathbf{b} + (\mathbf{I} - \mathbf{A}) \cdot \mathbf{x}^{(n)} \\ &= \mathbf{x}^{(n)} + (\mathbf{b} - \mathbf{A} \cdot \mathbf{x}^{(n)}). \end{aligned} \tag{3.78}$$

We can rewrite Eq. (3.78) as

$$\mathbf{x}^{(n+1)} = \mathbf{x}^{(n)} + \mathbf{r}^{(n)}, \tag{3.79}$$

where $\mathbf{r}^{(n)} \equiv (\mathbf{b} - \mathbf{A} \cdot \mathbf{x}^{(n)})$ is the *residual* after n iterations. Multiply Eq. (3.79) by $-\mathbf{A}$ and add $\mathbf{b}$ to obtain the difference equation the residual obeys

$$\mathbf{b} - \mathbf{A} \cdot \mathbf{x}^{(n+1)} = \mathbf{b} - \mathbf{A} \cdot \mathbf{x}^{(n)} - \mathbf{A} \cdot \mathbf{r}^{(n)} \tag{3.80}$$

or

$$\mathbf{r}^{(n+1)} = (\mathbf{I} - \mathbf{A}) \cdot \mathbf{r}^{(n)} = (\mathbf{I} - \mathbf{A})^{n+1} \cdot \mathbf{r}^{(0)}. \tag{3.81}$$

A more general form of (modified) Richardson iteration replaces Eq. (3.79) by

$$\mathbf{x}^{(n+1)} = \mathbf{x}^{(n)} + \alpha_{n+1} \mathbf{r}^{(n)}, \tag{3.82}$$

where the α_n are real constants that can be chosen to minimize some norm of the residual $\|\mathbf{r}^n\|$. This leads to the relation $\mathbf{r}^{(n+1)} = (\mathbf{I} - \alpha_{n+1}\mathbf{A}) \cdot \mathbf{r}^{(n)}$, which can also be written

$$\mathbf{r}^{(n+1)} = \mathbf{P}_{(n+1)}(\mathbf{A}) \cdot \mathbf{r}^{(0)}, \tag{3.83}$$

where we have introduced the matrix operator

$$\mathbf{P}_{(n)}(\mathbf{A}) \equiv \prod_{j=0}^{n} (\mathbf{I} - \alpha_j \mathbf{A}).$$

Different Krylov space methods correspond to different rules for selecting the α_j.

For simplicity in the following discussion, we will discuss the choice $\alpha_j = 1$. Let us also suppose that $\mathbf{x}^{(0)} = 0$. This is really no loss of generality, since for the situation $\mathbf{x}^{(0)} \neq 0$, we can transform through a simple linear transformation $\mathbf{z} = \mathbf{x} - \mathbf{x}^{(0)}$ so that the transformed system $\mathbf{A} \cdot \mathbf{z} = \mathbf{b} - \mathbf{A} \cdot \mathbf{x}^{(0)} = \tilde{\mathbf{b}}$, for which obviously $\mathbf{z}^{(0)} = 0$.

For the Richardson iteration introduced above, it follows from Eqs. (3.79) and (3.81) that

$$\mathbf{x}^{(n+1)} = \mathbf{r}^{(0)} + \mathbf{r}^{(1)} + \mathbf{r}^{(2)} + \dots + \mathbf{r}^{(n)} = \sum_{j=0}^{n} (\mathbf{I} - \mathbf{A})^j \cdot \mathbf{r}^{(0)}. \tag{3.84}$$

Thus, $\mathbf{x}^{(n+1)}$ lies in the space generated by

$$\{\mathbf{r}^{(0)}, \mathbf{A} \cdot \mathbf{r}^{(0)}, ..., \mathbf{A}^n \cdot \mathbf{r}^{(0)}\} \equiv \mathcal{K}_{n+1}(\mathbf{A}; \mathbf{r}^{(0)}),$$

where $\mathcal{K}_n(\mathbf{A}; \mathbf{r}^{(0)})$ is the *Krylov subspace* of dimension n generated by $\mathbf{A}$ and $\mathbf{r}^{(0)}$.

The Richardson iteration thereby generates elements of Krylov subspaces of increasing dimension. Including local iteration parameters in the iteration, i.e., allowing $\alpha_n \neq 1$, leads to other elements of the same Krylov subspaces, and hence to other polynomial expressions for the error and the residual. A natural question to ask is: "What is the best approximate solution that we can select from the Krylov subspace of a given dimension?" Posed mathematically, "best" might be defined as that which minimizes the norm of the residual or makes the residual orthogonal to the subspace.

In order to discuss the better approximations in the Krylov subspace, we need to define a basis for this subspace, one that can be extended naturally for subspaces of increasing dimension. The obvious basis

$$\mathbf{r}^{(0)}, \mathbf{A} \cdot \mathbf{r}^{(0)}, ..., \mathbf{A}^n \cdot \mathbf{r}^{(0)} \tag{3.85}$$

for $\mathcal{K}_{n+1}(\mathbf{A}; \mathbf{r}^{(0)})$ is not very attractive from a numerical point of view, since the vectors $\mathbf{A}^n \cdot \mathbf{r}^{(0)}$ point more and more in the direction of the dominant eigenvector for increasing n, and hence they lose their linear independence.

Instead of this standard basis, one usually works with an *orthonormal basis* that is constructed by a modified Gram–Schmidt procedure. Thus, if we assume that we already have an orthonormal basis $\mathbf{v}^1, ..., \mathbf{v}^n$ for $\mathcal{K}_n(\mathbf{A}; \mathbf{r}^{(0)})$, then this basis is expanded by computing $\mathbf{t} = \mathbf{A} \cdot \mathbf{v}^n$, and by orthonormalizing this vector with respect to $\mathbf{v}^1, ..., \mathbf{v}^n$. We start this process with $\mathbf{v}^1 \equiv \mathbf{r}^{(0)}/\|\mathbf{r}^{(0)}\|_2$, where we have normalized by the ℓ^2 norm of the vector.

There are several approaches to finding the optimal iteration strategy within the Krylov subspace. Two of the more popular methods are (1) the *Ritz–Galerkin* approach that leads to the *conjugate-gradient* (CG) method, requiring that $\mathbf{b} - \mathbf{A} \cdot \mathbf{x}^{(n)} \perp \mathcal{K}_n(\mathbf{A}, \mathbf{r}^{(0)})$ and (2) the *minimum residual* approach that leads to methods such as GMRES, requiring that $\|\mathbf{b} - \mathbf{A} \cdot \mathbf{x}^{(n)}\|_2$ be minimal over $\mathcal{K}_n(\mathbf{A}; \mathbf{r}^{(0)})$. We discuss a variational formulation of the conjugate gradient method in the next section and the GMRES approach in the section after that. Additional analysis and discussion of other Krylov space methods can be found in [34, 46].

3.7.1　Steepest Descent and Conjugate Gradient

If the matrix $\mathbf{A}$ is symmetric and positive definite, there is a variational derivation of the conjugate gradient method [50], which we present here. For a given $N \times N$ matrix $\mathbf{A}$ and N vector $\mathbf{b}$, we define the scalar function

$$\phi(\mathbf{x}) = \frac{1}{2}\mathbf{x}^T \cdot \mathbf{A} \cdot \mathbf{x} - \mathbf{b}^T \cdot \mathbf{x}. \tag{3.86}$$

Equation (3.76) is equivalent to finding the N-vector $\bar{\mathbf{x}}$ that minimizes $\phi(\mathbf{x})$, since this implies that $\nabla\phi(\bar{\mathbf{x}}) = \mathbf{A} \cdot \bar{\mathbf{x}} - \mathbf{b} = 0$, where ∇ is an N-dimensional gradient.

We first observe that for an arbitrary scalar α and vector $\mathbf{y}$, it follows from Eq. (3.86) that

$$\phi(\mathbf{x} + \alpha\mathbf{y}) = \phi(\mathbf{x}) + \alpha\mathbf{y}^T \cdot (\mathbf{A} \cdot \mathbf{x} - \mathbf{b}) + \frac{1}{2}\alpha^2\mathbf{y}^T \cdot \mathbf{A} \cdot \mathbf{y}. \tag{3.87}$$

If we impose the requirement that $\phi(\mathbf{x} + \alpha\mathbf{y})$ be stationary with respect to α, i.e., $\frac{\partial}{\partial\alpha}\phi(\mathbf{x} + \alpha\mathbf{y}) = 0$, then this implies

$$\mathbf{y}^T \cdot [\mathbf{A} \cdot (\mathbf{x} + \alpha\mathbf{y}) - \mathbf{b}] = 0, \tag{3.88}$$

where α is given by

$$\alpha = -\frac{\mathbf{y}^T \cdot (\mathbf{A} \cdot \mathbf{x} - \mathbf{b})}{\mathbf{y}^T \cdot \mathbf{A} \cdot \mathbf{y}}. \tag{3.89}$$

To specify an iteration, we choose an initial iterate, $\mathbf{x}_0$, and for $i = 0, 1, 2, \dots$ we choose a search direction $\mathbf{p}_i$, and define the iteration by

$$\mathbf{x}_{i+1} = \mathbf{x}_i + \alpha_i\mathbf{p}_i, \tag{3.90}$$

where α_i is chosen to minimize $\phi(\mathbf{x}_i + \alpha_i\mathbf{p}_i)$. It follows from Eq. (3.89) that

$$\alpha_i = \frac{\mathbf{p}_i^T \cdot \mathbf{r}_i}{\mathbf{p}_i^T \cdot \mathbf{A} \cdot \mathbf{p}_i}, \tag{3.91}$$

where $\mathbf{r}_i \equiv \mathbf{b} - \mathbf{A} \cdot \mathbf{x}_i$ is the residual.

The *steepest descent* method chooses the search direction $\mathbf{p}_i$ to be $\mathbf{r}_i$ (in the direction of the residual). The iteration method is then given by

$$\mathbf{x}_{i+1} = \mathbf{x}_i + \left(\frac{\mathbf{r}_i^T \cdot \mathbf{r}_i}{\mathbf{r}_i^T \cdot \mathbf{A} \cdot \mathbf{r}_i}\right)\mathbf{r}_i. \tag{3.92}$$

This will decrease the scalar function $\phi(\mathbf{x})$ defined in Eq. (3.86) every iteration, but it converges very slowly because it is always minimizing along the residual direction at the expense of the other directions.

The conjugate gradient method improves on this shortcoming of the steepest descent method by introducing a set of $\mathbf{A}$-*conjugate vectors* $\mathbf{p}_0, \mathbf{p}_1, \mathbf{p}_2, \dots$ with the orthogonality property:

$$\mathbf{p}_i^T \cdot \mathbf{A} \cdot \mathbf{p}_j = \begin{cases} = 0, & i \neq j \\ \neq 0, & i = j \end{cases}.$$

We note that these vectors are linearly independent, i.e., they have the property that if

$$\sum_{j=0}^{k} c_j\mathbf{p}_j = 0 \tag{3.93}$$

then this implies

$$c_i = 0; \quad i = 0, \cdots, k.$$

This can be seen by multiplying Eq. (3.93) on the left by $\mathbf{p}_i^T \cdot \mathbf{A}$ and using the orthogonality property.

$\mathbf{A}$-conjugate vectors have the property that

$$\phi(\mathbf{x}_{k+1}) \begin{array}{c} \\ \mathbf{x}_n \end{array} \begin{array}{c} = c_0, c_1, \overset{\min}{\cdots}, c_k \; \phi \left(\mathbf{x}_0 + \sum_{j=0}^{k} c_j \mathbf{p}_j \right) , \\ = \bar{\mathbf{x}} \end{array} \tag{3.94}$$

which can be verified directly using Eq. (3.88) and the orthogonality property of $\mathbf{p}_j$. This important property allows one to minimize sequentially with respect to the c_ks. The second equality in Eq. (3.94) follows from the fact that the vectors form a complete basis in this N-space $\mathcal{R}^N$.

We next consider how to construct the $\mathbf{p}$ vectors. A sequence of $\mathbf{A}$-conjugate $\mathbf{p}_i$s can be generated from any set of N independent vectors $\mathbf{u}_0, \mathbf{u}_1, \cdots$ through a Gram–Schmidt process. We begin with $\mathbf{p}_0 = \mathbf{u}_0$, an arbitrary N-vector. Then, for $i = 1, 2, \cdots$ define

$$\mathbf{p}_i = \mathbf{u}_i + \sum_{j=0}^{i-1} \beta_{i,j} \mathbf{p}_j. \tag{3.95}$$

Multiplying on the left by $\mathbf{p}_j^T \cdot \mathbf{A}$ and using the orthogonality property gives:

$$\beta_{i,j} = -\frac{\mathbf{p}_j^T \cdot \mathbf{A} \cdot \mathbf{u}_i}{\mathbf{p}_j^T \cdot \mathbf{A} \cdot \mathbf{p}_j}. \tag{3.96}$$

We note that the $\mathbf{p}_0, \cdots, \mathbf{p}_j$ are expressible as linear combinations of the $\mathbf{u}_0, \cdots, \mathbf{u}_j$ and vice versa. Let us take for the $\mathbf{u}_i$ to be the ith residual, $\mathbf{u}_i = \mathbf{r}_i = \mathbf{b} - \mathbf{A} \cdot \mathbf{x}_i$.

The residuals have the important property that the $\mathbf{r}_i$; $i = 0, \cdots, N-1$, form an orthogonal basis for $\mathcal{R}^n$. To see this, note that $\mathbf{x}_{i+1}$ minimizes $\phi(\mathbf{x}_0 + \sum_{j=0}^{i} \alpha_j \mathbf{p}_j)$ with respect to the $\alpha_0, \cdots, \alpha_i$. This is equivalent to minimizing $\phi(\mathbf{x}_0 + \sum_{j=0}^{i} d_j \mathbf{r}_j)$ with respect to the $d_0, \cdots, d_i$ since each $\mathbf{p}_j$ is a linear combination of $\mathbf{r}_0, \cdots, \mathbf{r}_j$. Thus, from Eq. (3.88), we have for $k = 0, \cdots, i$:

$$\frac{\partial}{\partial d_k} \phi \left(\mathbf{x}_0 + \sum_{j=0}^{i} d_j \mathbf{r}_j \right) = 0 \quad \Rightarrow \quad \mathbf{r}_k^T \cdot [\mathbf{A} \cdot \overbrace{(\mathbf{x}_0 + \underbrace{\sum_{j=0}^{i} d_j \mathbf{r}_j)}_{\mathbf{x}_{i+1}} - \mathbf{b}]}^{-\mathbf{r}_{i+1}} = 0. \tag{3.97}$$

We next note that the α_i in Eq. (3.91) can be expressed as:

$$\alpha_i = \frac{\mathbf{r}_i^T \cdot \mathbf{r}_i}{\mathbf{p}_i^T \cdot \mathbf{A} \cdot \mathbf{p}_i}. \tag{3.98}$$

To see this, start from Eq. (3.91) and note that

$$
\begin{aligned}
\mathbf{p}_i &= \mathbf{r}_i + \text{linear combination of } \mathbf{p}_0, ..., \mathbf{p}_{i-1} \\
&= \mathbf{r}_i + \text{linear combination of } \mathbf{r}_0, ..., \mathbf{r}_{i-1}.
\end{aligned}
\tag{3.99}
$$

Therefore, $\mathbf{p}_i^T \cdot \mathbf{r}_i = \mathbf{r}_i^T \cdot \mathbf{r}_i$, since $\mathbf{r}_i^T \cdot \mathbf{r}_j = 0$ for $i \neq j$.
It can also be shown that in Eq. (3.96),

$$
\beta_{i,j} =
\begin{cases}
0, & j < i - 1 \\
\dfrac{\mathbf{r}_i^T \cdot \mathbf{r}_i}{\mathbf{r}_{i-1}^T \cdot \mathbf{r}_{i-1}}, & j = i - 1.
\end{cases}
\tag{3.100}
$$

This can be seen by starting with Eq. (3.90) for the jth iterate, and premultiplying each side by $\mathbf{A}$ and subtracting $\mathbf{b}$ from each side

$$
\overbrace{\mathbf{A} \cdot \mathbf{x}_{j+1} - \mathbf{b}}^{-\mathbf{r}_{j+1}} = \overbrace{\mathbf{A} \cdot \mathbf{x}_j - \mathbf{b}}^{-\mathbf{r}_j} + \alpha_j \mathbf{A} \cdot \mathbf{p}_j,
\tag{3.101}
$$

thus,

$$
\mathbf{A} \cdot \mathbf{p}_j = \frac{1}{\alpha_j}(\mathbf{r}_j - \mathbf{r}_{j+1}).
\tag{3.102}
$$

Therefore, using the orthogonality property of the $\mathbf{r}_i$, we have

$$
\mathbf{r}_i^T \cdot \mathbf{A} \cdot \mathbf{p}_j =
\begin{cases}
0, & j < i - 1 \\
\dfrac{-\mathbf{r}_i^T \cdot \mathbf{r}_i}{\alpha_{i-1}}, & j = i - 1.
\end{cases}
\tag{3.103}
$$

Equation (3.100) follows from substitution of Eqs. (3.103) and (3.98) into Eq. (3.96), making use of the symmetry property of $\mathbf{A}$.

A stencil for the Conjugate Gradient method is as follows:

$$
\begin{aligned}
&\text{Choose } \mathbf{x}_0 \\
&\mathbf{r}_0 = \mathbf{b} - \mathbf{A} \cdot \mathbf{x}_0 \\
&\mathbf{p}_0 = \mathbf{r}_0 \\
&\text{for } i = 0, 1, 2, ... \text{ do} \\
&\quad \alpha_i = \frac{\mathbf{r}_i^T \cdot \mathbf{r}_i}{\mathbf{p}_i^T \cdot \mathbf{A} \cdot \mathbf{p}_i} \\
&\quad \mathbf{x}_{i+1} = \mathbf{x}_i + \alpha_i \mathbf{p}_i \\
&\quad \mathbf{r}_{i+1} = \mathbf{r}_i - \alpha_i \mathbf{A} \cdot \mathbf{p}_i \\
&\quad \beta_i = \frac{\|\mathbf{r}_{i+1}\|_2^2}{\|\mathbf{r}_i\|_2^2} \\
&\quad \mathbf{p}_{i+1} = \mathbf{r}_{i+1} + \beta_i \mathbf{p}_i \\
&\text{continue until convergence criteria is met.}
\end{aligned}
\tag{3.104}
$$

The $\mathbf{A}$-conjugate property of the $\mathbf{p}_j$ and the orthogonality property of the

$\mathbf{r}_j$ combine to make the Conjugate Gradient method a particularly elegant method which is fairly simple and efficient to implement.

It is shown in [50] that to reduce the error by a factor of ϵ will require a number of iterations given by

$$N \geq \frac{1}{2} \left| ln\frac{\epsilon}{2} \right| \left(\frac{\lambda_{\max}(\mathbf{A})}{\lambda_{\min}(\mathbf{A})} \right)^{\frac{1}{2}}. \tag{3.105}$$

Preconditioning should be employed to reduce this ratio of the largest to smallest eigenvalues of $\mathbf{A}$, known as the condition number, as is discussed in Section 3.7.3.

One can use the techniques of Section 3.4.5 to show that for Laplace's equation with homogeneous boundary conditions, Eq. 3.105 gives a convergence scaling for conjugate gradient (CG) similar to that found in Section 3.5.2 for dynamic relaxation (DR). Also, both CG and DR use two vectors to compute the new iterate value: in CG it is $\mathbf{x}_i$ and $\mathbf{p}_i$ and in DR it is $\mathbf{u}^n$ and $\mathbf{u}^{n+1}$. In fact, it has recently been shown [51] that there is a close connection between these seemingly totally different approaches and algorithms, and that the α_i and β_i in CG correspond to special choices (iteration dependent) for the δt and τ in the DR algorithm.

3.7.2   Generalized Minimum Residual (GMRES)

One of the most popular methods for solution of a matrix equation when the matrix is not necessarily symmetric is known as the generalized minimum residual method, or GMRES [52]. It is an iterative method that approximates the solution by the vector in a Krylov subspace with minimum residual.

Let $\mathbf{A}$ be an $m \times m$ matrix that satisfies the equation $\mathbf{A} \cdot \mathbf{x} = \mathbf{b}$. We start by constructing a set of n orthonormal vectors of length m that span the associated n-dimensional Krylov subspace, where $n \leq m$. Let $\mathbf{x}_0$ be the initial approximation to the solution (which could be zero). Define the initial residual as

$$\mathbf{r}_0 = \mathbf{b} - \mathbf{A} \cdot \mathbf{x}_0.$$

The corresponding Krylov subspace of dimension n is constructed as

$$\mathcal{K}_n = \left\{ \mathbf{r}_0, \mathbf{A} \cdot \mathbf{r}_0, \mathbf{A}^2 \cdot \mathbf{r}_0, \cdots, \mathbf{A}^{n-1} \cdot \mathbf{r}_0, \right\}. \tag{3.106}$$

The vectors in Eq. (3.106) are not suitable for use as basis vectors in $\mathcal{K}_n$ since they become increasingly colinear, approaching the eigenvector corresponding to the largest eigenvalue of $\mathbf{A}$. The Arnoldi iteration [53] performs a Gram–Schmidt process to produce a series of orthonormal vectors $\mathbf{q}_1, \mathbf{q}_2, \cdots, \mathbf{q}_n$, each of length m, that span the Krylov space $\mathcal{K}_n$. We start with

$$\mathbf{q}_1 = \frac{\mathbf{r}_0}{\|\mathbf{r}_0\|_2}.$$

We then perform the following process successively for each value of $k = 2, \cdots, n$. First define the m-vector:

$$\mathbf{z}_k = \mathbf{A} \cdot \mathbf{q}_{k-1} - \sum_{j=1}^{k-1} h_{j,k-1} \mathbf{q}_j. \qquad (3.107)$$

The coefficients $h_{j,k-1}$ are defined so that $\mathbf{z}_k$ is orthogonal to all the $\mathbf{q}_j$ for $j < k$. By taking the inner product of Eq. (3.107) with each of the $\mathbf{q}_j$ and using the orthonormality property, we have

$$h_{j,k-1} = \mathbf{q}_j^T \cdot \mathbf{A} \cdot \mathbf{q}_{k-1}; \qquad j = 1, k - 1. \qquad (3.108)$$

We then normalize the kth basis vector by defining

$$h_{k,k-1} = \|\mathbf{z}_k\|_2, \qquad (3.109)$$

$$\mathbf{q}_k = \frac{\mathbf{z}_k}{h_{k,k-1}}. \qquad (3.110)$$

The process is then repeated.

We can define a matrix $\mathbf{Q}_n$ with n columns and m rows consisting of the vectors as defined in Eq. (3.110).

$$\mathbf{Q}_n = \left[\begin{array}{c|c|c|c} \mathbf{q}_1 & \mathbf{q}_2 & \cdots & \mathbf{q}_n \end{array} \right]. \qquad (3.111)$$

One can verify that the coefficients $h_{j,k}$ as defined by Eqs. (3.108) and (3.109) form an upper Hessenburg matrix defined by

$$\mathbf{H}_n = \mathbf{Q}_n^T \cdot \mathbf{A} \cdot \mathbf{Q}_n = \begin{bmatrix} h_{1,1} & h_{1,2} & h_{1,3} & \cdots & h_{1,n} \\ h_{2,1} & h_{2,2} & h_{2,3} & \cdots & h_{2,n} \\ 0 & h_{3,2} & h_{3,3} & \cdots & h_{3,n} \\ \cdot & \cdot & \cdot & \cdot & \cdot \\ 0 & \cdots & 0 & h_{n,n-1} & h_{n,n} \end{bmatrix}. \qquad (3.112)$$

The matrix $\mathbf{H}_n$ can be viewed as the representation in the basis formed by the orthogonal vectors $\mathbf{q}_j$ of the orthogonal projection of $\mathbf{A}$ onto the Krylov subspace $\mathcal{K}_n$.

The relation between the $\mathbf{Q}_n$ matrices in subsequent iterations is given by

$$\mathbf{A} \cdot \mathbf{Q}_n = \mathbf{Q}_{n+1} \cdot \tilde{\mathbf{H}}_n. \qquad (3.113)$$

Here,

$$\tilde{\mathbf{H}}_n = \begin{bmatrix} h_{1,1} & h_{1,2} & h_{1,3} & \cdots & h_{1,n} \\ h_{2,1} & h_{2,2} & h_{2,3} & \cdots & h_{2,n} \\ 0 & h_{3,2} & h_{3,3} & \cdots & h_{3,n} \\ \cdot & \cdot & \cdot & & \cdot \\ 0 & \cdots & 0 & h_{n,n-1} & h_{n,n} \\ 0 & 0 & 0 & 0 & h_{n+1,n} \end{bmatrix} \qquad (3.114)$$

is an $(n+1) \times n$ matrix formed by adding an extra row to $\mathbf{H}_n$ with a single entry in the last column.

We now seek the vector $\mathbf{x}_n$ that lies within the subspace $\mathcal{K}_n$ that minimizes the norm of the residual $\mathbf{b} - \mathbf{A} \cdot \mathbf{x}_n$. Any vector within this subspace can be written as

$$\mathbf{x}_n = \mathbf{Q}_n \cdot \mathbf{y}_n \tag{3.115}$$

for some vector $\mathbf{y}_n$. Consider now the norm of the residual

$$
\begin{aligned}
\|\mathbf{b} - \mathbf{A} \cdot \mathbf{x}_n\|_2 &= \|\mathbf{b} - \mathbf{A} \cdot \mathbf{Q}_n \cdot \mathbf{y}_n\|_2 \\
&= \left\|\mathbf{b} - \mathbf{Q}_{n+1} \cdot \tilde{\mathbf{H}}_n \cdot \mathbf{y}_n\right\|_2 \\
&= \left\|\mathbf{Q}_{n+1} \cdot \left(\|\mathbf{r}_0\|_2 \mathbf{e}^1 - \tilde{\mathbf{H}}_n \cdot \mathbf{y}_n\right)\right\|_2 \\
&= \left\|\|\mathbf{r}_0\|_2 \mathbf{e}^1 - \tilde{\mathbf{H}}_n \cdot \mathbf{y}_n\right\|_2.
\end{aligned}
$$

The third step follows because $\mathbf{b} - \mathbf{A} \cdot \mathbf{x}_0$ is proportional to the first orthogonal vector, $\mathbf{e}^1 = (1, 0, 0, ..., 0)$ and $\|\mathbf{r}_0\|_2 = \|\mathbf{b} - \mathbf{A} \cdot \mathbf{x}_0\|_2$. The final step follows because $\mathbf{Q}_{n+1}$ is an orthonormal transformation. The minimizing vector $\mathbf{y}_n$ can then be found by solving a least squares problem.

The least squares problem is solved by computing a $\mathbf{QR}$ decomposition: find an $(n+1) \times (n+1)$ orthogonal matrix $\mathbf{\Omega}_n$ and an $(n+1) \times n$ upper triangular matrix $\tilde{\mathbf{R}}_n$ such that

$$\mathbf{\Omega}_n \cdot \tilde{\mathbf{H}}_n = \tilde{\mathbf{R}}_n.$$

The upper triangular matrix $\tilde{\mathbf{R}}_n$ has one more row than it has columns, so its bottom row consists of all zeros. Hence, it can be decomposed as

$$\tilde{\mathbf{R}}_n = \begin{bmatrix} \mathbf{R}_n \\ 0 \end{bmatrix}, \tag{3.116}$$

where $\mathbf{R}_n$ is an $n \times n$ triangular matrix and the "0" denotes a row of zeros. The $\mathbf{QR}$ decomposition is updated efficiently from one iteration to the next because the Hessenberg matrices differ only by a row of zeros and a column,

$$\tilde{\mathbf{H}}_{n+1} = \begin{bmatrix} \tilde{\mathbf{H}}_n & \mathbf{h}_{n+1} \\ 0 & h_{n+2,n+1} \end{bmatrix}, \tag{3.117}$$

where $\mathbf{h}_{n+1} \equiv (h_{1,n+1}, ..., h_{n+1,n+1})^T$. This implies that premultiplying the Hessenberg matrix with $\mathbf{\Omega}_n$, augmented with zeros and a row with multiplicative identity, yields almost a triangular matrix.

$$\begin{bmatrix} \mathbf{\Omega_n} & 0 \\ 0 & 1 \end{bmatrix} \cdot \tilde{\mathbf{H}}_{n+1} = \begin{bmatrix} \mathbf{R}_n & \mathbf{r}_n \\ 0 & \rho \\ 0 & \sigma \end{bmatrix}. \tag{3.118}$$

This equation determines $\mathbf{r}_n$ and ρ, and gives $\sigma = h_{n+2,n+1}$. The matrix on the right in Eq. (3.118) would be triangular, and thus of the desired form, if σ were zero. To remedy this, one applies the *Givens rotation*,

$$\mathbf{G}_n = \begin{bmatrix} \mathbf{I}_n & 0 & 0 \\ 0 & c_n & s_n \\ 0 & -s_n & c_n \end{bmatrix}, \qquad (3.119)$$

where $\mathbf{I}_n$ is the identity matrix of rank n and

$$c_n = \frac{\rho}{\sqrt{\rho^2 + \sigma^2}}, \qquad s_n = \frac{\sigma}{\sqrt{\rho^2 + \sigma^2}}. \qquad (3.120)$$

With this Givens rotation, we form

$$\mathbf{\Omega}_{n+1} = \mathbf{G}_n \cdot \begin{bmatrix} \mathbf{\Omega}_n & 0 \\ 0 & 1 \end{bmatrix}. \qquad (3.121)$$

Using Eqs. (3.118) and (3.121), we obtain the triangular matrix

$$\mathbf{\Omega}_{n+1} \cdot \tilde{\mathbf{H}}_{n+1} = \mathbf{G}_n \cdot \begin{bmatrix} \mathbf{R}_n & \mathbf{r}_n \\ 0 & \rho \\ 0 & \sigma \end{bmatrix} = \begin{bmatrix} \mathbf{R}_n & \mathbf{r}_n \\ 0 & r_{nn} \\ 0 & 0 \end{bmatrix} = \tilde{\mathbf{R}}_{n+1}, \qquad (3.122)$$

where $r_{nn} = \sqrt{\rho^2 + \sigma^2}$. This is the **QR** decomposition we were seeking. The minimization problem is now easily solved by noting

$$\begin{aligned} \left\| \|\mathbf{r}_0\|_2 \mathbf{e}^1 - \tilde{\mathbf{H}}_n \cdot \mathbf{y}_n \right\|_2 &= \left\| \mathbf{\Omega}_n \cdot \left(\|\mathbf{r}_0\|_2 \mathbf{e}^1 - \tilde{\mathbf{H}}_n \cdot \mathbf{y}_n \right) \right\|_2 \\ &= \left\| \|\mathbf{r}_0\|_2 \mathbf{\Omega}_n \cdot \mathbf{e}^1 - \tilde{\mathbf{R}}_n \cdot \mathbf{y}_n \right\|_2. \end{aligned} \qquad (3.123)$$

Now, denote the vector $\|\mathbf{r}_0\|_2 \mathbf{\Omega}_n \cdot \mathbf{e}^1$ by

$$\tilde{\mathbf{g}}_n = \begin{bmatrix} \mathbf{g}_n \\ \gamma_n \end{bmatrix}. \qquad (3.124)$$

We then have

$$\left\| \|\mathbf{r}_0\|_2 \mathbf{\Omega}_n \cdot \mathbf{e}^1 - \tilde{\mathbf{R}}_n \cdot \mathbf{y}_n \right\|_2 = \left\| \begin{bmatrix} \mathbf{R}_n \\ 0 \end{bmatrix} \cdot \mathbf{y}_n - \begin{bmatrix} \mathbf{g}_n \\ \gamma_n \end{bmatrix} \right\|_2. \qquad (3.125)$$

The vector that minimizes this is given by

$$\mathbf{R}_n \cdot \mathbf{y}_n = \mathbf{g}_n. \qquad (3.126)$$

This now is easily solved by back substitution since $\mathbf{R}_n$ is a triangular matrix. In summary, the GMRES method consists of the following steps:

   i Do one step of the Arnoldi method to compute a new column of the orthonormal matrix $\mathbf{Q}_n$ and thereby a new column and row of the matrix $\tilde{\mathbf{H}}_n$.

ii Find the $\mathbf{y}_n$ that minimizes the residual $\mathbf{r}_n$ by solving a least squares problem using a $\mathbf{QR}$ decomposition.

iii Compute $\mathbf{x}_n = \mathbf{Q}_n \cdot \mathbf{y}_n$.

iv Repeat if the residual is not small enough.

Because both the storage and the amount of computation increase with the number of steps, GMRES is typically restarted after a small number of iterations with the final iterate $\mathbf{x}_n$ serving as the initial guess of the next restart.

3.7.3   Preconditioning

It was stated in the discussion of Eq. (3.105) that the matrix $\mathbf{A}$ should have a low condition number in order for the iteration to converge rapidly. In order to achieve this, what is normally done is to construct a *preconditioning* matrix $\mathbf{K}$ with the property that $\mathbf{K}^{-1}$ approximates the inverse of $\mathbf{A}$. Then, Eq. (3.76) is essentially premultiplied by $\mathbf{K}^{-1}$ and we solve the system $\mathbf{K}^{-1} \cdot \mathbf{A} \cdot \mathbf{x} = \mathbf{K}^{-1} \cdot \mathbf{b}$ instead of the original system $\mathbf{A} \cdot \mathbf{x} = \mathbf{b}$.

There are different strategies for choosing the preconditioning matrix $\mathbf{K}$. In general, (1) $\mathbf{K}$ should be a good approximation to $\mathbf{A}$ in the sense that $\mathbf{K}^{-1} \cdot \mathbf{A}$ has a smaller condition number than does $\mathbf{A}$, (2) the cost of the construction of $\mathbf{K}$ must not be prohibitive, and (3) the system $\mathbf{K} \cdot \mathbf{x} = \mathbf{b}$ must be much easier to solve than the original system. This third property is because it is never necessary to actually compute either $\mathbf{K}^{-1}$ or $\mathbf{K}^{-1} \cdot \mathbf{A}$ explicitly. The Krylov subspace methods under discussion need the operator of the linear system only for computing matrix-vector products. Thus, a template for the Richardson iteration applied to the preconditioned system

$$\mathbf{K}^{-1} \cdot \mathbf{A} \cdot \mathbf{x} = \mathbf{K}^{-1} \cdot \mathbf{b}$$

would be the following: Given an initial guess $\mathbf{x}^0$, compute the residual $\mathbf{r}^0 = \mathbf{b} - \mathbf{A} \cdot \mathbf{x}^0$. Then, for $i = 0, 1, 2, \dots$ do

$$\begin{aligned}
\mathbf{K} \cdot \mathbf{z}^i &= \mathbf{r}^i \\
\mathbf{x}^{i+1} &= \mathbf{x}^i + \mathbf{z}^i \\
\mathbf{r}^{i+1} &= \mathbf{b} - \mathbf{A} \cdot \mathbf{x}^{i+1}.
\end{aligned}$$

We note that when solving the preconditioned system using a Krylov subspace method, we will get very different subspaces than for the original system. The idea is that approximations in the new sequence of subspaces will approach the solution more quickly than in the original subspaces.

The choice of $\mathbf{K}$ varies from purely "black box" algebraic techniques that can be applied to general matrices to "problem-dependent" preconditioners that exploit special features of a particular class of problems. Perhaps the simplest form of preconditioning is the *Jacobi preconditioner* in which the preconditioning matrix is chosen to be the matrix $\mathbf{D}$ in Eq. (3.44), the diagonal of the matrix $\mathbf{A}$.

A class of preconditioners follow from starting with a direct solution method for $\mathbf{A} \cdot \mathbf{x} = \mathbf{b}$ and making variations to keep the approximate solution from becoming too expensive. As we have seen in Section 3.3.2, a common direct technique is to factorize $\mathbf{A}$ as $\mathbf{A} = \mathbf{L} \cdot \mathbf{U}$. However, for a sparse matrix, the number of operations can be prohibitive and the number of entries in the factors is substantially greater than in the original matrix. In *incomplete LU factorization*, the factors are kept artificially sparse in order to save computer time and storage for the decomposition. The preconditioning matrix $\mathbf{K}$ is then given in the form $\mathbf{K} = \tilde{\mathbf{L}} \cdot \tilde{\mathbf{U}}$, where $\tilde{\mathbf{L}}$ and $\tilde{\mathbf{U}}$ are approximate incomplete factors of $\mathbf{A}$. In this example, $\mathbf{z} = \mathbf{K}^{-1} \cdot \mathbf{y}$ is computed by first solving for $\mathbf{w}$ from solving $\tilde{\mathbf{L}} \cdot \mathbf{w} = \mathbf{y}$ and then compute $\mathbf{z}$ from $\tilde{\mathbf{U}} \cdot \mathbf{z} = \mathbf{w}$.

Another popular preconditioner is the *incomplete Cholesky factorization* of a symmetric positive definite matrix. The Cholesky factorization of a positive definite matrix $\mathbf{A}$ is $\mathbf{A} = \mathbf{L}\mathbf{L}^T$ where $\mathbf{L}$ is a lower triangular matrix. An incomplete Cholesky factorization uses a sparse lower triangular matrix $\tilde{\mathbf{L}}$, where $\tilde{\mathbf{L}}$ is "close to" $\mathbf{L}$. The preconditioner is then $\mathbf{K} = \tilde{\mathbf{L}}\tilde{\mathbf{L}}^T$. To find $\tilde{\mathbf{L}}$, one uses the algorithm for finding the exact Cholesky decomposition, except that any entry is set to zero if the corresponding entry in $\mathbf{A}$ is also zero. This gives an incomplete Cholesky factorization matrix which is as sparse as the matrix $\mathbf{A}$ [47].

There are also a large class of *domain decomposition methods* that split the entire problem being solved into smaller problems on subdomains and iterate to coordinate the solution between adjacent subdomains. These are especially attractive when doing parallel computing since the problems on the subdomains are independent. The most popular of these is the *additive Schwarz method*. A detailed discussion of these methods is outside the scope of the present work, but "black box" routines are available and there is an extensive literature on their theory and applicability [48, 49].

In the case of the *conjugate gradient method with preconditioning*, PCG, a relatively simple modification of the algorithm presented in Section 3.7.1 is possible if the preconditioning matrix $\mathbf{K}$ is symmetric. Consider the following stencil, which is a modification of Eq. (3.104):

$$
\begin{aligned}
&\text{Choose } \mathbf{x}_0 \\
&\mathbf{r}_0 = \mathbf{b} - \mathbf{A} \cdot \mathbf{x}_0 \\
&\gamma_0 = \mathbf{K}^{-1} \cdot \mathbf{r}_0 \\
&\mathbf{p}_0 = \gamma_0 \\
&\text{for } i = 0, 1, 2, \dots \text{ do} \\
&\qquad \alpha_i = \frac{\gamma_i^T \cdot \mathbf{r}_i}{\mathbf{p}_i^T \cdot \mathbf{A} \cdot \mathbf{p}_i} \\
&\qquad \mathbf{x}_{i+1} = \mathbf{x}_i + \alpha_i \mathbf{p}_i \\
&\qquad \mathbf{r}_{i+1} = \mathbf{r}_i - \alpha_i \mathbf{A} \cdot \mathbf{p}_i \\
&\qquad \gamma_{i+1} = \mathbf{K}^{-1} \cdot \mathbf{r}_{i+1} \\
&\qquad \beta_i = \frac{\gamma_{i+1}^T \cdot \mathbf{r}_{i+1}}{\gamma_i^T \cdot \mathbf{r}_i} \\
&\qquad \mathbf{p}_{i+1} = \gamma_{i+1} + \beta_i \mathbf{p}_i \\
&\text{continue until convergence criteria is met.}
\end{aligned}
\tag{3.127}
$$

This algorithm can be derived similarly to that in Section 3.7.1, except that in the equivalent of Eq. (3.97) we make note that each $\mathbf{p}_j$ is a linear combination of $\gamma_0, \cdots, \gamma_j$, and thus it follows from Eq. (3.97) that $\gamma_i^T \cdot \mathbf{r}_j = 0$ for $i \neq j$. With preconditioning, the error estimates given in Eq. (3.105) to reduce the error by a factor of ϵ now become [50]

$$
N \geq \frac{1}{2} \left| \ln \frac{\epsilon}{2} \right| \left(\frac{\lambda_{\max}\left(\mathbf{K}^{-1} \cdot \mathbf{A}\right)}{\lambda_{\min}\left(\mathbf{K}^{-1} \cdot \mathbf{A}\right)} \right)^{\frac{1}{2}}.
\tag{3.128}
$$

There is an extensive and evolving literature on preconditioning that the reader is referred to [34, 47, 54, 55].

3.8 Finite Fourier Transform

We have seen in Section 2.5 that there exists a finite (or discrete) Fourier transform pair defined by

$$
g_k = \frac{1}{N} \sum_{n=0}^{N-1} G_n \exp(2\pi i k n / N); \qquad k = 0, \cdots, N-1,
\tag{3.129}
$$

$$
G_n = \sum_{k=0}^{N-1} g_k \exp[-(2\pi i k n / N)]; \qquad n = 0, \cdots, N-1,
\tag{3.130}
$$

and that this is an exact transformation, so that transforming back and forth does not represent any loss of information.

There are many variations of the transform pair given by Eqs. (3.129) and (3.130) that are equivalent but are more convenient for certain applications. We used one, the 2D discrete sine transform, in Section 3.4.5, as it was well suited for examining the properties of a particular finite difference equation applied to a problem with homogeneous boundary conditions. When concerned with real arrays with non-zero boundary conditions, it is often convenient to use the real finite Fourier transform relations:

$$G_n = \frac{1}{2}g_0^c + \frac{1}{2}g_{N/2}^c(-1)^n + \sum_{k=1}^{N/2-1}\left\{g_k^c\cos\frac{2\pi nk}{N} + g_k^s\sin\frac{2\pi nk}{N}\right\},$$

$$g_k^c = \frac{2}{N}\sum_{n=0}^{N-1}G_n\cos\frac{2\pi nk}{N}; \qquad k = 0,\cdots,N/2,$$

$$g_k^s = \frac{2}{N}\sum_{n=0}^{N-1}G_n\sin\frac{2\pi nk}{N}; \qquad k = 1,\cdots,N/2-1, \qquad (3.131)$$

where now the G_n, g_k^c, and g_k^s are all real. Note that if the G_n were purely real in Eq. (3.129), then we would have, for $k = 0,\cdots,N/2$,

$$g_k^c = 2Re\{g_k\},$$
$$g_k^s = 2Im\{g_k\}.$$

Returning now to the notation of Eq. (3.130), we again observe that Eq. (3.130) is equivalent to a matrix multiplication. If we define the quantity

$$W \equiv e^{-i2\pi/N}, \qquad (3.132)$$

then the elements of the matrix are W^{nk}, where the superscript now means "raise to the power $n \times k$." Using this convention and the implied summation notation, we can rewrite Eq. (3.130) compactly as:

$$G_n = W^{nk}g_k; \qquad n = 0,\cdots,N-1, \qquad (3.133)$$

where the index k is summed from 0 to $N-1$. We note that W as defined in Eq. (3.132) has the following properties, for any integers j and p:

$$W^{N+j} = W^j,$$
$$W^{N/2+p} = -W^p. \qquad (3.134)$$

3.8.1 Fast Fourier Transform

The matrix multiplication needed to perform the finite transform, as defined in Eq. (3.133) can be expensive for N large, requiring N^2 multiplications and $N(N-1)$ additions since W is a full matrix. The fast Fourier transform algorithm (FFT) [27, 56] is essentially a *clever trick* which allows one, if N is equal to a power of 2, to factor W^{nk} into a product of $\log_2 N$ sparse matrices of a particular form. The total work involved in multiplying all of the sparse factor matrices turns out to be much less than that required to multiply by the original matrix.

For a specific example, let us consider explicitly the case $N = 4$. Using the properties in Eq. (3.134) and the fact that $W^o = 1$, Eq. (3.133) becomes

$$
\begin{bmatrix} G_0 \\ G_1 \\ G_2 \\ G_3 \end{bmatrix} = \begin{bmatrix} 1 & 1 & 1 & 1 \\ 1 & W^1 & W^2 & W^3 \\ 1 & W^2 & W^0 & W^2 \\ 1 & W^3 & W^2 & W^1 \end{bmatrix} \cdot \begin{bmatrix} g_0 \\ g_1 \\ g_2 \\ g_3 \end{bmatrix}.
\tag{3.135}
$$

The matrix multiplication in Eq. (3.135) requires $N^2 = 16$ multiplications and $N(N-1) = 12$ additions to evaluate. Alternatively, now consider the pair of matrix multiplications

$$
\begin{bmatrix} y_0 \\ y_1 \\ y_2 \\ y_3 \end{bmatrix} = \begin{bmatrix} 1 & 0 & W^0 & 0 \\ 0 & 1 & 0 & W^0 \\ 1 & 0 & W^2 & 0 \\ 0 & 1 & 0 & W^2 \end{bmatrix} \cdot \begin{bmatrix} g_0 \\ g_1 \\ g_2 \\ g_3 \end{bmatrix},
$$

$$
\begin{bmatrix} G_0 \\ G_2 \\ G_1 \\ G_3 \end{bmatrix} = \begin{bmatrix} 1 & W^0 & 0 & 0 \\ 1 & W^2 & 0 & 0 \\ 0 & 0 & 1 & W^1 \\ 0 & 0 & 1 & W^3 \end{bmatrix} \cdot \begin{bmatrix} y_0 \\ y_1 \\ y_2 \\ y_3 \end{bmatrix}.
\tag{3.136}
$$

The pair of matrix multiplications defined in Eq. (3.136) requires only a total of 8 additions and 4 multiplications if we make use of the fact that $W^2 = -W^0$, as per Eq. (3.134). This fits a general pattern where if N is a power of 2 such that $N = 2^\gamma$, we factor the $N \times N$ full matrix into $\gamma N \times N$ sparse matrices which have only 2 non-zero entries in each row. Evaluating these then requires $N\gamma/2$ multiplications and $N\gamma$ additions.

Developing the general pattern for the FFT involves utilizing a binary representation for the integers k and n in the transform relation of Eq. (3.133). Again, let us consider the case $N = 4$. We write the integers k and n in terms of binary numbers k_1, k_0, n_1, n_0 that only take on the values of 0 or 1:

$$
\begin{aligned}
k &= 2k_1 + k_0; & k_1 = 0, 1; & \quad k_0 = 0, 1, \\
n &= 2n_1 + n_0; & n_1 = 0, 1; & \quad n_0 = 0, 1.
\end{aligned}
\tag{3.137}
$$

With this notation, specifying the values of k_1, k_0, n_1, n_0 implies values for k and n through Eq. (3.137). Thus, the transform relation of Eq. (3.133) can be written

$$
G_{n_1,n_0} = \sum_{k_0=0}^{1} \sum_{k_1=0}^{1} g_{k_1,k_0} W^{(2n_1+n_0)(2k_1+k_0)}.
\tag{3.138}
$$

Now, we can rewrite

$$
\begin{aligned}
W^{(2n_1+n_0)(2k_1+k_0)} &= W^{(2n_1+n_0)2k_1} W^{(2n_1+n_0)k_0}, \\
&= \left[W^{4n_1 k_1} \right] W^{2n_0 k_1} W^{(2n_1+n_0)k_0}, \\
&= W^{2n_0 k_1} W^{(2n_1+n_0)k_0},
\end{aligned}
$$

where the term in brackets is unity from Eq. (3.134). Equation (3.138) becomes

$$
G_{n_1,n_0} = \sum_{k_0=0}^{1} \left[\sum_{k_1=0}^{1} g_{k_1,k_0} W^{2n_0 k_1} \right] W^{(2n_1+n_0)k_0}. \tag{3.139}
$$

Equation (3.139) can be written as three sequential steps, namely

$$
y_{n_0,k_0} = \sum_{k_1=0}^{1} g_{k_1,k_0} W^{2n_0 k_1},
$$

$$
Z_{n_0,n_1} = \sum_{k_0=0}^{1} y_{n_0,k_0} W^{(2n_1+n_0)k_0}, \tag{3.140}
$$

$$
G_{n_1,n_0} = Z_{n_0,n_1}.
$$

Note that in matrix form, Eq. (3.140) can be written as

$$
\begin{bmatrix} y_{0,0} \\ y_{0,1} \\ y_{1,0} \\ y_{1,1} \end{bmatrix} = \begin{bmatrix} 1 & 0 & W^0 & 0 \\ 0 & 1 & 0 & W^0 \\ 1 & 0 & W^2 & 0 \\ 0 & 1 & 0 & W^2 \end{bmatrix} \cdot \begin{bmatrix} g_{0,0} \\ g_{0,1} \\ g_{1,0} \\ g_{1,1} \end{bmatrix},
$$

$$
\begin{bmatrix} Z_{0,0} \\ Z_{0,1} \\ Z_{1,0} \\ Z_{1,1} \end{bmatrix} = \begin{bmatrix} 1 & W^0 & 0 & 0 \\ 1 & W^2 & 0 & 0 \\ 0 & 0 & 1 & W^1 \\ 0 & 0 & 1 & W^3 \end{bmatrix} \cdot \begin{bmatrix} y_{0,0} \\ y_{0,1} \\ y_{1,0} \\ y_{1,1} \end{bmatrix},
$$

$$
\begin{bmatrix} G_{0,0} \\ G_{1,0} \\ G_{0,1} \\ G_{1,1} \end{bmatrix} = \begin{bmatrix} Z_{0,0} \\ Z_{0,1} \\ Z_{1,0} \\ Z_{1,1} \end{bmatrix},
$$

which are equivalent to the matrix relations in Eq. (3.136).

The general FFT algorithm [27] for $N = 2^\gamma$, where γ is any integer, is a straightforward generalization of the $N = 4$ example presented here. When $N = 2^\gamma$, n and k are represented in binary form as

$$n = 2^{\gamma-1}n_{\gamma-1} + 2^{\gamma-2}n_{\gamma-2} + \cdots + n_0,$$
$$k = 2^{\gamma-1}k_{\gamma-1} + 2^{\gamma-2}n_{\gamma-2} + \cdots + k_0.$$

The FFT algorithm can be written by introducing intermediate results in a generalization of Eq. (3.140):

$$x_1(n_0, k_{\gamma-2}, \cdots, k_0) = \sum_{k_{\gamma-1}=0}^{1} g(k_{\gamma-1}, k_{\gamma-2}, \cdots, k_0) W^{n_0 2^{\gamma-1} k_{\gamma-1}},$$

$$x_2(n_0, n_1, k_{\gamma-3}, \cdots, k_0) = \sum_{k_{\gamma-2}=0}^{1} x_1(n_0, k_{\gamma-2}, \cdots, k_0) W^{(2n_1+n_0)2^{\gamma-2}k_{\gamma-2}},$$

$$\vdots$$

$$x_\gamma(n_0, n_1, \cdots, n_{\gamma-1}) = \sum_{k_0=0}^{1} x_{\gamma-1}(n_0, n_1, \cdots, k_0)$$
$$\times W^{(2^{\gamma-1}n_{\gamma-1}+2^{\gamma-2}n_{\gamma-2}+\cdots+n_0)k_0},$$

$$G(n_{\gamma-1}, n_{\gamma-2}, \cdots, n_0) = x_\gamma(n_0, n_1, \cdots, n_{\gamma-1}).$$

A careful counting of operations shows that $N\gamma/2$ complex multiplications and $N\gamma$ complex additions are required to perform the transform. However, the important thing is not the numerical coefficient but the scaling as $\sim N \log_2 N$ for an exact solution. This algorithm is due to Cooley and Tukey [56].

The FFT based on the binary representation is the most efficient, but it is also restrictive, limiting applicability to mesh sizes that are powers of 2. The methods presented here can be straightforwardly extended to numbers that are powers of other prime numbers. For example, if $N = 3^\gamma$, we would represent an index as

$$n = 3^{\gamma-1}n_{\gamma-1} + 3^{\gamma-2}n_{\gamma-2} + \cdots + n_0,$$

where now each of the n_i takes on the value of $0, 1, 2$. It is also possible to construct a fast algorithm for any factorization where $N = r_1^{m_1} r_2^{m_2} \cdots r_n^{m_n}$ [27].

3.8.2 Application to 2D Elliptic Equations

Let us consider the application of transform techniques to the solution of the finite difference approximation to common elliptic equations in two dimensions (R, Z) [57]. Consider an equation of the form

$$R^{-m}\frac{\partial}{\partial R}R^m\frac{\partial u}{\partial R} + \frac{\partial^2 u}{\partial Z^2} = S(R, Z). \tag{3.141}$$

We are assuming that the function $S(R, Z)$ is known and that Dirichlet boundary conditions are given. For $m = 0(1)$ this is Poisson's equation in rectangular (cylindrical) geometry. For $m = -1$ it is the toroidal elliptic operator Δ^* that will be discussed in Chapter 4. For simplicity, consider a rectangular domain with $N + 1$ equally spaced grid points in each dimension with grid spacing h. A conservative, second-order finite difference approximation to Eq. (3.141) is given by

$$\left(1 + \frac{h}{2R_i}\right)^m (u_{i+1,j} - u_{i,j}) - \left(1 - \frac{h}{2R_i}\right)^m (u_{i,j} - u_{i-1,j})$$
$$+ u_{i,j+1} - 2u_{i,j} + u_{i,j-1} = h^2 S_{i,j}. \qquad (3.142)$$

The boundary conditions are assumed to be in the locations $u_{0,j}$, $u_{N,j}$, $u_{i,0}$, $u_{i,N}$ for $i, j = 0, \cdots N$. We perform the following five steps to obtain an exact solution of Eq. (3.142) based on the 1D sine transform introduced in Section 3.4.5.

Step (1): Replace the difference equation (3.142) with an equivalent difference equation for discrete variables $\bar{u}_{i,j} = u_{i,j} - (A_i + jB_i)$ and an appropriately defined $\bar{S}_{i,j}$ such that $\bar{u}_{i,j}$ has homogeneous (top/bottom) boundary conditions at $j = 0$ and $j = N$. Define the arrays for $i = 0, \cdots, N$:

$$\bar{a}_i \equiv \left(1 + \frac{h}{2R_i}\right)^m, \qquad \bar{c}_i \equiv \left(1 - \frac{h}{2R_i}\right)^m, \qquad \bar{b}_i \equiv \bar{a}_i + \bar{c}_i,$$

$$A_i \equiv u_{i,0}, \qquad\qquad B_i \equiv (u_{i,N} - u_{i,0})/N. \qquad (3.143)$$

The new left-right boundary conditions become, for $j = 0, \cdots, N$,

$$\bar{u}_{i,j} = u_{i,j} - (A_i + jB_i), \qquad\qquad i = 0 \,\&\, N.$$

The new source function becomes, for $i, j = 1, \cdots, N - 1$,

$$\bar{S}_{i,j} = S_{i,j} \quad - \quad h^2 \left[\bar{a}_i A_{i+1} - \bar{b}_i A_i + \bar{c}_i A_{i-1}\right]$$
$$- \quad jh^2 \left[\bar{a}_i B_{i+1} - \bar{b}_i B_i + \bar{c}_i B_{i-1}\right].$$

Step (2): Transform the source function $\bar{S}_{i,j}$ for $i, j = 1, \cdots, N - 1$ using a sine transform. For $k = 0, \cdots, N - 1$ we define:

$$\tilde{S}_{i,k} = \frac{2}{N} \sum_{j=1}^{N-1} \sin \frac{jk\pi}{N} \bar{S}_{i,j}, \qquad i = 1, \cdots, N - 1.$$

Similarly for the (left/right) boundary values $\bar{u}_{0,j}$ and $\bar{u}_{N,j}$, for $k = 1, \cdots, N - 1$ we have:

$$\tilde{u}_{0,k} = \frac{2}{N} \sum_{j=1}^{N-1} \sin \frac{jk\pi}{N} \bar{u}_{0,j}, \qquad \tilde{u}_{N,k} = \frac{2}{N} \sum_{j=1}^{N-1} \sin \frac{jk\pi}{N} \bar{u}_{N,j}.$$

With these, the sine transform of the equivalent difference equation obtained by substituting the variables defined in Step (1) into Eq. (3.142) can be written:

$$\sum_{k=0}^{N-1} \sin \frac{jk\pi}{N} \left[\bar{a}_i \tilde{u}_{i+1,k} - b_i \tilde{u}_{i,k} + \bar{c}_i \tilde{u}_{i-1,k} - \tilde{S}_{i,k} \right] = 0. \tag{3.144}$$

Here we have introduced the coefficient array $b_i = \bar{a}_i + \bar{c}_i + 2 - 2\cos\frac{k\pi}{N}$.

Step (3): In order to satisfy Eq. (3.144), each term in the bracket must vanish since it is multiplied by a different linearly independent function. Since each term corresponds to a different k value, and they are uncoupled, each is a standard tridiagonal equation in the index i. We may solve them using the method of Section 3.3.3 using $\tilde{u}_{0,k}$ and $\tilde{u}_{N,k}$ as boundary conditions to obtain $\tilde{u}_{i,k}$ for $i = 1, \cdots, N-1$.

Step (4): Transform $\tilde{u}_{i,k}$ back to $\bar{u}_{i,j}$ using the inverse sine transform. For $i, j = 1, \cdots N-1$, we have

$$\bar{u}_{i,j} = \sum_{k=0}^{N-1} \tilde{u}_{i,k} \sin \frac{jk\pi}{N}. \tag{3.145}$$

Step (5): Finally, construct $u_{i,j}$ from $\bar{u}_{i,j}$ by adding back the term subtracted in Step (1):

$$u_{i,j} = \bar{u}_{i,j} + (A_i + jB_i); \qquad i, j = 1, \cdots, N-1. \tag{3.146}$$

The computationally intensive work involved transforming the source function and boundary conditions, solving N tridiagonal equations, and transforming the solution back. Therefore, the total number of operations required to obtain a solution is about $N^2(4\ln N + 5)$. This is for an *exact solution* of the difference equations! The fast direct methods described here are generally more efficient than the methods discussed earlier in this chapter and should be used whenever applicable. In general, they can be used whenever the coefficient of the derivatives in the original differential equation are independent of at least one coordinate.

3.9   Summary

When elliptic equations are transformed into finite difference equations, they take the form of sparse matrix equations. There are special solvability constraints associated with Neumann boundary conditions. In 1D, the difference equations can be solved directly using the tridiagonal algorithm. In 2D or 3D there are many approaches to solving these sparse matrix systems. If the equations and the geometry are sufficiently simple, then transform techniques utilizing the FFT can and should be applied. If not, there are a number

of different options. In 2D, direct methods based on Gauss elimination may
be competitive with iterative methods, but not normally in 3D. Basic iterative methods can be defined by decomposing the matrix into diagonal, upper
triangular, and lower triangular parts and using these to apply the Jacobi,
Gauss–Seidel, or SOR algorithms. Each of these has an analogue in the physical approach of adding time derivatives to the original elliptic equation and
solving to steady-state. However, other methods such as dynamic relaxation
are more naturally derived using the physical approach. When these iterative
methods are combined with the multigrid algorithm they can be exceptionally
effective. Methods based on Krylov space projections, especially the preconditioned conjugate gradient method (if the coefficient matrix is symmetric)
and the preconditioned generalized minimum residual method, are also very
effective, especially for large systems.

Problems

3.1: Show that a derivative boundary condition at the origin, $u'(0)$, can be
incorporated in the matrix equation (3.4) to second order in h by specifying:

$$
\begin{aligned}
B_0 &= 1 \\
A_0 &= 1 \\
D_0 &= hu'(0) + \frac{1}{2}h^2 R(0).
\end{aligned}
$$

3.2: Tridiagonal matrix with corner elements: Suppose we wish to solve the
finite difference equation in Eq. (3.3) but with periodic boundary conditions.
(a) Show that the corresponding sparse matrix is tridiagonal in form but with
additional non-zero corner elements.
(b) Show that if you apply Gauss elimination to this matrix by first taking
linear combinations of rows to zero the band below the main diagonal (going
from top left to bottom right), and then taking linear combinations to zero
the band above the main diagonal (going from bottom right to top left), you
will be left with a matrix with only a main diagonal band and one or two
additional non-zero columns.
(c) Show that a 2×2 autonomous system splits off and can be solved for two
unknowns independent of the rest of the matrix.
(d) Once the 2×2 system is solved, show that the rest of the unknowns can
be solved by a form of back substitution.

3.3: Non-constant coefficients: Consider the general 1D elliptic equation:

$$
\frac{\partial}{\partial x} D(x) \frac{\partial u}{\partial x} = R(x),
$$

where $D(x)$ and $R(x)$ are given, and $u(x)$ is the unknown function to be solved for.

(a) Using centered finite differences, derive the matrix coefficients A_j, B_j, C_j, and D_j appearing in Eq. (3.5) for the *conservative differencing* form of the equation. (See Section 2.6.)

(b) Repeat for the *non-conservative* form of differencing this equation.

(c) What are the possible advantages of the conservative form of the difference equations over the non-conservative form?

3.4: Relation of tridiagonal algorithm to Gauss elimination: Show that defining the vectors E_j and F_j by Eq. (3.9) is equivalent to applying Gauss elimination to convert the matrix in Eq. (3.4) to an upper triangular matrix with 1s on the main diagonal, and a single upper diagonal band with $-E_j$ on row j and F_j in the right-side vector.

3.5: Convergence of the Gauss–Seidel method:

(a) Show that Eq. (3.147) is the eigenvalue equation that the Gauss–Seidel operator satisfies:

$$\frac{1}{4}\left(u^n_{j+1,k} + \lambda u^n_{j-1,k} + u^n_{j,k+1} + \lambda u^n_{j,k-1}\right) = \lambda u^n_{j,k}. \tag{3.147}$$

(b) Make the substitution

$$u^n_{j,k} = \lambda^{(j+k)/2} v_{j,k} \tag{3.148}$$

to obtain an eigenvalue equation for $v_{j,k}$ similar in form to Eq. (3.54). Solve this to obtain an expression for the lowest eigenvalue, and use this to estimate the convergence rate of the Gauss–Seidel method and compare it to that of the Jacobi method.

3.6: Successive over-relaxation: Consider the partial differential equation:

$$2\alpha\frac{\partial u}{\partial t} = \frac{\partial^2 u}{\partial x^2} + \frac{\partial^2 u}{\partial y^2} - \frac{\partial^2 u}{\partial x \partial t} - \frac{\partial^2 u}{\partial y \partial t} - R(x,y). \tag{3.149}$$

(a) Let $u^n_{j,k}$ be the finite difference approximation to $u(t^n, x_j, y_k)$ on an $N \times N$ grid of domain size $0 < x < L$ and $0 < y < L$, where $t^n = n\delta t$, $x_j = j\delta x$, and $y_k = k\delta y$. Write down a second-order finite difference approximation to Eq. (3.149) that is centered in space about the point x_j, y_k and is centered in time halfway between time t^n and t^{n+1}:

$$\frac{\partial^2 u}{\partial y \partial t} = \frac{1}{2\delta y \delta t}\left[u^{n+1}_{j,k+1} - u^{n+1}_{j,k-1} - u^n_{j,k+1} + u^n_{j,k-1}\right],$$

$$\frac{\partial^2 u}{\partial y^2} = \frac{1}{2\delta y^2}\left[u^{n+1}_{j,k\,|\,1} - 2u^{n+1}_{j,k} + u^{n+1}_{j,k-1} + u^n_{j,k\,|\,1} - 2u^n_{j,k} + u^n_{j,k-1}\right],$$

$$\frac{\partial u}{\partial t} = \frac{1}{\delta t}\left[u^{n+1}_{j,k} - u^n_{j,k}\right],$$

etc.

(b) Let $\delta t = \delta x = \delta y \equiv h = L/N$, and combine terms to show that the finite difference equation reduces to the successive over-relaxation formula, Eq. (3.49), if we identify

$$\omega = \frac{2}{\alpha h + 1}. \tag{3.150}$$

(c) Use von Neumann stability analysis to analyze the numerical stability of Eq. (3.49), and find the conditions on ω for stability by examining the critical points of the exponential functions.

(d) Examine Eq. (3.149) analytically by looking for a separable solution of the form:

$$u(x, y, t) \sim u_{\widetilde{m,n}} exp\left[-\lambda t + i\left(m\pi x/L + n\pi y/L\right)\right]. \tag{3.151}$$

Solve for the real part of λ for the slowest decaying mode and use this to estimate the number of iterations required for convergence as a function of the parameter α.

(e) Find the value of α (and hence ω) that minimizes the number of iterations, and estimate the number of iterations required at the optimal value.

3.7: Show how the stencil for the conjugate gradient method with preconditioning, Eq. (3.128), follows from the derivation given in Section 3.7.1 when we take the $\mathbf{p}_j$ to be a linear combination of $\gamma_0, \cdots, \gamma_j$, and thus it follows from Eq. (3.97) that $\gamma_i^T \cdot \mathbf{r}_j = 0$ for $i \neq j$.

3.8: Write down the equations for the FFT algorithm for $N = 3^\gamma$.

Chapter 4

Plasma Equilibrium

4.1 Introduction

The equilibrium equation,

$$\nabla p = \mathbf{J} \times \mathbf{B}, \tag{4.1}$$

plays a central role in MHD, on both the ideal and the transport time scales. On the ideal time scales, if the pressure tensor terms are negligible, this is the time independent solution of the ideal MHD equations with $\mathbf{u} = 0$. Linearized ideal MHD stability problems require these equilibrium solutions to perturb about. On the transport time scales, over long times when inertia is unimportant, this is the force balance equation. It constrains the pressure, magnetic field, and current density profiles, dictates what external fields are required to confine the plasma, and determines the plasma shape. In this chapter, we start by deriving the basic scalar partial differential equation that is equivalent to the vector equation (4.1). This equation was first derived independently by three groups [58, 59, 60], and is now known as the Grad–Shafranov–Schlüter (GSS) equation, or more commonly, the Grad–Shafranov equation.

4.2 Derivation of the Grad–Shafranov Equation

We will work in axisymmetric geometry and use standard cylindrical coordinates (R, ϕ, Z); see Figure 4.1. We seek axisymmetric solutions so that all derivatives with respect to ϕ are set to zero:

$$\frac{\partial}{\partial \phi} = 0. \tag{4.2}$$

The magnetic field $\mathbf{B}$ can be written as the curl of the vector potential,

$$\mathbf{A} = (A_R, A_\phi, A_Z).$$

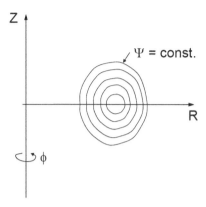

FIGURE 4.1: Cylindrical coordinates (R, ϕ, Z).

Using axisymmetry, Eq. (4.2), we have

$$\mathbf{B} = \nabla \times (A_\phi \hat{\phi}) + \left(\frac{\partial A_R}{\partial Z} - \frac{\partial A_Z}{\partial R} \right) \hat{\phi}, \tag{4.3}$$

where $\hat{\phi}$ is a unit vector in the ϕ direction. We define two new variables, the *poloidal flux function*,

$$\Psi(R, Z) = -R A_\phi,$$

and the *toroidal field function*,

$$g(R, Z) = R\left(\frac{\partial A_R}{\partial Z} - \frac{\partial A_Z}{\partial R} \right),$$

which at this point in the derivation is a general 2D function of R and Z. Also, recalling that the gradient of the coordinate ϕ is given by

$$\nabla \phi = \frac{1}{R} \hat{\phi},$$

we can write Eq. (4.3) in the form

$$\mathbf{B} = \nabla \phi \times \nabla \Psi(R, Z) + g(R, Z) \nabla \phi. \tag{4.4}$$

This is a general result for axisymmetric systems in that the magnetic field can always be expressed in terms of two scalar functions in this form, even if it is not an equilibrium field. The current density is calculated by taking the curl of Eq. (4.4),

$$\mu_0 \mathbf{J} = \Delta^* \Psi \nabla \phi + \nabla g \times \nabla \phi. \tag{4.5}$$

Here we have defined the toroidal elliptic operator,

$$\Delta^* \Psi \equiv R^2 \nabla \cdot \frac{1}{R^2} \nabla \Psi = R \frac{\partial}{\partial R} \left(\frac{1}{R} \frac{\partial \Psi}{\partial R} \right) + \frac{\partial^2 \Psi}{\partial Z^2}.$$

We substitute Eq. (4.4) for **B** and Eq. (4.5) for **J** into the equilibrium equation, Eq. (4.1). Taking the projection, or dot product, of Eq. (4.1) successively with the three vectors $\nabla\phi$, **B**, and $\nabla\Psi$, we obtain the three conditions

$$g = g(\Psi), \tag{4.6}$$

$$p = p(\Psi), \tag{4.7}$$

$$\Delta^*\Psi + \mu_0 R^2 \frac{dp}{d\Psi} + g\frac{dg}{d\Psi} = 0, \tag{4.8}$$

respectively. The plasma pressure, p, and the toroidal field function, g, are therefore both functions only of the poloidal flux, Ψ, in equilibrium, and the three scalar functions $\Psi(R, Z)$, $p(\Psi)$, and $g(\Psi)$ are related by the Grad–Shafranov equation, Eq. (4.8). Note that Eq. (4.8) is not of a standard form in that the functions $p(\Psi)$ and $g(\Psi)$ are differentiated by Ψ, which is the function being solved for. To solve these equations, we must prescribe the functional dependence of p and g on Ψ and also supply boundary conditions for Ψ.

4.2.1 Equilibrium with Toroidal Flow

Suppose the steady-state flow is non-negligible. It is still possible to obtain a scalar equilibrium equation similar to the Grad–Shafranov equation, but that allows for toroidal flow of arbitrary strength [61, 62]. Consider the steady-state scalar pressure force balance equation with flow included:

$$\rho\mathbf{u} \cdot \nabla\mathbf{u} + \nabla p = \mathbf{J} \times \mathbf{B}. \tag{4.9}$$

The most general toroidal velocity field that satisfies the ideal MHD steady-state condition which follows from Eq. (1.61),

$$\nabla \times (\mathbf{u} \times \mathbf{B}) = 0, \tag{4.10}$$

is given by

$$\mathbf{u} = R^2\omega(\Psi)\nabla\phi. \tag{4.11}$$

Here, we have introduced the toroidal angular velocity, $\omega(\Psi)$, which can be prescribed as an arbitrary function of the poloidal magnetic flux, Ψ.

In the presence of equilibrium flow, the plasma pressure will no longer be a function only of Ψ, but can depend also on the coordinate R. However, the temperature will remain almost constant on a magnetic surface and thus close to a function of Ψ because of the large parallel thermal conductivities. We can still express the pressure as the product of the density and temperature:

$$p(R, \Psi) = 2n_e(R, \Psi)k_B T(\Psi). \tag{4.12}$$

Here, we have defined a species average temperature for convenience: $T(\Psi) = .5\left[T_e(\Psi) + T_i(\Psi)\right]$, and have made use of the quasineutrality condition,

Eq. (1.27), to set $n_i = n_e$. The mass density and number density are related by $\rho(R, \Psi) = m_i n_e(R, \Psi)$, where m_i is the ion mass.

The $\nabla\phi$ component of Eq. (4.9) still gives Eq. (4.6), $g = g(\Psi)$. However, the other projections have an additional term present. We use the result that with the velocity of the form of Eq. (4.11), we have

$$\rho \mathbf{u} \cdot \nabla \mathbf{u} = -\rho \omega^2(\Psi) R \hat{R}. \tag{4.13}$$

Taking the dot product of Eq. (4.9) with $\mathbf{B}$ now gives, instead of Eq. (4.7), the constraint

$$\mathbf{B} \cdot \nabla \left[2k_B T(\Psi) \log n_e(R, \Psi) - \frac{1}{2} m_i \omega^2(\Psi) R^2 \right] = 0.$$

This constraint will be satisfied if we take the pressure and density to be of the form

$$p(R, \Psi) = 2n_e(R, \Psi)k_B T(\Psi) = p_0(\Psi) \exp \left[\frac{P_\omega(\Psi)}{p_0(\Psi)} \frac{(R^2 - R_0^2)}{R_0^2} \right]. \tag{4.14}$$

Here, R_0 is a nominal major radius introduced for convenience. We also introduce nominal number densities and mass densities defined as:

$$n_{e0}(\Psi) = n_e(R_0, \Psi), \tag{4.15}$$
$$\rho_0(\Psi) = m_i n_{e0}(\Psi) = m_i n_e(R_0, \Psi). \tag{4.16}$$

Using these definitions, the nominal fluid and rotation pressures are

$$p_0(\Psi) = 2n_{e0}(\Psi)k_B T(\Psi), \tag{4.17}$$
$$P_\omega(\Psi) = \frac{1}{2} R_0^2 \rho_0(\Psi)\omega^2(\Psi). \tag{4.18}$$

It follows that the partial derivatives of the pressure are given by

$$\frac{\partial}{\partial \Psi} p(R, \Psi) = \frac{p(R, \Psi)}{p_0(\Psi)} \tag{4.19}$$

$$\times \left[p_0'(\Psi) + \frac{R^2 - R_0^2}{R_0^2} \left(P_\omega'(\Psi) - \frac{P_\omega}{p_0} p_0'(\Psi) \right) \right], \tag{4.20}$$

$$\frac{\partial}{\partial R} p(R, \Psi) = \rho(R, \Psi)\omega^2(\Psi)R, \tag{4.21}$$

where we have used "prime" to denote differentiation with respect to Ψ, i.e. $()' \equiv \partial()/\partial\Psi$. This form for the pressure manifestly satisfies force balance in the $\hat{R}$ direction and allows us to write the component of Eq. (4.9) in the $\nabla\Psi$ direction as

$$\Delta^* \Psi + \mu_0 R^2 \frac{\partial}{\partial \Psi} p(R, \Psi) + g \frac{dg}{d\Psi} = 0. \tag{4.22}$$

This is the desired generalization of the Grad–Shafranov equation, Eq. (4.8), to include the effect of toroidal equilibrium flow. Problem specification requires the three functions: $g(\Psi)$, $P_\omega(\Psi)$, and $p_0(\Psi)$. The temperatures $T_e(\Psi)$ and $T_i(\Psi)$ are then needed to obtain the density from the pressure using Eq. (4.12) and the definitions that follow.

There have been several efforts in formulating and solving the equilibrium equation including an arbitrary *poloidal* flow as well as a toroidal flow [63, 64, 65]. This leads to a coupled non-linear system involving five free functions of Ψ. The coupled system involves an (algebraic) Bernoulli equation and a generalization of the Grad–Shafranov equation. Since the poloidal flow is strongly damped in most toroidal systems [66, 67], the conventional reasoning is that because of the additional complexity involved and the lack of physical guidance in selecting the five free functions, including poloidal flow in the Grad–Shafranov form of the equilibrium problem is not justified for most applications. However, there may be strongly driven systems where this is required.

An alternative to this approach is to solve the full two-dimensional (axisymmetric) resistive or two-fluid forms of the time-dependent MHD equations to steady state [68, 69]. This approach requires specifying the forms of sources of particles, energy, and current, as well as any externally supplied parallel electric field (or loop voltage).

4.2.2   Tensor Pressure Equilibrium

There are many equilibrium applications of interest where the plasma pressure is not well described by a single scalar function. Examples are a driven system where an anisotropy is present and sustained by intense neutral beam injection, or a system that preferentially loses a particular class of particles such as the loss cone in a magnetic mirror device. Here we consider the equilibrium described by a particular form of pressure tensor,

$$\mathbf{P} = p_\perp \mathbf{I} + (p_\parallel - p_\perp)\frac{\mathbf{BB}}{B^2}, \tag{4.23}$$

where $p_\parallel$ and $p_\perp$ are scalar functions representing the pressure parallel and perpendicular to the magnetic field direction. The equilibrium equation with this form for the pressure tensor becomes

$$\nabla \cdot \mathbf{P} = \mathbf{J} \times \mathbf{B}. \tag{4.24}$$

To proceed, we substitute Eq. (4.23) into Eq. (4.24) and make use of the identity that for an arbitrary scalar quantity α:

$$\begin{aligned}
\nabla \cdot [\alpha \mathbf{BB}] &= \mathbf{B}(\mathbf{B} \cdot \nabla \alpha) + \alpha \mathbf{B} \cdot \nabla \mathbf{B} \\
&= \mathbf{B}(\mathbf{B} \cdot \nabla \alpha) + \alpha \left[\frac{1}{2}\nabla B^2 + (\nabla \times \mathbf{B}) \times \mathbf{B}\right].
\end{aligned}$$

Here, we have defined the magnitude of the magnetic field as $B \equiv |\mathbf{B}|$. Using this identity with

$$\alpha = \frac{p_\parallel - p_\perp}{B^2},$$

the tensor pressure equilibrium equation becomes

$$\nabla p_\perp + \mathbf{B}(\mathbf{B} \cdot \nabla \alpha) + \frac{1}{2}\alpha \nabla B^2 = \left(\frac{1}{\mu_0} - \alpha\right)(\nabla \times \mathbf{B}) \times \mathbf{B}. \tag{4.25}$$

We proceed to take projections of this equation. Taking the dot product with the magnetic field $\mathbf{B}$ gives, after several cancellations,

$$\mathbf{B} \cdot \nabla p_\parallel = \frac{p_\parallel - p_\perp}{B}\mathbf{B} \cdot \nabla B,$$

or,

$$\left.\frac{\partial p_\parallel}{\partial B}\right|_\Psi = \frac{p_\parallel - p_\perp}{B}. \tag{4.26}$$

Upon defining the scalar quantity

$$\sigma = 1 - \mu_0 \alpha = 1 + \mu_0 \frac{p_\perp - p_\parallel}{B^2}, \tag{4.27}$$

the dot product of Eq. (4.25) with $\nabla \phi$ gives the condition that the product of σg be constant on a flux surface, i.e.,

$$\sigma g = g^*(\Psi), \tag{4.28}$$

for some function $g^*(\Psi)$. Using these, the component of the equilibrium equation in the $\nabla \Psi$ direction gives the tensor pressure analogue of the Grad–Shafranov equation [70]:

$$\Delta^*\Psi + \frac{\mu_0}{\sigma}R^2 \frac{\partial}{\partial \Psi}p_\parallel(B, \Psi) + \frac{1}{\sigma}\nabla \Psi \cdot \nabla \sigma + \frac{1}{\sigma^2}g^* \frac{d}{d\Psi}g^* = 0. \tag{4.29}$$

This now requires that the two functions $g^*(\Psi)$ and $p_\parallel(B, \Psi)$ be specified. From these, the function $p_\perp$ is determined from Eq. (4.26) and the function g is determined from Eqs. (4.28) and (4.27).

Analytic forms for $p_\parallel(B, \Psi)$ and $p_\perp(B, \Psi)$ have been proposed [71] that correspond to a neutral beam slowing down distribution function (where the pitch-angle scattering operator has been ignored) in an otherwise isotropic pressure plasma. If $\mu_B(\psi)$ is the value of the magnetic moment $\mu = 1/2m_i V_\perp^2/B$ of the beam particles at the outer midplane point of the flux surface Ψ, normalized to the total particle energy $(E = 1/2m_i V_\parallel^2 + 1/2m_i V_\perp^2)$ there, then the pressure can be written as

$$\begin{aligned}
p_\parallel(B, \Psi) &= p_0(\Psi) + P_B(\Psi)B\sqrt{1 - \mu_B(\Psi)B}, \\
p_\perp(B, \Psi) &= p_0(\Psi) + \frac{\frac{1}{2}P_B(\Psi)\mu_B(\Psi)B^2}{\sqrt{1 - \mu_B(\Psi)B}}.
\end{aligned} \tag{4.30}$$

Here, $p_0(\Psi)$ is the isotropic background pressure and $P_B(\Psi)$ is proportional to the beam deposition profile. One can readily verify that Eq. (4.26) is satisfied with these functions. Note that because of the singularity at the turning points, these functional forms are only appropriate for passing beam particles, i.e., we require $\mu_B(\Psi)B < 1$ everywhere in the plasma.

The form of Eq. (4.30) shows that the beam component of the pressure, $p_\perp$ in particular, will be larger on the inner edge (small major radius side) of the torus where the magnetic field is highest. From a kinetic standpoint, this effect is due to the fact that the hot ions move more slowly along the field lines on the inner edge of the torus and thus spend more time there on the average.

4.3   The Meaning of Ψ

The contours $\Psi(R, Z) = $ constant are closed within the plasma region, forming nested toroidal surfaces as shown in Figure 4.1. In the scalar pressure zero rotation case, these constant Ψ contours are also constant pressure contours since $p = p(\Psi)$. Both the magnetic field and the current lie inside these $\Psi = $ const surfaces since it follows from Eqs. (4.4), (4.5), and (4.6) that

$$\mathbf{B} \cdot \nabla\Psi = 0 \qquad (4.31)$$

and

$$\mathbf{J} \cdot \nabla\Psi = 0. \qquad (4.32)$$

We can calculate the magnetic flux through a disk lying in the $z = 0$ plane, as shown in Figure 4.2. By direct computation, using the form of $\mathbf{B}$ from Eq. (4.4) and $d\mathbf{A} = 2\pi R dR \hat{z}$, the poloidal flux is the integral over the disk shown

$$
\begin{aligned}
\Psi_{PF} &= \int \mathbf{B} \cdot d\mathbf{A} \\
&= -\int_0^R \frac{1}{R}\frac{\partial\Psi}{\partial R} 2\pi R dR \\
&= -2\pi\Psi.
\end{aligned}
$$

Thus, Ψ is the negative of the poloidal magnetic flux per radian.

It is interesting and instructive to look at the topology of the constant Ψ contours in both the plasma and the vacuum region at different times in a typical tokamak discharge as illustrated in Figure 4.3. We plot the contours in the upper half plane for an up-down symmetric configuration with only two sets of external coils: *vertical field coils* located at $R = 5m$ and $Z = \pm 2m$, and an *OH solenoid* located at $R = 1m$ and extending from $Z = $

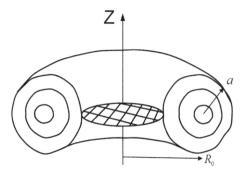

FIGURE 4.2: Calculate magnetic flux associated with disk in the $z = 0$ plane as shown.

$-2m$ to $Z = +2m$. This is the minimum set of coils needed for a tokamak. The vertical field coils produce the $\hat{z}$-directed external magnetic field required for a toroidal equilibrium. The required strength can be estimated from the Shafranov formula [72]:

$$B_Z^{ert} = -\frac{\mu_0 I_p}{4\pi R_0}\left[\log\frac{8R_0}{a} - \frac{3}{2} + \frac{\ell_i}{2} + \beta_\theta\right]. \tag{4.33}$$

Here I_P is the plasma current, R_0 is the plasma major radius, a is the approximate minor radius (see Figure 4.2), and the internal inductance and poloidal beta are defined in terms of integrals over the plasma volume as follows:

$$\beta_\theta \equiv \frac{4}{I_P^2\mu_0 R_0}\int p\,dV, \tag{4.34}$$

$$\ell_i \equiv \frac{2}{I_P^2\mu_0^2 R_0}\int \frac{|\nabla\Psi|^2}{R^2}\,dV. \tag{4.35}$$

Because these coils do not produce exactly a uniform vertical field, the field produced by them will also modify the plasma shape. This shape modification can be compensated for by additional coils as discussed below. The OH solenoid has toroidal currents that are programmed to change in time in order to produce and sustain the plasma current. Early in time, these currents are positive (in the same direction as the plasma current) as shown in Figure 4.3a. As the OH solenoid currents are decreased in magnitude, through zero as in Figure 4.3b and to a negative value as in Figure 4.3c, the contours in the plasma remain approximately similar in shape, but uniformly increase in value due to resistive diffusion as will be discussed in Chapter 6. The flux increase from the OH solenoid compensates for the flux increase in the plasma due to the resistive diffusion. As with the vertical field coils, the OH solenoid will also normally produce stray fields that need to be compensated for by

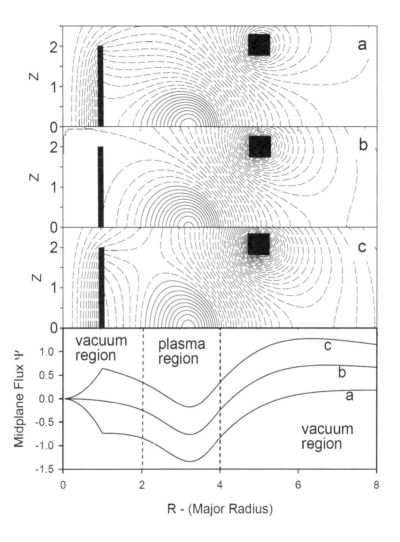

FIGURE 4.3: Parts a, b, c show the poloidal magnetic flux contours, $\Psi = const.$ in the upper half plane for a typical up-down symmetric toka-mak discharge at three different times. The bottom plot is of the midplane ($Z = 0$) value of the poloidal flux for the three times. The three snapshots correspond to (a) a positive toroidal current in the OH solenoidal coil located at R=1, (b) zero current in that coil, and (c) a negative current in the OH coil. Note that the flux contours in the plasma (solid lines) remain approximately similar in shape, but uniformly increase in value as time progresses, as shown in the bottom graph. This increase is due to resistive dissipation in the plasma as is described in Chapter 6. At the magnetic axis (O-point), this increase in time is described by the equation $\dot{\Psi} = (\eta/\mu_0)\Delta^*\Psi$, where η is the resistivity at the magnetic axis.

shaping coils. Note that not shown in Figure 4.3 is the current initiation and buildup phase, which occurs in the time before Figure 4.3a, and the current decay phase, which occurs in the time after Figure 4.3c.

A modern tokamak will have additional sets of coils as well as those shown in Figure 4.3. A number of *shaping coils* will be present to control the shape of the plasma-vacuum interface. If the plasma cross section is elongated, a vertical feedback system must also be present that can stabilize the vertical motion of the plasma [73, 74]. This will normally involve a pair of coils constrained to have up-down antisymmetric currents as discussed further in Section 4.6.3.

4.4   Exact Solutions

There are a restricted class of exact solutions to the Grad–Shafranov equilibrium equation, Eq. (4.8). We discuss next some useful solutions for axisymmetric vacuum fields, and in the following section a class of exact solutions for a particular set of tokamak profiles.

4.4.1   Vacuum Solution

In the vacuum region, both the pressure p and the current $\mathbf{J}$ are zero. From Eq. (4.5), the flux function Ψ satisfies the homogeneous elliptic equation

$$\Delta^* \Psi = 0, \tag{4.36}$$

subject to appropriate boundary conditions. Even when a plasma is present, the total magnetic field can be thought of as being the sum of the field that arises from the plasma currents, and the *external field* that is produced by nearby coils. We discuss Green's function methods for calculating the poloidal field produced by external coils in Section 4.6.3. However, if the coils are somewhat far away, it can be a useful approximation to represent their fields in terms of a multipolar expansion for Ψ, where each term in the expansion satisfies Eq. (4.36). The first few terms in such an expansion about the point $(R, Z) = (R_0, 0)$, constrained to be even about the $Z = 0$ midplane, are [75,

76]:

$$\Psi_0 = R_0^2,$$

$$\Psi_2 = \frac{1}{2}\left[R^2 - R_0^2\right],$$

$$\Psi_4 = \frac{1}{8R_0^2}\left[(R^2 - R_0^2)^2 - 4R^2 Z^2\right],$$

$$\Psi_6 = \frac{1}{24R_0^4}\left[(R^2 - R_0^2)^3 - 12R^2 Z^2(R^2 - R_0^2) + 8R^2 Z^4\right],$$

$$\Psi_8 = \frac{1}{320R_0^6}\left[5(R^2 - R_0^2)^4 - 120R^2 Z^2(R^2 - R_0^2)^2\right.$$
$$+ \left. 80Z^4 R^2(3R^2 - 2R_0^2) - 64R^2 Z^6\right].$$

These correspond to the even nullapole, dipole, quadrupole, hexapole, and octopole components of the external fields expanded about that point. The first few odd multipoles are:

$$\Psi_1 = 0,$$

$$\Psi_3 = \frac{1}{R_0}R^2 Z,$$

$$\Psi_5 = \frac{1}{6R_0^3}R^2 Z\left[3\left(R^2 - R_0^2\right) - 4Z^2\right],$$

$$\Psi_7 = \frac{1}{60R_0^5}\left[15R^2 Z\left(R^2 - R_0^2\right)^2 + 20R^2 Z^3\left(2R_0^2 - 3R^2\right) + 24R^2 Z^5\right],$$

$$\Psi_9 = \frac{1}{280R_0^7}R^2 Z\left[35\left(R^2 - R_0^2\right)^3 + 140Z^2\left(-2R^4 + 3R^2 R_0^2 - R_0^4\right)\right.$$
$$+ \left. 168Z^4\left(2R^2 - R_0^2\right) - 64Z^6\right].$$

For some studies, it may be advantageous to represent the external field as a superposition of these components rather than use the field produced by actual coils.

In an axisymmetric system, the toroidal field function g in vacuum must be constant, $g = g_0$, so that the toroidal field in vacuum is given by

$$\mathbf{B}_T = g_0\nabla\phi = \frac{g_0}{R}\hat{\phi}. \qquad (4.37)$$

It is straightforward to see the physical significance of the constant g_0. If we apply Ampere's law to a circle of radius R that is enclosed by the toroidal field (TF) magnets as shown in Figure 4.4, we find

$$g_0 = \frac{\mu_0 I_{TF}}{2\pi},$$

where I_{TF} is the total current in all the toroidal field magnets. In the region topologically exterior to the TF magnets, a similar argument shows $g_0 = 0$.

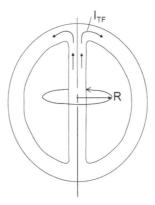

FIGURE 4.4: For an axisymmetric system, the vacuum toroidal field constant g_0 is proportional to the total current in the toroidal field coils, I_{TF}.

4.4.2 Shafranov–Solovév Solution

An exact solution [77] can be found for the particular choice of profiles satisfying

$$\frac{dp}{d\Psi} = -c_1, \qquad g\frac{dg}{d\Psi} = -c_2 R_0^2.$$

For real constants c_1, c_2, and R_0, the solution is given by

$$\Psi = \frac{1}{2}\left(c_2 R_0^2 + c_0 R^2\right)Z^2 + \frac{1}{8}\left(c_1 - c_0\right)\left(R^2 - R_0^2\right)^2, \qquad (4.38)$$

where c_0 is an arbitrary constant. A particularly useful choice for tokamak applications is to define new constants [78] such that $c_2 = 0$, $c_1 = B_0(\kappa_0^2 + 1)/\mu_0 R_0^2 \kappa_0 q_0$, and $c_0 = B_0/R_0^2 \kappa_0 q_0$. This gives, for $0 < \Psi < \Psi_B$:

$$\Psi = \frac{B_0}{2R_0^2 \kappa_0 q_0}\left[R^2 Z^2 + \frac{\kappa_0^2}{4}\left(R^2 - R_0^2\right)^2\right],$$

$$p(\Psi) = \frac{B_0\left(\kappa_0^2 + 1\right)}{\mu_0 R_0^2 \kappa_0 q_0}\left[\Psi_B - \Psi\right],$$

$$g = R_0 B_0. \qquad (4.39)$$

We can identify the constants as $R_0 = $ major radius of the magnetic axis, $B_0 = $ toroidal field strength at R_0, $\kappa_0 = $ ellipticity of the surfaces at the magnetic axis, and $q_0 = $ safety factor at magnetic axis (see Section 5.3). The constant Ψ_B is related to the approximate inverse aspect ratio, ϵ, by [78]

$$\epsilon \simeq \left[\frac{2q_0 \Psi_B}{\kappa_0 R_0^2 B_0}\right]^{\frac{1}{2}}. \qquad (4.40)$$

The solution for this class of profiles has also been extended by adding a homogeneous solution to this particular solution so that a wider variety of shapes can be represented [79, 80, 81, 82].

Other analytic solutions to the Grad–Shafranov equation have been found, but none as simple as the ones presented here. In particular, we note that Atanasiu et al. [83] and Guazzotto et al. [84] have published solutions where $p(\Psi)$ and $g^2(\Psi)$ are quadratic functions of Ψ, but they involve the evaluation of sums of Whittaker functions to fit boundary conditions, making them inconvenient for most applications.

4.5   Variational Forms of the Equilibrium Equation

It is often desirable to have a variational statement of a problem. One value of such a form is that it can be cast in new variables and the principle will still hold. To this end, we seek an energy functional W such that the condition that the variation of W vanish is equivalent to the plasma equilibrium equation, Eq. (4.1). It would suffice to find a functional W such that its variation is given by

$$\delta W = \int \boldsymbol{\xi} \cdot [\nabla p - \mathbf{J} \times \mathbf{B}] d^3 \mathbf{x}, \qquad (4.41)$$

for an arbitrary displacement field $\boldsymbol{\xi}(\mathbf{x})$. For, if Eq. (4.41) vanishes for all $\boldsymbol{\xi}$, then the equilibrium equation is satisfied everywhere. The form of W will depend on the constraints imposed during the variation.

Let us first consider the *ideal MHD* constraints that flux and entropy are conserved. This is the most physical of possible constraints, and the second variation determines the stability of the system (see Chapter 8). Since the displacement field $\boldsymbol{\xi}(\mathbf{x})$ is the time integral of the velocity field $\mathbf{u}$, we can obtain the ideal MHD constraints from Eqs. (1.61) and (1.63). Thus, we examine the functional

$$W = \int \left[\frac{B^2}{2\mu_0} + \frac{p}{\gamma - 1} \right] d^3 \mathbf{x}, \qquad (4.42)$$

subject to magnetic flux conservation

$$\delta \mathbf{B} = \nabla \times (\boldsymbol{\xi} \times \mathbf{B}), \qquad (4.43)$$

and entropy conservation

$$\begin{aligned} \delta p &= -\boldsymbol{\xi} \cdot \nabla p - \gamma p \nabla \cdot \boldsymbol{\xi}, \\ &= (\gamma - 1)\boldsymbol{\xi} \cdot \nabla p - \gamma \nabla \cdot p \boldsymbol{\xi}. \end{aligned} \qquad (4.44)$$

The first variation of Eq. (4.42) gives

$$\delta W = \int \left[\frac{1}{\mu_0} \mathbf{B} \cdot \delta \mathbf{B} + \frac{\delta p}{\gamma - 1} \right] d^3 \mathbf{x}. \qquad (4.45)$$

Using the constraints from Eqs. (4.43) and (4.44), Eq. (4.45) becomes

$$\delta W = \int \boldsymbol{\xi} \cdot (\nabla p - \mathbf{J} \times \mathbf{B}) \, d^3 x - \int dS \cdot \left\{ \frac{1}{\mu_0} \boldsymbol{\xi}_\perp B^2 + \frac{\gamma}{\gamma - 1} p \boldsymbol{\xi} \right\}. \quad (4.46)$$

If we have the additional condition that $dS \cdot \boldsymbol{\xi} = 0$ on the boundary, which also follows from flux and entropy conservation, then Eq. (4.46) is of the desired form of Eq. (4.41).

The ideal MHD variational form given by Eqs. (4.42) to (4.44) is valid in up to three-dimensional space. There is another useful variational statement due to Grad and Hirshman [85, 86], but it is only valid in two dimensions (axisymmetry). It is to vary the Lagrangian functional

$$L = \int \left[\frac{B_P^2}{2\mu_0} - \frac{B_T^2}{2\mu_0} - p(\Psi) \right] R \, dR \, dZ, \quad (4.47)$$

subject to the constraints

$$\begin{aligned} \mathbf{B}_P &= \nabla \phi \times \nabla \Psi, \\ \mathbf{B}_T &= g(\Psi) \nabla \phi, \\ p &= p(\Psi). \end{aligned} \quad (4.48)$$

Varying L in Eq. (4.47) with respect to $\Psi(R, Z)$ gives

$$\delta L = \quad - \int \delta \Psi \left[\nabla \cdot \frac{1}{\mu_0 R^2} \nabla \Psi + \frac{1}{\mu_0 R^2} g \frac{dg}{d\Psi} + \frac{dp}{d\Psi} \right] R dR dZ \quad (4.49)$$

$$+ \int dS \cdot \frac{1}{\mu_0 R^2} \nabla \Psi \delta \Psi = 0, \quad (4.50)$$

which implies that the Grad–Shafranov equation, Eq. (4.8), will be satisfied everywhere if we constrain $\delta \Psi = 0$ on the boundary. We return to these forms in Section 5.5.

4.6   Free Boundary Grad–Shafranov Equation

Here we describe an efficient method [87] for solving the Grad–Shafranov equation, Eq. (4.8), given a prescribed set of external currents and possibly some shape and position control feedback systems as described in Section 4.6.6. The partial differential equation we are considering is

$$\Delta^* \Psi + \mu_0 R^2 \frac{dp}{d\Psi} + g \frac{dg}{d\Psi} = 0, \quad (4.51)$$

or, in cylindrical coordinates,

$$R\frac{\partial}{\partial R}\frac{1}{R}\frac{\partial \Psi}{\partial R} + \frac{\partial^2 \Psi}{\partial Z^2} = \mu_0 R J_\phi(R, \Psi), \qquad (4.52)$$

where

$$\mu_0 R J_\phi(R, \Psi) = -(\mu_0 R^2 \frac{dp}{d\Psi} + g\frac{dg}{d\Psi}). \qquad (4.53)$$

In Eq. (4.51), the functions $p(\Psi)$ and $g(\Psi)$ are prescribed in a way which will be described shortly. We will describe the three major steps in solving Eq. (4.52): (i) inverting the elliptic operator in Eq. (4.52) for Ψ assuming the right-hand side (RHS) is known, (ii) iterating on the Ψ dependence in the RHS function $J_\phi(R, \Psi)$, and (iii) determining Ψ on the boundary.

4.6.1 Inverting the Elliptic Operator

We use the superscript n to denote the iteration level for the iteration on the non-linearity in the source function $J_\phi(R, \Psi)$. Thus, denote by J_ϕ^n the "old" value of this function which is assumed known. The 2D elliptic finite difference equation satisfied by $\Psi_{i,j}^{n+1} = \Psi(i\delta R, j\delta Z)$ at the new iteration level is given by

$$\frac{R_i}{\delta R}\left[\frac{1}{R}\frac{\partial \Psi}{\partial R}\bigg|_{i+1/2,j}^{n+1} - \frac{1}{R}\frac{\partial \Psi}{\partial R}\bigg|_{i-1/2,j}^{n+1}\right] + \frac{1}{\delta Z}\left[\frac{\partial \Psi}{\partial Z}\bigg|_{i,j+1/2}^{n+1} - \frac{\partial \Psi}{\partial Z}\bigg|_{i,j-1/2}^{n+1}\right]$$
$$= \mu_0 R_i J_{\phi i,j}^n.$$

or

$$\frac{R_i}{(\delta R)^2}\left[\frac{1}{R_{i+1/2}}\left(\Psi_{i+1,j}^{n+1} - \Psi_{i,j}^{n+1}\right) - \frac{1}{R_{i-1/2}}\left(\Psi_{i,j}^{n+1} - \Psi_{i-1,j}^{n+1}\right)\right]$$
$$+\frac{1}{(\delta Z)^2}\left[\Psi_{i,j+1}^{n+1} - 2\Psi_{i,j}^{n+1} + \Psi_{i,j-1}^{n+1}\right] = \mu_0 R_i J_{\phi i,j}^n. \qquad (4.54)$$

Note that the finite difference equation, Eq. (4.54), is in conservative form, and will thus satisfy the discrete form of Ampere's law exactly.

Equation (4.54) is of the form that can be solved by the transform method described in Section 3.8.2. Because of the factors involving the coordinate R multiplying some of the differences, the transform must be performed in the Z direction (index j), and the tridiagonal matrix equation is solved in the R direction (index i).

4.6.2 Iterating on $J_\phi(R, \Psi)$

The iteration we are considering here is due to the non-linear dependence of the right-hand side of Eq. (4.52) on the solution Ψ itself. The most straight-forward iteration scheme, which is called *Picard iteration*, just amounts to evaluating J_ϕ at the "old" value, as already discussed. Most often used is a slight generalization of Picard iteration that incorporates *blending* or *backaveraging*. It is defined as follows. We first calculate a provisional new Ψ value from the old current:

$$\Delta^* \widetilde{\Psi^{n+1}} = \mu_0 R J_\phi(R, \Psi^n). \tag{4.55}$$

The value of Ψ at the new time is then taken as a linear combination of the provisional new value and the old value. Thus, for some blending parameter $1 \geq \alpha_B > 0$, we define the new value as:

$$\Psi^{n+1} = \alpha_B \widetilde{\Psi^{n+1}} + (1 - \alpha_B)\Psi^n. \tag{4.56}$$

It has been found quasi-empirically that choosing a value of $\alpha_B \sim 0.5$ helps in convergence and eliminates the possibility of the solution converging to an odd-even limit cycle.

As a further aid in convergence, and to prevent converging to the trivial solution, we define the source function in such a way so as to keep certain physical quantities constant from one iteration to the next. A common constraint is to keep the total plasma toroidal current and the central value of the plasma pressure fixed. To this end, we define a normalized flux function,

$$\tilde{\Psi} = \frac{\Psi_\ell - \Psi}{\Psi_\ell - \Psi_0},$$

where Ψ_ℓ is the value of Ψ at the limiter (plasma edge) and Ψ_0 is the value at the magnetic axis. The normalized flux $\tilde{\Psi}$ is between 0 and 1 for all Ψ values in the interior of the plasma.

We then define the free functions $p(\Psi)$ and $g(\Psi)$ in terms of $\tilde{\Psi}$ and some constants that can be adjusted during the iteration to keep the prescribed global quantities fixed. One such prescription would be

$$p(\Psi) = p_0 \hat{p}(\tilde{\Psi}), \tag{4.57}$$

$$\frac{1}{2}g^2(\Psi) = \frac{1}{2}g_0^2[1 + \alpha_g \hat{g}(\tilde{\Psi})], \tag{4.58}$$

where p_0 is the fixed value of the central pressure, g_0 is the fixed vacuum value of the toroidal field function, and $\hat{p}(\Psi)$ and $\hat{g}(\Psi)$ are functions of the normalized flux, $\tilde{\Psi}$. For example,

$$\hat{p}(\tilde{\Psi}) = \tilde{\Psi}^{n_1},$$

$$\hat{g}(\tilde{\Psi}) = \tilde{\Psi}^{n_2}.$$

The central pressure is thereby held fixed by the form of Eq. (4.57). The constant α_g, introduced in Eq. (4.58), is adjusted during the iteration to keep the total plasma current fixed from one interaction to the next. We have introduced the integer exponents n_1 and n_2 which will determine the peakedness of the profiles. To implement the constraint on the total plasma current, we compute the total plasma current by summing the current density times the zone area over all grid points. From the equality in Eq. (4.53), we have:

$$
\begin{aligned}
I_P &= \sum_{i,j} \delta R \delta Z J^n_{\phi i,j} \\
&= \sum_{i,j} \left\{ \frac{R_i p_0}{\Delta \Psi} \hat{p}'(\tilde{\Psi}^n_{i,j}) + \frac{g_0^2 \alpha_g}{2\mu_0 R_i \Delta \Psi} \hat{g}'(\tilde{\Psi}^n_{i,j}) \right\} \delta R \delta Z,
\end{aligned}
\tag{4.59}
$$

where $\Delta \Psi = \Psi_\ell - \Psi_0$. We then solve Eq. (4.59) for α_g, keeping I_P fixed, to obtain

$$
\alpha_g = \mu_0 \left[\frac{-p_0 \sum_{i,j} R_i \hat{p}'(\tilde{\Psi}^n_{i,j}) + I_P \Delta \Psi / \delta R \delta Z}{\frac{1}{2} g_0^2 \sum_{i,j} \hat{g}'(\tilde{\Psi}^n_{i,j})/R_i} \right].
\tag{4.60}
$$

When this value for α_g is inserted back into Eq. (4.58), and J_ϕ is evaluated from Eq. (4.53), the total current will come out to be I_P.

It is often desirable to fix additional global parameters besides the central pressure and total plasma current. For example, one may wish to constrain the central value of the toroidal current density, or the central safety factor, q_0 (to be defined in Section 5.3). This is readily accomplished by including additional constants and functions in the definition of the pressure and toroidal field function. For example, Eq. (4.58) could be generalized as follows:

$$
\frac{1}{2} g^2(\Psi) = \frac{1}{2} g_0^2 [1 + \alpha_1 \hat{g}_1(\tilde{\Psi}) + \alpha_2 \hat{g}_2(\tilde{\Psi})].
$$

Here, $\hat{g}_1(\tilde{\Psi})$ and $\hat{g}_2(\tilde{\Psi})$ need to be linearly independent functions and α_1 and α_2 are determined in a way analogous to Eq. (4.60) to enforce the constraints.

For many equilibrium applications, it is not the normalized pressure and form for the toroidal field function that are to be held fixed, but rather other quantities involving averages over the magnetic surfaces. In particular, the parallel (to the magnetic field) current inside a given surface, and the safety factor profile (inverse rotational transform) have special significance and importance. We will return to techniques for solving for equilibria that hold these profiles fixed in Section 5.5.

4.6.3 Determining Ψ on the Boundary

Up until this point, we have been assuming that the value of Ψ is known on the boundary of the computational domain. If this is the case, we have seen how to invert the elliptic operation $\Delta^* \Psi$, and how to iterate on the nonlinearity in the source function $J_\phi(R, \Psi)$. Here we discuss how to determine the value of Ψ on the boundary if what is given is the values of current I_i in N_c discrete conductors located at coordinates R^c_i, Z^c_i as shown in Figure 4.5.

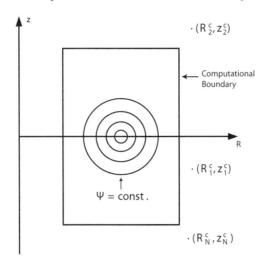

FIGURE 4.5: Poloidal flux Ψ on the boundary of the computational domain is obtained from a Green's function. External coils are represented as discrete circular current loops.

We use the fact that there is a Green's function for the toroidal elliptic operator Δ^*. It is given by

$$G(\mathbf{R};\mathbf{R}') = \frac{1}{2\pi}\frac{\sqrt{RR'}}{k}[(2-k^2)K(k) - 2E(k)], \qquad (4.61)$$

where $K(k)$ and $E(k)$ are complete elliptic integrals of the first and second kind, and the argument is given by

$$k^2 = \frac{4RR'}{[(R+R')^2 + (Z-Z')^2]}. \qquad (4.62)$$

This is derived in Jackson [88] as the vector potential due to an axisymmetric current source. The Green's function, $G(\mathbf{R};\mathbf{R}')$, and the flux function, $\Psi(\mathbf{R})$, satisfy

$$
\begin{aligned}
\Delta^* G(\mathbf{R};\mathbf{R}') &= R\delta(R-R')\delta(Z-Z'), \\
\Delta^* \Psi(\mathbf{R}) &= \mu_0 R J_\phi(\mathbf{R}), \quad \text{in the plasma region} \\
&= \mu_0 \sum_{i=1}^{N_c} R I_i \delta(R - R_i^c)\delta(Z - Z_i^c) \quad \text{in vacuum.} \quad (4.63)
\end{aligned}
$$

The form of Green's theorem we will need is

$$
\begin{aligned}
\nabla \cdot &\left[\Psi \frac{1}{R^2}\nabla G(\mathbf{R};\mathbf{R}') - G(\mathbf{R};\mathbf{R}')\frac{1}{R^2}\nabla\Psi\right] \\
&= \frac{1}{R^2}\Psi\Delta^* G(\mathbf{R};\mathbf{R}') - \frac{1}{R^2}G(\mathbf{R};\mathbf{R}')\Delta^*\Psi. \quad (4.64)
\end{aligned}
$$

We integrate Eq. (4.64) over all space, letting the observation point $\mathbf{R}'$ lie on the boundary of the computational grid; see Figure 4.5. We use Gauss's theorem to convert the divergences to surface integrals which vanish at infinity. This gives the result that Ψ at a boundary point (R', Z') is given by

$$\Psi_b(R', Z') = \int_{\mathrm{P}} G(R, Z; R', Z') J_\phi(R, Z) dR dZ$$

$$+ \sum_{i=1}^{N_c} G(R_i^c, Z_i^c; R', Z') I_i, \tag{4.65}$$

where the integral is over the plasma area. The right side of Eq. (4.65) consists of two parts, the contribution from the plasma currents and the contribution from the external currents. Let us consider how to evaluate the first part. The integral in Eq. (4.65) can be approximated by a sum. If the computational grid has N grid points in each direction, we would approximate the two-dimensional integral over the plasma by

$$\Psi_b(R', Z') = \int_{\mathrm{P}} G(R, Z; R', Z') J_\phi(R, Z) dR dZ$$

$$\simeq \sum_{i=1}^{N} \sum_{j=1}^{N} G(R_i, Z_j; R', Z') J_\phi(R_i, Z_j) \delta R \delta Z. \tag{4.66}$$

The double sum appearing in Eq. (4.66) is expensive to evaluate, requiring N^2 evaluations of the Green's function for each of the $4N$ boundary points. There is another way to evaluate the contribution from the plasma currents, due to von Hagenow and Lackner [89], which we now discuss.

4.6.4 Von Hagenow's Method

Consider now a function $U(R, Z)$ that satisfies the same differential equation that Ψ does in the interior of the computational boundary, but which vanishes on the boundary:

$$\Delta^* U = \Delta^* \Psi = \mu_0 R J_\phi,$$
$$U = 0 \qquad \text{on boundary.} \tag{4.67}$$

We use the following form of Green's theorem:

$$\nabla \cdot \left[U \frac{1}{R^2} \nabla G(\mathbf{R}; \mathbf{R}') \right] - \nabla \cdot \left[G(\mathbf{R}; \mathbf{R}') \frac{1}{R^2} \nabla U \right]$$

$$= \frac{1}{R^2} U \Delta^* G(\mathbf{R}; \mathbf{R}') - \frac{1}{R^2} G(\mathbf{R}; \mathbf{R}') \Delta^* U. \tag{4.68}$$

As shown in Figure 4.6, we take the observation point $\mathbf{R}'$ a small distance ϵ outside the computational boundary, and take the limit as $\epsilon \to 0$. Integrating

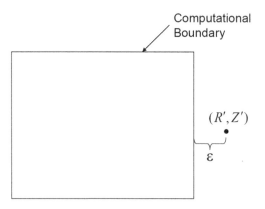

FIGURE 4.6: Singularity due to self-field term is resolved by taking the limit as ϵ approaches zero from the outside

Eq. (4.68) over the *computational domain only*, using Gauss's theorem, and evaluating the differential volume element as $dV = 2\pi R dR dZ$ gives

$$\int \frac{dS}{R^2} U \frac{\partial G(\mathbf{R};\mathbf{R}')}{\partial n} - \int \frac{dS}{R^2} G(\mathbf{R};\mathbf{R}') \frac{\partial U}{\partial n}$$
$$= -2\pi\mu_0 \int G(\mathbf{R};\mathbf{R}') J_\phi(R) dR dZ, \qquad (4.69)$$

where the surface integrals are over the computational boundary. Now, let the differential surface element be related to the differential line element by $dS = 2\pi R d\ell$ and note that the first term in Eq. (4.69) vanishes due to the fact that $U = 0$ on the boundary, Eq. (4.67). We are left with the identity

$$\int_P G(\mathbf{R};\mathbf{R}') J_\phi(R) dR dZ = \int_b \frac{d\ell}{R} G(\mathbf{R};\mathbf{R}') \frac{\partial U}{\partial n}, \qquad (4.70)$$

where the line integral is now over the boundary of the computational domain. The right side of Eq. (4.70) requires considerable less work to evaluate than does the left side as it is a line integral rather than an area integral. We have seen that the solution U can be obtained using a fast direct elliptic solver in $\mathcal{O}(N \ln N)$ machine operations. Add this to the $\mathcal{O}(N^2)$ calculations of the Green's function needed to evaluate the line integral in Eq. (4.70) for each boundary point, and it is still much less expensive than the $\mathcal{O}(N^3)$ evaluations needed to evaluate the area sum in Eq. (4.66). The final expression to obtain Ψ on the computational boundary is

$$\Psi_b(R',Z') = \int_b \frac{d\ell}{R} G(R,Z;R',Z') \frac{\partial U}{\partial n} + \sum_{i=1}^{N_c} G(R_i^c, Z_i^c; R', Z') I_i. \qquad (4.71)$$

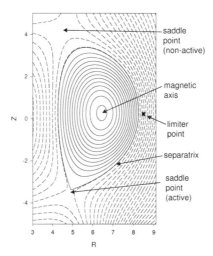

FIGURE 4.7: Poloidal flux at magnetic axis, Ψ_0, is a local minimum. Limiter value of flux, Ψ_ℓ, is the minimum of the value of the flux at the limiter points and at the saddle points.

Note that the derivation of this formula assumed that $\Psi(R', Z')$ lies outside or on the computational boundary, and not interior to it. Were the observation point to lie inside the computational boundary, the first term on the right in Eq. (4.68) would have given a contribution when integrated over the domain, and the resulting expression for the flux at a point interior to the computational boundary would be

$$\Psi_{interior}(R', Z') = \int_b \frac{d\ell}{R} G(R, Z; R', Z') \frac{\partial U}{\partial n} + U(R', Z')$$

$$+ \sum_{i=1}^{N_c} G(R_i^c, Z_i^c; R', Z') I_i. \qquad (4.72)$$

4.6.5 Calculation of the Critical Points

In Section 4.6.2 we referred to the value of Ψ at the magnetic axis as Ψ_0 and the value of Ψ at the plasma-vacuum boundary as Ψ_ℓ. The calculation of these during the iteration involves searching for local extremum values of Ψ: minima and saddle points.

The basic technique is to apply a multidimensional Newton iteration to look for points where $\nabla\Psi = 0$. If we use a crude search of mesh values to locate a point (R, Z) near a critical point, we can calculate the approximate location of the critical point as $(R + \delta R, Z + \delta Z)$, where:

$$\begin{bmatrix} \delta R \\ \delta Z \end{bmatrix} = \frac{1}{D} \begin{bmatrix} -\Psi_{ZZ} & \Psi_{RZ} \\ \Psi_{RZ} & -\Psi_{RR} \end{bmatrix} \cdot \begin{bmatrix} \Psi_R \\ \Psi_Z \end{bmatrix}. \tag{4.73}$$

Here, we have used subscripts to denote partial differentiation, and have defined:

$$D = \Psi_{RR}\Psi_{ZZ} - \Psi_{RZ}^2. \tag{4.74}$$

This can be iterated until $|\nabla\Psi| = 0$ to some tolerance. If the critical point so found has $D > 0$, it is a minimum, and if it has $D < 0$, it corresponds to a saddle point and an associated separatrix. The first and second derivatives can be calculated using a local bivariate interpolation routine that enforces continuous first derivatives [90].

In most free boundary equilibrium calculations, there are a number of fixed *limiter points* as well, corresponding to the location of nearby material structures. The outermost flux value in the plasma, Ψ_ℓ, is then defined as the minimum Ψ contour that either passes through a limiter point or a saddle point; see Figure 4.7.

4.6.6   Magnetic Feedback Systems

The boundary value iteration described in Section 4.6.3 normally involves the implementation of one or more magnetic feedback systems. These can be classified as (i) radial position control, (ii) vertical position control, and (iii) shape control.

The radial position feedback system is needed because Eq. (4.33) only gives an approximate condition for the vertical component of the external magnetic field required for a plasma of minor radius a and major radius R_0. It is common to use this formula as the initial guess for the vertical field, and to add to this an additional field determined from a feedback system designed to locate the plasma major radius at a particular position.

The basic magnetic feedback system strives to make the difference in the poloidal flux at two locations zero. Consider the feedback algorithm that increases the current in a coil located at (R_i^c, Z_i^c) for each iteration in proportion to the difference in the measured poloidal flux at two locations (R_1, Z_1) and (R_2, Z_2). We can write this as

$$\delta I_i = \alpha \times [\Psi(R_1, Z_1) - \Psi(R_2, Z_2)], \tag{4.75}$$

for some proportionality constant α. It is convenient to define the normalized proportionality constant α^* so that in the absence of the plasma response, Eq. (4.75) would be satisfied exactly for $\alpha^* = 1$. Using the Green's functions from Section 4.6.3, we have:

$$\alpha = -\frac{\alpha^*}{[G(R_1, Z_1; R_i^c, Z_i^c) - G(R_2, Z_2; R_i^c, Z_i^c)]}. \tag{4.76}$$

For $\alpha^* > 0$, the vacuum field response will be such as to decrease the flux difference. In general, we would choose α^* as large as possible in order to reduce the flux difference to near zero, but if the gain is chosen to be too large, the feedback system itself can become unstable.

A vertical position feedback system is required to keep the plasma from drifting off in the Z direction if the system is physically unstable to a vertical displacement and if the solution is not constrained to be symmetric about the midplane [73, 74]. Basic physics considerations show that in order to make a shaped plasma, the external "vertical" field will have some curvature to it. The externally produced magnetic field curvature index n is defined as

$$n = - \frac{R}{B_Z} \frac{\partial B_Z}{\partial R}\bigg|_{R_0}. \tag{4.77}$$

If n defined this way is negative, the system will be unstable to a rigid vertical displacement and a vertical position feedback system must be present, both in the equilibrium calculation and in the experiment. This can be stabilized in the equilibrium calculation in a manner similar to what was described above, but with the observation points chosen symmetrically above and below the desired Z location of the plasma and with the feedback coil system chosen to be one that is effective in producing a radial magnetic field. Such a system would be a a pair of coils located some distance $\pm L_Z$ with respect to the midplane, and with currents constrained to be of opposite polarity.

Shape control systems can be defined by taking superpositions of the basic feedback system defined above. However, once one goes beyond the basic vertical and radial feedback systems, it becomes increasingly important that the different feedback systems be nearly orthogonal, otherwise they will interfere with one another and potentially cause instabilities.

If enough shape control points are present, the coil currents are best determined by solving the overdetermined problem of finding the least squares error [87, 89] in the poloidal flux at the M plasma boundary points (R_j^b, Z_j^b) produced by N coils at set locations $I_i(R_i^c, Z_i^c)$ with $M > N$. Assuming that the plasma boundary points lie interior to the computational boundary, we use Eq. (4.72) to express the function we seek to minimize as

$$\epsilon = \sum_{j=1}^{M} \left[\int \frac{d\ell}{R} G(\mathbf{R}; \mathbf{R}_j^b) \frac{\partial U}{\partial n} + U(\mathbf{R}_j^b) + \sum_{i=1}^{N} G(\mathbf{R}_i^c; \mathbf{R}_j^b) I_i - \Psi^b \right]^2 \tag{4.78}$$

$$+ \gamma \sum_{i=1}^{N} I_i^2,$$

where Ψ^b is the desired value of flux at the boundary points (R_j^b, Z_j^b), $j = 1, M$. Differentiating ϵ with respect to each of the coil currents and setting to zero leads to a linear matrix equation that can be solved for the coil currents I_i. Note that the last term, with a small positive value for γ, is needed to keep the system from becoming ill posed [89].

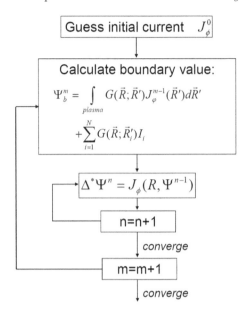

FIGURE 4.8: A double-nested iteration loop is used to converge the interior and boundary values.

4.6.7   Summary of Numerical Solution

Finally, we summarize the equilibrium algorithm in Figure 4.8. Initially, we guess a distribution for the plasma current. We use this to give a first guess for the flux on the boundary using Eq. (4.71). With this boundary flux held fixed, we solve for a consistent solution in the interior. When this converges, we use the current distribution so computed to update the boundary flux. This whole process is repeated until the flux stops changing between successive iterations.

4.7   Experimental Equilibrium Reconstruction

An important application of the equilibrium equation is to interpret magnetic and other diagnostic measurements in order to reconstruct the plasma equilibrium [62, 91, 92]. This is an essential capability that is part of all modern magnetic fusion experiments. For this application, instead of the plasma pressure and toroidal field functions being prescribed, they are expanded in basis functions, and the coefficients in the expansion are determined as best fits to whatever magnetic and other measurements are available.

In the most straightforward equilibrium reconstruction problem, the flux,

Ψ, is measured at N locations close to the plasma by an array of axisymmetric *flux loops* by time integrating the voltage induced in the loop using Eq. (6.13). Thus, the data consist of an array of measured values Ψ_i^M corresponding to locations (R_i, Z_i). The pressure and toroidal field functions are expanded in terms of known basis functions:

$$p'(\Psi) = \sum_j \alpha_j y_j(\tilde{\psi}), \qquad gg'(\Psi) = \sum_j \beta_j y_j(\tilde{\psi}). \qquad (4.79)$$

Here, $\tilde{\psi} = (\Psi_l - \Psi)/(\Psi_l - \Psi_0)$ is the normalized poloidal magnetic flux, with Ψ_0 and Ψ_l the poloidal magnetic flux at the magnetic axis and the plasma boundary, and $0 < \tilde{\psi} < 1$. Simple polynomial basis functions would correspond to $y_j(\tilde{\psi}) = \tilde{\psi}^j$, although it is generally found to be desirable to choose a basis that incorporates the boundary conditions and is more nearly orthogonal.

We then use the Green's function formula, Eq. (4.65), to express the computed flux at the flux loop locations as a function of the α_j and β_j, as well as the measured coil currents in the external coils:

$$\begin{aligned} \Psi_i^C(R_i, Z_i) &= \int_P G(R, Z; R_i, Z_i) J_\phi(R, \psi, \alpha_j, \beta_j) dRdZ \\ &+ \sum_{l=1}^{N_c} G(R_l^c, Z_l^c; R_i, Z_i) I_l. \end{aligned} \qquad (4.80)$$

The problem is then best formulated as a *least squares* problem where at each iteration, we choose the α_j and β_j to minimize the weighted sum of the squares of the error. For each measurement Ψ_i^M we define an experimental uncertainty weighting value σ_i and use this to define the error function,

$$\chi^2 = \Sigma_i \left(\frac{\Psi_i^M - \Psi_i^C}{\sigma_i} \right)^2. \qquad (4.81)$$

Differentiating with respect to the α_j and β_j leads to the matrix equation:

$$\mathbf{R} \cdot \mathbf{x} = \mathbf{M}, \qquad (4.82)$$

which is solved for the unknown $\mathbf{x} = (\alpha_j, \beta_j)$ at each step in the iteration for the plasma current. This equilibrium reconstruction technique can be readily extended to include toroidal flow [62] or anisotropic pressure [93].

4.8 Summary

The plasma equilibrium equation is probably the most widely used equation in plasma physics. It is extensively used in theoretical studies and in the design and interpretation of experiments. The basic Grad–Shafranov equation is readily extended to handle equilibrium states with flow and with a particular type of anisotropic pressure tensor. It is seen that the poloidal flux function Ψ is of central importance in equilibrium studies. There are some limited analytic solutions for $\Psi(R, Z)$, but generally it needs to be solved for numerically. We described the standard technique for solving for Ψ if a given set of coils are present, and also how to use this equation to reconstruct many features of an equilibrium from magnetic measurements. These applications are normally called the *free boundary* solution and equilibrium reconstruction methods. We will return to this equation in the next chapter to discuss *fixed boundary* techniques using magnetic flux coordinates to produce an equilibrium with a given shape for the last closed flux surface.

Problems

4.1: Verify that the components of the equilibrium equation with flow, Eq. (4.9), in the ∇R and the $\nabla \Psi$ directions are satisfied for the velocity **u** given by Eq. (4.11) and the pressure p given by Eq. (4.14), if Eq. (4.22) is satisfied.

4.2: Verify that the forms of the parallel and perpendicular pressure profiles given by Eq. (4.30) satisfy the constraint given in Eq. (4.26).

4.3: Evaluate ℓ_i from Eq. (4.35) in the *large aspect ratio, circular cross section* limit where the major radius $R \to R_0$, a constant, and the minor radius is a, where $a \ll R_0$. Using circular, cylindrical coordinates, assume the current density has the form:

$$J(r) = \nabla^2 \Psi = \frac{1}{r}\frac{d}{dr} r \frac{d\Psi}{dr} = J_0 \left(1 - \left(\frac{r}{a}\right)^\alpha\right),$$

where α is a constant.

4.4: Show that at large aspect ratio, $(R^2 + Z^2)/R_0^2 \gg 1$, substitution of $R = R_0 + X$ into the expressions in Section 4.4.1 for multipolar fields yields to lowest order

$$\Psi_{2m} + i\Psi_{2m+1} = \frac{(X + iZ)^m}{mR_0^{m-2}}.$$

4.5: Verify that the analytic functions given in Eq. (4.39) satisfy the Grad–Shafranov equation, Eq. (4.8).

4.6: Ampere's law is that for any closed curve $\delta\Omega$ surrounding an area Ω, we have the integral relation:

$$\oint_{\delta\Omega} \mathbf{B} \cdot \mathbf{d}\ell = \mu_0 \int_{\Omega} \mathbf{J} \cdot \mathbf{dA}.$$

Show that the discrete form of this equation is exactly satisfied if the conservative difference formula Eq. (4.54) is used and the integration contour is chosen correctly.

4.7: How is Eq. (4.69) modified if we take the limit that as $\epsilon \to 0$, we approach the boundary from *inside* the computational domain?

4.8: Derive the Newton iteration formula, Eq. (4.73), by Taylor expanding Ψ_R and Ψ_Z about the point (R, Z) and solving for the point where they both vanish.

4.9: Since the equilibrium equation, Eq. (4.8), only depends on the toroidal field function $g(\Psi)$ through $gdg/d\Psi$, show that once a single solution for $\Psi(R, Z)$ is obtained, a family of related solutions can be generated by varying the boundary value g_0, Eq. (4.37). This is called *Bateman scaling* [94]. What similarities do the different solutions in this family have? What differences?

Chapter 5

Magnetic Flux Coordinates in a Torus

5.1 Introduction

A strongly magnetized high-temperature plasma is very anisotropic due to the presence of the magnetic field itself. Physically, this anisotropy is due to the fact that particles are relatively free to stream in the direction along the field, but exhibit gyro-orbits in the directions perpendicular to the field, and also due to the fact that two out of three of the wave characteristics are aligned with the field direction. Mathematically, this anisotropy manifests itself in the fact that the MHD equations are higher order in the direction along the magnetic field than in the directions perpendicular to it. Because of this anisotropy, there is often considerable simplification to be gained in performing analysis or numerical solutions using a coordinate system aligned with the magnetic field. Here, we lay some of the groundwork for these coordinate systems in toroidal geometry [95, 96, 97, 98].

5.2 Preliminaries

To define a coordinate system, we need to introduce three spatial functions

$$\psi(\mathbf{x}), \theta(\mathbf{x}), \phi(\mathbf{x}),$$

which have *non-coplanar gradients*. Mathematically, this is a condition on the triple vector product of the gradients, or the *Jacobian J*,

$$\nabla\psi \times \nabla\theta \cdot \nabla\phi \equiv J^{-1} \geq 0, \tag{5.1}$$

where the equality can hold only at isolated singular points, for example at the origin of the coordinate system (or at the magnetic axis).

The infinitesimal volume element associated with the differentials $d\psi, d\theta, d\phi$ is given by

$$d\tau = Jd\psi d\theta d\phi = \frac{d\psi d\theta d\phi}{[\nabla\psi \times \nabla\theta \cdot \nabla\phi]} . \tag{5.2}$$

To make this a magnetic flux coordinate system, we choose ψ to be a flux function, i.e.,

$$\mathbf{B} \cdot \nabla \psi = 0. \tag{5.3}$$

We take θ to be a poloidal angle (increasing the short way around the torus) and ϕ to be a toroidal angle (increasing the long way around the torus), each varying between 0 and 2π:

$$0 \leq \theta < 2\pi , \qquad\qquad 0 \leq \phi < 2\pi .$$

Here we will choose ϕ to be the same angle as that in a cylindrical coordinate system, making these *axisymmetric* magnetic flux coordinates. Thus, we have $\partial / \partial \phi = 0$ for axisymmetric quantities, and $|\nabla \phi|^2 = 1/R^2$. This is not the only choice possible for the toroidal angle ϕ, but it is a convenient one for calculating if the geometry is axisymmetric such as in a tokamak or reversed field pinch. This restriction is relaxed in Section 5.4.2.

All physical quantities must be periodic functions of the toroidal angle ϕ, and of the poloidal angle θ if the $\psi = $ constant contours form closed nested surfaces, as is normally the case of interest. These coordinates are illustrated in Figure 5.1.

5.2.1 Jacobian

The magnetic flux coordinate ψ need not be the poloidal flux function Ψ, it need only be some function of the poloidal flux so that $\mathbf{B} \cdot \nabla \psi = 0$. For example, ψ could be identified with the toroidal flux, or with the volume inside a $\psi = $ constant surface, or with the pressure, if it is strictly a function of ψ. Also, the poloidal angle θ could vary between 0 and 2π in many different ways.

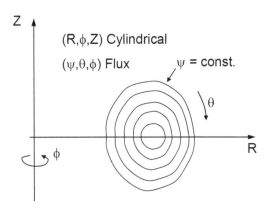

FIGURE 5.1: Flux coordinates use a magnetic flux ψ and a poloidal angle θ.

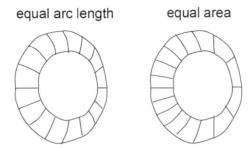

FIGURE 5.2: Two definitions of the angle θ are illustrated. Many more are possible.

For example, it could increase in such a way that the *arc length* $d\ell$ along a constant ψ contour is everywhere proportional to $d\theta$ or in such a way that the area associated with an infinitesimal area element $dA = (J/R)d\theta d\psi$ is equal to a constant as the constant ψ contour is transversed. These two possibilities are illustrated in Figure 5.2, but many other definitions of the angle θ are, of course, possible. We will return to this in Section 5.3.

Note that in general, $\nabla\theta \cdot \nabla\psi \neq 0$, so that we will have a *non-orthogonal* coordinate system. The complication due to this is more than compensated for by the flexibility which can be used to achieve additional simplifications. In axisymmetric magnetic flux coordinates, both $\nabla\psi$ and $\nabla\theta$ lie in constant ϕ planes so that $\nabla\theta \cdot \nabla\phi = 0$ and $\nabla\psi \cdot \nabla\phi = 0$.

It is usual to complete the definition of the ψ and the θ coordinates by specifying the functional form of the Jacobian J. To see how this uniquely defines the coordinate ψ, we go back to the definition of the differential volume element in cylindrical and flux coordinates and equate these:

$$dR\,dZ\,Rd\phi \;=\; Jd\psi d\theta d\phi.$$

Noting that the angle ϕ is the same in both coordinate systems, and dividing by the Jacobian gives

$$\frac{dR\,dZ\,R}{J} \;=\; d\psi d\theta. \tag{5.4}$$

Next integrate both sides of Eq. (5.4) over the area interior to the contour ψ = const as shown in Figure 5.3a. Noting that the θ integral on the right side of Eq. (5.4) is 2π, we obtain

$$\iint_A \frac{R\,dR\,dZ}{J} = 2\pi \int_{\psi_0}^{\psi} d\psi', \tag{5.5}$$

or,

$$\psi - \psi_0 = \frac{1}{2\pi} \iint_A \frac{R\,dR\,dZ}{J}, \tag{5.6}$$

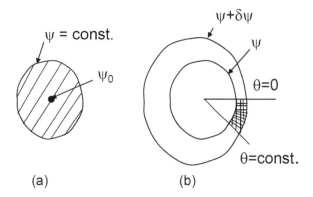

FIGURE 5.3: Choosing the Jacobian determines both the (a) ψ and the (b) θ coordinate.

where the integral is over the area inside the contour $\psi = $ const. Similarly, if we integrate Eq. (5.4) over the shaded area in Figure 5.3b, which is defined by

$$\psi < \psi' < \psi + \delta\psi \; , \; 0 \le \theta' < \theta,$$

we obtain the defining relation for θ, where A is the shaded area in Figure 5.3b. Therefore, both θ and ψ are determined (up to a constant) if the form of the Jacobian is given.

$$\theta = \frac{1}{\delta\psi} \int\!\!\int_A \frac{dR dZ\, R}{J},$$

5.2.2   Basis Vectors

There are two separate systems of basis vectors formed from the gradients of the coordinate functions ψ, θ, ϕ. These are the *covariant* basis

$$\nabla\psi, \qquad \nabla\theta, \qquad \nabla\phi,$$

and the *contravariant* basis

$$\nabla\theta \times \nabla\phi J, \qquad \nabla\phi \times \nabla\psi J, \qquad \nabla\psi \times \nabla\theta J.$$

We note that the scalar product between a vector in either of these two bases with its corresponding vector in the other basis gives unity, while it is orthogonal to the remaining members of the other basis. For example,

$$\nabla\psi \cdot \nabla\theta \times \nabla\phi J = 1,$$
$$\nabla\psi \cdot \nabla\phi \times \nabla\psi J = 0,$$

etc. Both of these bases are useful for representing vectors, and we need to be able to transform from one representation to the other. If $\mathbf{A}$ is an arbitrary vector quantity, we can represent it in terms of the covariant vectors

$$\mathbf{A} = A_1 \nabla\psi + A_2 \nabla\theta + A_3 \nabla\phi, \tag{5.7}$$

where

$$\begin{aligned}
A_1 &= \mathbf{A} \cdot \nabla\theta \times \nabla\phi J, \\
A_2 &= \mathbf{A} \cdot \nabla\phi \times \nabla\psi J, \\
A_3 &= \mathbf{A} \cdot \nabla\psi \times \nabla\theta J,
\end{aligned}$$

or, alternatively, in terms of the contravariant vectors

$$\mathbf{A} = A^1 \nabla\theta \times \nabla\phi J + A^2 \nabla\phi \times \nabla\psi J + A^3 \nabla\psi \times \nabla\theta J, \tag{5.8}$$

where

$$\begin{aligned}
A^1 &= \mathbf{A} \cdot \nabla\psi, \\
A^2 &= \mathbf{A} \cdot \nabla\theta, \\
A^3 &= \mathbf{A} \cdot \nabla\phi.
\end{aligned}$$

5.2.3 Grad, Div, Curl

The gradient of a scalar function $f(\psi, \theta, \phi)$ is readily calculated from the chain rule of differentiation,

$$\nabla f = \frac{\partial f}{\partial \psi}\nabla\psi + \frac{\partial f}{\partial \theta}\nabla\theta + \frac{\partial f}{\partial \phi}\nabla\phi. \tag{5.9}$$

Note that the gradient operation produces a vector in the covariant representation. The inverse to Eq. (5.9) is the calculation of the derivative of f with respect to one of the coordinates. By taking the scalar product of Eq. (5.9) with each of the contravariant basis vectors, we obtain

$$J\nabla\theta \times \nabla\phi \cdot \nabla f = \frac{\partial f}{\partial \psi},$$

$$J\nabla\phi \times \nabla\psi \cdot \nabla f = \frac{\partial f}{\partial \theta},$$

$$J\nabla\psi \times \nabla\theta \cdot \nabla f = \frac{\partial f}{\partial \phi}.$$

It is useful to keep in mind that $\partial f/\partial \psi$ is proportional to the projection of ∇f in the direction orthogonal to the other remaining coordinates, and not in the $\nabla\psi$ direction, as illustrated in Figure 5.4.

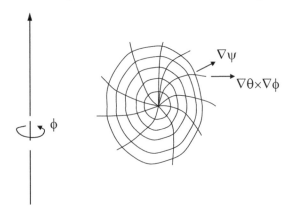

FIGURE 5.4: The vector $\frac{\partial f}{\partial \psi}$ is proportional to the projection of ∇f in the direction orthogonal to the other remaining coordinates, and not in the $\nabla \psi$ direction.

To take the divergence of a vector, it should first be in the contravariant representation since we can use the fact that the divergence of the cross product of any two gradients is zero, i.e.,

$$\nabla \cdot (\nabla \alpha \times \nabla \beta) = 0 \tag{5.10}$$

for any scalar quantities α and β. Using this, Eq. (5.9), and the orthogonality properties, we can easily compute

$$
\begin{aligned}
\nabla \cdot \mathbf{A} &= \nabla \cdot \left[A^1 \nabla \theta \times \nabla \phi J + A^2 \nabla \phi \times \nabla \psi J + A^3 \nabla \psi \times \nabla \theta J \right] \\
&= \frac{1}{J} \left[\frac{\partial}{\partial \psi} \left(A^1 J \right) + \frac{\partial}{\partial \theta} (A^2 J) + \frac{\partial}{\partial \phi} (A^3 J) \right].
\end{aligned} \tag{5.11}
$$

To take the curl of a vector, it should be in the covariant representation since we can make use of the fact that the curl of a gradient is zero, i.e.,

$$\nabla \times (\nabla \alpha) = 0, \tag{5.12}$$

for any scalar α. This, Eq. (5.9), and the orthogonality properties give

$$
\begin{aligned}
\nabla \times \mathbf{A} &= \nabla \times [A_1 \nabla \psi + A_2 \nabla \theta + A_3 \nabla \phi] \\
&= \frac{1}{J} \left[\frac{\partial A_3}{\partial \theta} - \frac{\partial A_2}{\partial \phi} \right] \nabla \theta \times \nabla \phi J + \frac{1}{J} \left[\frac{\partial A_1}{\partial \phi} - \frac{\partial A_3}{\partial \psi} \right] \nabla \phi \times \nabla \psi J \\
&\quad + \frac{1}{J} \left[\frac{\partial A_2}{\partial \psi} - \frac{\partial A_1}{\partial \theta} \right] \nabla \psi \times \nabla \theta J.
\end{aligned} \tag{5.13}
$$

Note that taking the curl leaves a vector in the contravariant basis.

5.2.4 Metric Tensor

In carrying out vector manipulations, we sometimes need to convert a vector in the covariant representation to one in the contravariant, or vice versa. To do this, we need relations of the form

$$\nabla\psi = a^1 \nabla\theta \times \nabla\phi J + a^2 \nabla\phi \times \nabla\psi J + a^3 \nabla\psi \times \nabla\theta J,$$

etc. Taking the scalar product successively with $\nabla\psi$, $\nabla\theta$, $\nabla\phi$, and using the orthogonality property gives

$$
\begin{aligned}
a^1 &= |\nabla\psi|^2 \\
a^2 &= \nabla\theta \cdot \nabla\psi \\
a^3 &= 0.
\end{aligned}
$$

Repeating for the other basis vectors gives the metric tensor

$$
\begin{bmatrix} \nabla\psi \\ \nabla\theta \\ \nabla\phi \end{bmatrix} = \begin{bmatrix} |\nabla\psi|^2 & \nabla\theta \cdot \nabla\psi & 0 \\ \nabla\theta \cdot \nabla\psi & |\nabla\theta|^2 & 0 \\ 0 & 0 & 1/R^2 \end{bmatrix} \cdot \begin{bmatrix} \nabla\theta \times \nabla\phi J \\ \nabla\phi \times \nabla\psi J \\ \nabla\psi \times \nabla\theta J \end{bmatrix}. \tag{5.14}
$$

Similarly, to transform from contravariant to covariant, we need relations of the form

$$\nabla\theta \times \nabla\phi J = a_1 \nabla\psi + a_2 \nabla\theta + a_3 \nabla\phi.$$

Taking the scalar product with the other contravariant basis vectors gives expressions for a_1, a_2, and a_3. Again, we can summarize this in terms of the inverse metric tensor

$$
\begin{bmatrix} \nabla\theta \times \nabla\phi J \\ \nabla\phi \times \nabla\psi J \\ \nabla\psi \times \nabla\theta J \end{bmatrix} = \begin{bmatrix} |\nabla\theta|^2 \frac{J^2}{R^2} & -\nabla\theta \cdot \nabla\psi \frac{J^2}{R^2} & 0 \\ -\nabla\theta \cdot \nabla\psi \frac{J^2}{R^2} & |\nabla\psi|^2 \frac{J^2}{R^2} & 0 \\ 0 & 0 & R^2 \end{bmatrix} \cdot \begin{bmatrix} \nabla\psi \\ \nabla\theta \\ \nabla\phi \end{bmatrix}.
$$

$$\tag{5.15}$$

The matrices in (5.14) and (5.15) are the inverse of one another. This can be used to give the identity

$$|\nabla\psi|^2 |\nabla\theta|^2 - (\nabla\theta \cdot \nabla\psi)^2 = R^2/J^2. \tag{5.16}$$

5.2.5 Metric Elements

In the last section, we introduced the coordinate functions $\psi(R, Z)$ and $\theta(R, Z)$ and proceeded to construct a coordinate system from them. However, computationally it is often more convenient to work in the inverse representation where we consider the cylindrical coordinates R and Z to be functions of ψ and θ:

$$
\begin{aligned}
R &= R(\psi, \theta), \\
Z &= Z(\psi, \theta). \tag{5.17}
\end{aligned}
$$

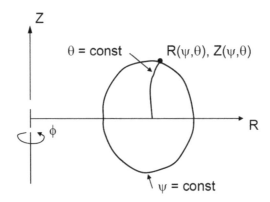

FIGURE 5.5: In the inverse representation we consider the cylindrical co-ordinates R and Z to be functions of ψ and θ.

Note that for ψ fixed, and θ varying between 0 and 2π, the curve defined by $[R(\psi,\theta), Z(\psi,\theta)]$ will trace out a closed flux surface as shown in Figure 5.5.

We begin by computing the gradients of R and Z, which are orthogonal unit vectors. Using Eq. (5.9), we have

$$\begin{aligned} \nabla R &= \hat{\mathbf{R}} = R_\psi \nabla\psi + R_\theta \nabla\theta \\ \nabla Z &= \hat{\mathbf{Z}} = Z_\psi \nabla\psi + Z_\theta \nabla\theta. \end{aligned} \tag{5.18}$$

Here and elsewhere in this section, we will denote partial differentiation with subscripts. Using Eq. (5.18), we can compute the triple product

$$\nabla R \times \nabla\phi \cdot \nabla Z = \frac{1}{R} = (R_\theta Z_\psi - R_\psi Z_\theta)\, \nabla\psi \times \nabla\theta \cdot \nabla\phi$$

or, using the definition of the Jacobian, Eq. (5.1),

$$J = R\,(R_\theta Z_\psi - R_\psi Z_\theta). \tag{5.19}$$

We can write the relations in Eq. (5.18) in matrix form:

$$\left[\begin{array}{c} \hat{\mathbf{R}} \\ \hat{\mathbf{Z}} \end{array}\right] = \left[\begin{array}{cc} R_\psi & R_\theta \\ Z_\psi & Z_\theta \end{array}\right] \cdot \left[\begin{array}{c} \nabla\psi \\ \nabla\theta \end{array}\right]. \tag{5.20}$$

Inverting this gives

$$\left[\begin{array}{c} \nabla\psi \\ \nabla\theta \end{array}\right] = \frac{R}{J}\left[\begin{array}{cc} -Z_\theta & R_\theta \\ Z_\psi & -R_\psi \end{array}\right] \cdot \left[\begin{array}{c} \hat{\mathbf{R}} \\ \hat{\mathbf{Z}} \end{array}\right]. \tag{5.21}$$

From Eq. (5.21), we can now compute the metric elements that appear in

Eq. (5.14):

$$|\nabla\psi|^2 = \frac{R^2}{J^2}\left(R_\theta^2 + Z_\theta^2\right),$$

$$|\nabla\theta|^2 = \frac{R^2}{J^2}\left(R_\psi^2 + Z_\psi^2\right),$$

$$\nabla\theta \cdot \nabla\psi = -\frac{R^2}{J^2}(R_\psi R_\theta + Z_\psi Z_\theta). \tag{5.22}$$

The first of these provides a useful relation between the arc length along a constant ψ surface, $d\ell$, and the angle differential $d\theta$. Using the relation $d\ell^2 = (R_\theta^2 + Z_\theta^2)d\theta^2$, we have

$$d\ell = \frac{|\nabla\psi|J}{R}d\theta. \tag{5.23}$$

5.3 Magnetic Field, Current, and Surface Functions

Recall that the equilibrium magnetic field is represented as

$$\mathbf{B} = \nabla\phi \times \nabla\Psi + g(\Psi)\nabla\phi,$$

$$= \Psi'\nabla\phi \times \nabla\psi + g(\Psi)\nabla\phi, \tag{5.24}$$

where we have denoted $d\Psi/d\psi$ by Ψ'. This is a mixed representation. We can convert to a covariant form by using the inverse metric tensor, Eq. (5.15):

$$\mathbf{B} = \left[-\Psi'\nabla\theta \cdot \nabla\psi\frac{J}{R^2}\right]\nabla\psi + \left[\Psi'|\nabla\psi|^2\frac{J}{R^2}\right]\nabla\theta + g\nabla\phi. \tag{5.25}$$

The current density is calculated directly by taking the curl of Eq. (5.25), using the identity in Eq. (5.13):

$$\mu_0\mathbf{J} = \nabla \times \mathbf{B}$$

$$= \left\{\left[\Psi'|\nabla\psi|^2\frac{J}{R^2}\right]_\psi + \left[\Psi'\nabla\theta \cdot \nabla\psi\frac{J}{R^2}\right]_\theta\right\}\nabla\psi \times \nabla\theta - g'\nabla\phi \times \nabla\psi. \tag{5.26}$$

Another useful expression related to the current parallel to the magnetic field is obtained by taking the dot product of Eqs. (5.25) and (5.26),

$$\mu_0\mathbf{J} \cdot \mathbf{B} = \frac{g^2}{J}\left\{\left[\frac{\Psi'}{g}|\nabla\psi|^2\frac{J}{R^2}\right]_\psi + \left[\frac{\Psi'}{g}\nabla\theta \cdot \nabla\psi\frac{J}{R^2}\right]_\theta\right\}. \tag{5.27}$$

We can now compute some surface functions of interest. The volume within a flux surface is given simply by

$$V(\psi) = \int d\tau = 2\pi \int_{\psi_0}^{\psi} d\psi \int_0^{2\pi} d\theta J. \tag{5.28}$$

This allows us to define the *differential volume* as

$$V'(\psi) \equiv \frac{dV}{d\psi} = 2\pi \int_0^{2\pi} d\theta J = 2\pi \oint \frac{R d\ell}{|\nabla \psi|}, \tag{5.29}$$

where the second equality follows from Eq. (5.23). For any scalar quantity $a(\psi, \theta)$ we define the *surface average* of a as the following integral over a constant ψ contour:

$$\langle a \rangle = \frac{2\pi}{V'} \int_0^{2\pi} d\theta J a. \tag{5.30}$$

Following Kruskal and Kulsrud [95], we compute magnetic and current fluxes that are best expressed as volume integrals interior to a given magnetic surface. Since both $\nabla \cdot \mathbf{B} = 0$ and $\nabla \cdot \mathbf{J} = 0$, the following integrands can be written as divergences of a scalar times a divergence-free vector. Using Gauss's theorem, and noting that since θ and ϕ are not single valued, it is necessary to cut the region of integration at $\theta = 0$ or $\phi = 0$. Since the field and current lie within surfaces, the boundary contribution vanishes except at the cut where it is seen to reduce to area integrals over the surface area. The area integral at $\theta = 0$ or $\phi = 0$ vanishes since it is multiplied by the angle, and only the contribution from the cut at 2π survives.

Using this technique and the expression for the volume element in Eq. (5.2), the *toroidal flux* inside a magnetic surface is

$$\Phi(\psi) = \frac{1}{2\pi} \int \mathbf{B} \cdot \nabla \phi \, d\tau = \frac{1}{2\pi} \int_{\psi_0}^{\psi} d\psi g(\psi) V' \left\langle R^{-2} \right\rangle. \tag{5.31}$$

The *poloidal flux* inside a magnetic surface is

$$\Psi_p(\psi) = \frac{1}{2\pi} \int \mathbf{B} \cdot \nabla \theta \, d\tau = 2\pi \left(\Psi - \Psi_0 \right). \tag{5.32}$$

The *poloidal current* within a magnetic surface is

$$K(\psi) = \frac{1}{2\pi} \int \mathbf{J} \cdot \nabla \theta \, d\tau = \frac{2\pi}{\mu_0} \left[g(0) - g(\psi) \right]. \tag{5.33}$$

And the *toroidal current* within a magnetic surface is

$$I(\psi) = \frac{1}{2\pi} \int \mathbf{J} \cdot \nabla \phi \, d\tau = \frac{\Psi'}{2\pi \mu_0} V' \left\langle \frac{|\nabla \psi|^2}{R^2} \right\rangle. \tag{5.34}$$

We can now compute the *safety factor* as

$$q(\psi) = d\Phi/d\Psi_p = \Phi'/\Psi_p' = \frac{gV'}{(2\pi)^2\Psi'}\left\langle R^{-2}\right\rangle. \qquad (5.35)$$

Another quantity of interest is related to the surface-averaged parallel current density, obtained by using Eq. (5.27),

$$J_{\parallel}(\psi) \equiv \frac{\langle \mathbf{J} \cdot \mathbf{B}\rangle}{\langle \mathbf{B} \cdot \nabla\phi\rangle} = \frac{g}{\mu_0 V'\langle R^{-2}\rangle}\left[\frac{\Psi'V'}{g}\left\langle\frac{|\nabla\psi|^2}{R^2}\right\rangle\right]_\psi. \qquad (5.36)$$

5.4 Constructing Flux Coordinates from $\Psi(R, Z)$

Suppose we are given the poloidal flux on an (R, Z) grid such as we would obtain from the equilibrium solution method discussed in Section 4.6. Here we discuss how to go about calculating an associated (ψ, θ) flux coordinate system with the desired Jacobian. The first step is to divide the flux in the plasma into N intervals with interval boundaries given by

$$\Psi_j = \Psi_0 + j\Delta\Psi; \qquad j = 0, \cdots, N. \qquad (5.37)$$

Here, the flux increment is defined by $\Delta\Psi = (\Psi_l - \Psi_0)/N$, where the limiter and axis fluxes, Ψ_l and Ψ_0, are those defined in Section 4.6. Next, calculate the differential volume with respect to Ψ, $V'(\Psi_j)$, by contouring each Ψ_j level as follows. Define a sequence of M rays extending outward from the magnetic axis that intersect each contour $\Psi(R, Z) = \Psi_j$. Denote the intersection point of ray i and contour j by $\mathbf{x}_{i,j} = (R_{i,j}, Z_{i,j})$. These can be equally spaced in polar angle, and should be periodic such that ray $i = M+1$ would be the same as ray $i = 1$. Using a suitable 2D interpolation method such as Akima [90] or cubic spline, solve for the intersection points $\mathbf{x}_{i,j}$ using standard root-finding techniques such as binary search or Newton's method.

Once the four points denoted $i - 1$, i, $i + 1$, and $i + 2$ illustrated in Figure 5.6 have been located in this fashion, a simple and accurate method for approximating the section of the arc length $\delta\ell$ between points denoted i and $i + 1$ is to use points $i - 1$ and $i + 1$ to define a tangent value at point i, and to use points $i + 2$ and i to define a tangent value at point $i + 1$. We construct a cubic polynomial that goes through the points i and $i + 1$ with the desired tangent values at those endpoints. This can be integrated analytically to give a high-order approximation to the arc length for that segment. If we construct a local coordinate system (x', z') with the origin at $\mathbf{x}_{i,j}$ such that the x' axis passes through the point $\mathbf{x}_{i+1,j}$ and require that the interpolating cubic have slopes m_1 and m_2 in this local coordinate system at the two points, then a good approximation to the arc length of the curve passing through $\mathbf{x}_{i,j}$ and

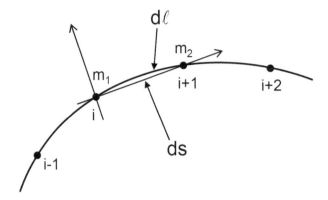

FIGURE 5.6: Points on a constant Ψ contour are found using root-finding techniques. The two slopes m_1 and m_2 relative to the line segment passing through points i and $i+1$ are used to define the local arc length $d\ell$.

$\mathbf{x}_{i+1,j}$ with the required endpoint slopes is given by

$$d\ell_{i+1/2} = ds_{i+1/2} \left[1 + (2m_1^2 + 2m_2^2 - m_1 m_2)/30 \right]. \tag{5.38}$$

Here $ds_{i+1/2}$ is the length of the straight-line segment passing through $\mathbf{x}_{i,j}$ and $\mathbf{x}_{i+1,j}$.

Approximating the integral in Eq. (5.29) by a finite sum, we have the following expression for the derivative of the volume within a flux surface with respect to the poloidal flux function:

$$\frac{dV}{d\Psi} = 2\pi \sum_{i=0}^{M} \left(\frac{R}{|\nabla\Psi|} \right)_{i+1/2} d\ell_{i+1/2}. \tag{5.39}$$

Quantities at the half-integer points are best evaluated as averages over the two neighboring integer points that lie on the contour,

$$\left(\frac{R}{|\nabla\Psi|} \right)_{i+1/2} \equiv \frac{1}{2} \left[\left(\frac{R}{|\nabla\Psi|} \right)_i + \left(\frac{R}{|\nabla\Psi|} \right)_{i+1} \right].$$

As discussed in Section 5.2.1, choosing a form for the Jacobian determines both the properties of the θ coordinate and the physical meaning of the ψ coordinate. Because of this, it is useful to prescribe the Jacobian as the product of two functions:

$$J(\psi, \theta) = f(\psi)h(\psi, \theta). \tag{5.40}$$

The function $h(\psi, \theta)$ will be prescribed to fix the properties of the θ coordinate on each surface, and the function $f(\psi)$ will determine the physical meaning of

TABLE 5.1: The poloidal angle θ is determined by the function $h(\psi, \theta)$ in the Jacobian as defined in Eq. (5.40).

$h(\psi, \theta)$	θ
$R/|\nabla\psi|$	constant arc length
R^2	straight field lines
R	constant area
1	constant volume

TABLE 5.2: Once the θ coordinate is determined, the ψ coordinate can be determined by the function $f(\psi)$ in the Jacobian definition, Eq. (5.40).

$f(\psi)$	ψ
$\left[2\pi \int_0^{2\pi} d\theta h(\psi, \theta)\right]^{-1}$	volume within surface
$\left[\int_0^{2\pi} d\theta h(\psi, \theta)g/R^2\right]^{-1}$	toroidal flux within surface

the ψ coordinate. Once $h(\psi, \theta)$ is chosen, we can use the relation in Eq. (5.23) to calculate the value of θ at a given point along the contour Ψ_j by performing the line integral:

$$\theta_i = 2\pi \left[\oint \frac{R d\ell}{h(\psi, \theta)|\nabla\psi|}\right]^{-1} \int_0^{\mathbf{x}_{i,j}} \frac{R d\ell}{h(\psi, \theta)|\nabla\psi|}. \tag{5.41}$$

Common choices for $h(\psi, \theta)$ and the corresponding name of the θ coordinate are given in Table 5.1. Once a function $h(\psi, \theta)$ is chosen, if we need to switch from the poloidal flux function to another physical meaning for the coordinate ψ, the function $f(\psi)$ can be defined appropriately. Two additional common choices for the surface coordinate are given in Table 5.2. These follow from Eqs. (5.29) and (5.31).

5.4.1 Axisymmetric Straight Field Line Coordinates

We call a flux coordinate system *straight field line* if it allows a magnetic field vector to be written in the form

$$\mathbf{B} = \nabla\phi \times \nabla F(\psi) + \nabla H(\psi) \times \nabla\theta, \tag{5.42}$$

where $F(\psi)$ and $H(\psi)$ are only functions of the coordinate ψ. When plotted as a function of the angles θ and ϕ, magnetic field lines on a surface appear

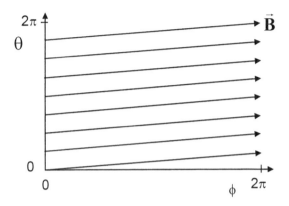

FIGURE 5.7: With straight field line angular coordinates, magnetic field lines on a surface ψ=const. appear as straight lines when plotted in (θ, ϕ) space.

straight as illustrated in Figure 5.7 since the ratio $\mathbf{B} \cdot \nabla\theta / \mathbf{B} \cdot \nabla\phi$ depends only on the coordinate ψ. We will see in Chapter 8 that a straight field line coordinate system can greatly simplify some analysis and make certain linear stability calculations much more efficient.

An axisymmetric equilibrium magnetic field is straight if we choose the Jacobian function $h(\psi, \theta)$ in Eq. (5.40) to be proportional to R^2, as indicated in Table 5.1. In a straight field line system, the equilibrium magnetic field can be written as

$$
\begin{aligned}
\mathbf{B} &= \nabla\phi \times \nabla\Psi + g(\psi)\nabla\phi, \\
&= \Psi' \left[\nabla\phi \times \nabla\psi + \frac{gJ}{\Psi'R^2}\nabla\psi \times \nabla\theta \right], \\
&= \Psi' \left[\nabla\phi \times \nabla\psi + q(\psi)\nabla\psi \times \nabla\theta \right],
\end{aligned} \tag{5.43}
$$

where the second line follows from Eq. (5.14) and $q(\psi)$ is the safety factor defined in Eq. (5.35).

It follows from this form that we can write the gradient along the field line of an arbitrary scalar variable α as

$$
\mathbf{B} \cdot \nabla\alpha = \frac{\Psi'}{J} \left[\alpha_\theta + q(\psi)\alpha_\phi \right]. \tag{5.44}
$$

Straight field line coordinates are especially useful for linear stability calculations where one often uses a spectral representation in the θ and ϕ directions. If the $(m, n)^{th}$ harmonic varies like $\exp i(m\theta - n\phi)$, the above operator becomes

$$
\mathbf{B} \cdot \nabla\alpha = \frac{\Psi'}{J} i \left[m - nq(\psi) \right] \alpha. \tag{5.45}
$$

It is seen that if the bracket vanishes anywhere, it will vanish over the entire flux surface. Therefore, in a straight field line system, all the resonant behavior of a given rational q value will be captured in the rational harmonics that satisfy the resonance condition $m/n = q(\psi)$. Some implications of this are discussed in Chapter 8.

5.4.2 Generalized Straight Field Line Coordinates

For some applications that require straight field line coordinates, the restriction that $J \sim R^2$ is too restrictive as this effectively determines the spacing and hence relative resolution of the θ coordinate. It has been shown [99] that one can obtain straight field line coordinates for any θ coordinate if we are willing to generalize what choice we make for the toroidal angle. Let us make the substitution $\phi \to \zeta$ in the toroidal angle, where ϕ is the cylindrical coordinate angle, and ζ is a generalized toroidal angle related to it by

$$\zeta \equiv \phi - q(\psi)\delta(\psi, \theta), \tag{5.46}$$

for some function $\delta(\psi, \theta)$. The requirement that the field lines be straight in the (ψ, θ, ζ) coordinate system is that we can write the magnetic field in the form

$$\mathbf{B} = \Psi' \left[\nabla\zeta \times \nabla\psi + q(\psi)\nabla\psi \times \nabla\theta \right]. \tag{5.47}$$

Now, since $\nabla\zeta \times \nabla\psi = \left(\nabla\phi - q\frac{\partial\delta}{\partial\theta}\nabla\theta \right) \times \nabla\psi$, we have

$$\mathbf{B} = \Psi' \left[\nabla\phi \times \nabla\psi + q(\psi)\left(1 + \frac{\partial\delta}{\partial\theta} \right) \nabla\psi \times \nabla\theta \right]. \tag{5.48}$$

However, if $\mathbf{B}$ is an equilibrium magnetic field, we know that it can also be written in the form of Eq. (5.24), i.e.,

$$\mathbf{B} = \Psi'\nabla\phi \times \nabla\psi + g(\psi)\nabla\phi. \tag{5.49}$$

Equating the $\nabla\phi$ components of Eqs. (5.48) and (5.49) gives the relation

$$\frac{\partial\delta}{\partial\theta} = \left(\frac{J}{R^2} \frac{g(\psi)}{\Psi'(\psi)q(\psi)} - 1 \right), \tag{5.50}$$

where J is the Jacobian of both the (ψ, θ, ϕ) and the (ψ, θ, ζ) coordinate systems. Note the requirement, that comes from periodicity, that the Jacobian satisfy

$$\frac{1}{2\pi} \int_0^{2\pi} \frac{J}{R^2} d\theta = \frac{\Psi'(\psi)q(\psi)}{g(\psi)}.$$

Since it is the form of the Jacobian that determines the spacing of the θ coordinate [for example, see Eq. (5.23)], this new flexibility allows us to obtain a straight field line system for any choice of θ coordinate by defining δ appropriately and using ζ as the toroidal angle instead of ϕ. This is at the expense of some additional complexity since the the metric quantities $\nabla\psi \cdot \nabla\zeta$ and $\nabla\theta \cdot \nabla\zeta$ will no longer be zero, and $|\nabla\zeta|^2 \neq 1/R^2$. However, the coordinate ζ is still ignorable with respect to axisymmetric equilibrium quantities, and the linear stability analysis (Chapter 8) can still be reduced to the study of a single mode varying as $\exp(in\zeta)$.

A particularly useful straight field line coordinate system, known as Boozer coordinates [100], is obtained if we take the Jacobian to be proportional to the inverse of the square of the magnetic field,

$$J \equiv [\nabla\psi \times \nabla\theta \cdot \nabla\zeta]^{-1} = \frac{f(\psi)}{B^2}, \tag{5.51}$$

for some function $f(\psi)$. It follows that an equilibrium magnetic field can then also be written in the covariant form

$$\mathbf{B} = g(\psi)\nabla\zeta + I(\psi)\nabla\theta + \beta_*\nabla\psi. \tag{5.52}$$

From Eqs. (5.47)–(5.52), we have the relation $f(\psi) = \Psi'[g(\psi)q(\psi) + I(\psi)]$. These coordinates have been shown to be particularly useful for calculating guiding center drift orbits in magnetic confinement devices [100, 101, 102].

5.5 Inverse Equilibrium Equation

We now are in a position to construct the inverse equilibrium equation

$$\nabla p = \mathbf{J} \times \mathbf{B}.$$

Substituting from Eqs. (5.25) and (5.26), we have

$$\Psi'\frac{R^2}{J}\left\{\left[\Psi'|\nabla\psi|^2\frac{J}{R^2}\right]_\psi + \left[\Psi'\nabla\theta \cdot \nabla\psi\frac{J}{R^2}\right]_\theta\right\} + gg' + R^2\mu_0 p' = 0. \tag{5.53}$$

Equation (5.53) is equivalent to the Grad–Shafranov equation, Eq. (4.8). If we prescribe the two functions $p(\psi)$ and $g(\psi)$, and the form of the Jacobian J, this determines the coordinate functions $R(\psi,\theta), Z(\psi,\theta)$ and the poloidal flux function $\Psi(\psi)$.

A technique for solving this equation, known as the *iterative metric method*, has been developed [103] that uses approximate flux coordinates (ψ,θ) to iteratively solve the general coordinate form of the Grad–Shafranov equation, Eq. (4.8), for the poloidal flux function $\Psi(\psi,\theta)$ where the toroidal elliptic operator takes the form

$$\Delta^*\Psi = \frac{R^2}{J} \left\{ \left[\Psi_\psi |\nabla\psi|^2 \frac{J}{R^2} \right]_\psi + \left[\Psi_\psi \nabla\theta \cdot \nabla\psi \frac{J}{R^2} \right]_\theta \right.$$
$$\left. + \left[\Psi_\theta \nabla\theta \cdot \nabla\psi \frac{J}{R^2} \right]_\psi + \left[\Psi_\theta |\nabla\theta|^2 \frac{J}{R^2} \right]_\theta \right\}.$$

The coordinates $R(\psi,\theta)$ and $Z(\psi,\theta)$ are adjusted each metric iteration cycle so that constant ψ surfaces lie on the computed constant Ψ contours, and so that the Jacobian integrals are satisfied. Once the iterations converge to a tolerance, the computed Ψ is only a function of the single coordinate ψ, and Eq. (5.53) is then satisfied.

5.5.1 q-Solver

An advantage of working in the flux coordinates is that one or both of the free functions needed for solution of the Grad–Shafranov equation can be specified in terms of quantities requiring surface averages to evaluate. Since stability studies depend sensitively on the q-profile, it is often desired to prescribe the functions $p(\Psi)$ and $q(\Psi)$ for an equilibrium solution. We can then eliminate the term gg' in Eqs. (4.8) and (5.53) using Eq. (5.35),

$$gg' = \left[\frac{(2\pi)^2 \Psi' q}{V' \langle R^{-2} \rangle} \right] \left[\frac{(2\pi)^2 \Psi' q}{V' \langle R^{-2} \rangle} \right]_\psi . \tag{5.54}$$

Since this involves two ψ derivatives acting on the poloidal flux function Ψ, direct substitution into Eq. (5.53) or (4.8) does not lead to an elliptic equation of standard form. Instead, we can use the surface average of R^{-2} times Eq. (5.53), with the gg' eliminated using Eq. (5.54),

$$\Psi' \left[\Psi' \left\langle \frac{|\nabla\psi|^2}{R^2} \right\rangle V' \right]_\psi = -\mu_0 V' p' - (2\pi)^4 \Psi' q \left[\frac{\Psi' q}{V' \langle R^{-2} \rangle} \right]_\psi , \tag{5.55}$$

to eliminate Ψ'' from Eq. (5.54). Insertion into Eq. (4.8) or (5.53) then yields the desired form of the Grad–Shafranov equation with the q-profile held fixed,

$$\left(\Delta^* - A \frac{\partial}{\partial\psi} \right) \Psi = -B\mu_0 \frac{dp}{d\Psi} . \tag{5.56}$$

Here,

$$A = \frac{(2\pi)^4}{D} \left[\frac{q}{V' \langle R^{-2} \rangle} \right]^3 \left[\left\langle \frac{|\nabla\psi|^2}{R^2} \right\rangle \langle R^{-2} \rangle \frac{V'}{q} \right]_\psi , \tag{5.57}$$

$$B = \frac{1}{D} \left[R^2 \left\langle \frac{|\nabla\psi|^2}{R^2} \right\rangle + \frac{(2\pi)^4}{\langle R^{-2} \rangle} \left(\frac{q}{V'} \right)^2 \left(R^2 - \frac{1}{\langle R^{-2} \rangle} \right) \right] , \tag{5.58}$$

$$D = \left[\left\langle \frac{|\nabla\psi|^2}{R^2} \right\rangle + \frac{(2\pi)^4}{\langle R^{-2} \rangle} \left(\frac{q}{V'} \right)^2 \right] . \tag{5.59}$$

Note that these expressions depend on q and V' only through the combination q/V' and are thus well behaved in the vicinity of a magnetic separatrix where $q \to \infty$ and $V' \to \infty$, but the ratio q/V' remains finite [104, 105]. Numerical codes using finite differences to solve this form of the equation using iteration methods have been developed [103].

5.5.2   *J*-Solver

Another function that is sometimes specified is the surface-averaged parallel current density, $J_\parallel$, defined in Eq. (5.36). This is useful in characterizing the stationary state of a resistive plasma in which the magnetic field is time-independent. It follows from Eq. (1.56) that the electric field can be written in the form

$$\mathbf{E} = -\nabla \Phi_P + \frac{V_L}{2\pi} \nabla \phi = -\mathbf{u} \times \mathbf{B} + \eta \mathbf{J}. \qquad (5.60)$$

Here, Φ_P is a single-valued electrical potential and the term involving $\nabla \phi$ is then needed to represent the most general axisymmetric curl-free electric field. We have introduced the constant V_L, which will be seen to be the *loop voltage* the long way around the torus.

It follows from taking the dot product of Eq. (5.60) with the magnetic field vector $\mathbf{B}$ and then surface averaging that the surface average parallel current in a stationary state is related to the plasma parallel resistivity profile $\eta_\parallel$ and the loop voltage by the relation

$$J_\parallel(\psi) \equiv \frac{\langle \mathbf{J} \cdot \mathbf{B} \rangle}{\langle \mathbf{B} \cdot \nabla \phi \rangle} = \frac{V_L}{2\pi \eta_\parallel(\psi)}. \qquad (5.61)$$

Since $\eta_\parallel$ is related to the plasma temperature in a known way, for example by Eq. (1.44), and the temperature can be measured accurately, this gives a powerful constraint on the current profile in a steady-state axisymmetric fusion device.

As in the discussion of the *q*-solver option above, Eq. (5.36) is not in a form to be solved directly to eliminate the g function that appears in the Grad–Shafranov equation because of the term involving Ψ''. We again use the surface average of R^{-2} times Eq. (5.53) to eliminate this term in order to obtain the following relation involving the function $G(\psi) \equiv \frac{1}{2} g^2$,

$$\left[\mu_0 J_\parallel \langle R^{-2} \rangle \Psi' + \mu_0 p' \right] G + \langle R^{-2} \rangle G G' + \frac{1}{2} \Psi'^2 \left\langle \frac{|\nabla \psi|^2}{R^2} \right\rangle G' = 0. \qquad (5.62)$$

We finite difference and let $G_j = G(\psi_j)$ with $\psi_j = j \delta \psi$ and evaluate the following two coefficients at the half-integer grid points

$$a_{j+1/2} = \delta\psi \times \left[\frac{\mu_0 J_\|(\psi)\langle R^{-2}\rangle \Psi' + \mu_0 p'}{\langle R^{-2}\rangle}\right]_{j+1/2},$$

$$b_{j+1/2} = \left[\frac{\Psi'^2 \langle\frac{|\nabla\psi|^2}{R^2}\rangle}{\langle R^{-2}\rangle}\right]_{j+1/2}.$$

By using centered differences and averaging appropriately, we can determine G_j from the following quadratic equation once G_{j+1} and the metric terms $a_{j+1/2}$ and $b_{j+1/2}$ are known,

$$G_j^2 - (a-b)_{j+1/2} \, G_j - \left[G_{j+1}^2 + (a+b)_{j+1/2} \, G_{j+1}\right] = 0. \qquad (5.63)$$

This is used to integrate G inward from the plasma boundary to the magnetic axis. Differentiating G then gives the function $gg\prime$ that appears in Eq. (4.8) or (5.53).

5.5.3   Expansion Solution

We can obtain an asymptotic analytic solution to the inverse equilibrium equation, Eq. (5.53), by performing an expansion in the inverse aspect ratio a/R_0 [106]. We introduce an expansion tag ϵ which does not have a value but merely indicates the order in the inverse aspect ratio expansion. We look for a solution with circular cross section with low β ordered such that $\mu_0 p/B^2 \sim \epsilon^2$. It is convenient to prescribe the form of the Jacobian (which is ordered ϵ^0) as

$$J = R\left(R_\theta Z_\psi - R_\psi Z_\theta\right) = \frac{a_0^2}{2R_0} R^2. \qquad (5.64)$$

This is consistent with $R(\psi, \theta)$ and $Z(\psi, \theta)$ being parameterized as

$$R(\psi, \theta) = R_0 + \epsilon a_0 \sqrt{\psi}\cos\theta - \epsilon^2\Delta + \epsilon^2\psi\left(\Delta_\psi + \frac{a_0^2}{2R_0}\right)[\cos 2\theta - 1] + \cdots,$$

$$Z(\psi, \theta) = -\epsilon a_0 \sqrt{\psi}\sin\theta - \epsilon^2\psi\left(\Delta_\psi + \frac{a_0^2}{2R_0}\right)\sin 2\theta + \cdots. \qquad (5.65)$$

Here the magnetic axis shift, $\Delta(\psi)$, is a function yet to be determined. Prescribing $R(\psi, \theta)$ and $Z(\psi, \theta)$ of this form is consistent with the prescribed form of the Jacobian in Eq. (5.64) to order ϵ^2. We can now compute the metric elements needed to evaluate Eq. (5.53). Direct computation yields, to lowest order,

$$|\nabla\psi|^2\frac{J}{R^2} = \frac{1}{J}\left(R_\theta^2 + Z_\theta^2\right) = \frac{2\psi}{R_0}\epsilon^2\left(1 + \frac{4\epsilon}{a_0}\sqrt{\psi}\Delta_\psi\cos\theta\right),$$

$$\nabla\theta\cdot\nabla\psi\frac{J}{R^2} = \frac{-1}{J}\left(R_\theta R_\psi + Z_\theta Z_\psi\right) = \frac{-4\epsilon^3}{R_0 a_0}\sqrt{\psi}\sin\theta\left[(\psi\Delta_\psi)_\psi + \frac{a_0^2}{4R_0}\right],$$

$$R^2 = R_0^2\left(1 + 2\frac{\epsilon a_0}{R_0}\sqrt{\psi}\cos\theta\right).$$

We now specify the functional forms of the pressure and the current density:

$$p(\psi) = \epsilon^2\frac{1}{2}\beta_0(1-\psi),$$

$$\Psi'(\psi) = \frac{1}{2}\frac{a_0^2}{q_0}.$$

The equilibrium equation, Eq. (5.53), then gives, to the first two orders,

$$g(\psi) = R_0\left[1 + \epsilon^2 g^{(2)}(\psi)\right],$$

$$g^{(2)}(\psi) = \frac{a_0^2}{q_0^2 R_0^2}(1-\beta_\theta)(1-\psi), \qquad (5.66)$$

$$\Delta(\psi) = -\left(a_0^2/8R\right)(1+4\beta_\theta)(1-\psi). \qquad (5.67)$$

Here, the poloidal beta is defined as

$$\beta_\theta = q_0^2 R_0^2 \mu_0 \beta_0/2a_0^2.$$

Equation (5.66) illustrates that a tokamak plasma switches from being paramagnetic ($g^{(2)} > 0$) to being diamagnetic ($g^{(2)} < 0$) when $\beta_\theta > 1$. The result in Eq. (5.67) is known as the Shafranov shift [72] of the magnetic axis. These relations provide a useful check on numerical solutions to the inverse equilibrium equation.

5.5.4   Grad–Hirshman Variational Equilibrium

Here we demonstrate the usefulness of a variational form for calculating axisymmetric inverse equilibrium. Recall the *Grad–Hirshman* variational principle from Section 4.5 in which we define the Lagrangian functional:

$$L = \int_V d\tau \left[\frac{B_P^2}{2\mu_0} - \frac{B_T^2}{2\mu_0} - p(\Psi)\right].$$

Here, the integral is over the plasma volume and from Eq. (5.24), the square of the poloidal and toroidal magnetic fields are given by $B_P^2 = |\nabla\Psi|^2/R^2$, and $B_T^2 = g^2(\Psi)/R^2$. The functional L is stationary with respect to variations of Ψ (with $\delta\Psi = 0$ on the boundary) for Ψ satisfying the Grad–Shafranov equation. This variational statement is coordinate system independent, so we can use it to compute inverse equilibrium. We introduce the normalized flux coordinate ψ which is constrained to be 0 at the magnetic axis and 1 at the plasma boundary. Upon making the coordinate transformation

$$(R, \phi, Z) \rightarrow (\psi, \theta, \phi)$$

the action integral becomes

$$L = \int_0^1 d\psi \int_0^{2\pi} d\theta J \left[\Psi'^2 \frac{(R_\theta^2 + Z_\theta^2)}{2J^2} - \frac{g^2(\Psi)}{2R^2} - \mu_0 p(\Psi) \right], \qquad (5.68)$$

where J is defined by Eq. (5.19).

The variational statement now is that δL as defined by Eq. (5.68) is stationary with respect to variations of the dependent variables $R(\psi, \theta)$, $Z(\psi, \theta)$, and $\Psi(\psi)$ for those functions satisfying the inverse equilibrium equation, Eq. (5.53). It can be verified by direct, although lengthy, calculation that the Ψ, R, and Z variations of the functional L in Eq. (5.68) yield

$$\delta L = \int_0^1 d\psi \delta\Psi \int_0^{2\pi} d\theta J G, \qquad (5.69)$$

$$\delta L = \int_0^1 d\psi \int_0^{2\pi} d\theta R Z_\theta G \delta R, \qquad (5.70)$$

$$\delta L = -\int_0^1 d\psi \int_0^{2\pi} d\theta R R_\theta G \delta Z, \qquad (5.71)$$

respectively, where

$$G = \frac{\Psi'}{J} \left\{ \frac{\partial}{\partial\psi} \left[\frac{R_\theta^2 + Z_\theta^2}{J} \Psi' \right] - \frac{\partial}{\partial\theta} \left[\frac{R_\theta R_\psi + Z_\theta Z_\psi}{J} \right] \Psi' \right\} + \mu_0 p' + \frac{1}{R^2} g g'. \qquad (5.72)$$

Since $\delta R, \delta Z, \delta\Psi$ are arbitrary, the fact that δL must vanish for each of the variations in Eq. (5.68) implies that $G = 0$.

A technique has been developed [86] that makes use of the variational statement of the inverse equilibrium problem by expanding $R(\psi, \theta)$ and $Z(\psi, \theta)$ in a truncated set of basis functions, and determining the amplitudes in this expansion by the appropriate Euler-Lagrange equations. Many different choices for the basis functions are possible. Efficiency dictates that a small number of terms be able to adequately approximate shapes of interest. The basis functions should be nearly orthogonal (or linearly independent) so that two or more parameters do not have nearly the same effect on the shape of the surfaces. The representation should be compatible with the singularity at the origin so that the basis functions remain regular there.

For up-down symmetric configurations, a good representation in light of these considerations is given by

$$R(\psi,\theta) \;=\; R_A^0(\psi) + \sqrt{\psi T}\, E(\psi)\cos(\theta) + \sum_{m=2}^{M} \psi^{\frac{m}{2}-1} R_A^m(\psi)\cos m\theta,$$

$$Z(\psi,\theta) \;=\; \sqrt{\psi T}\, E^{-1}(\psi)\sin(\theta) + \sum_{m=2}^{M} \psi^{\frac{m}{2}-1} R_A^m(\psi)\sin m\theta. \qquad (5.73)$$

Here, the area of the ψ contour is seen to be $\pi T\psi$, where T is a normalization constant that can be fixed. In this way, the remaining spectral amplitudes will vary the shape of a constant-ψ curve of fixed area. The basis functions $R_A^0(\psi)$, $E(\psi)$, and $R_A^m(\psi)$ are defined such that they will each be linear in ψ near $\psi = 0$. We can identify R_A^0 with the axis shift, E^{-2} with the ellipticity, R_A^2 with the triangularity of the surface, R_A^3 with the squareness, and so on. In the representation of Eq. (5.73) the poloidal angle θ and the Jacobian are implicitly defined.

Substituting these expansions into the action integral, Eq. (5.68), and performing the variation, we obtain an Euler equation for each harmonic amplitude. First, consider the Ψ variation. This just gives

$$\frac{\partial L}{\partial \Psi} - \frac{d}{d\psi}\frac{\partial L}{\partial \Psi'} = 0,$$

or, from Eq. (5.69),

$$\langle G \rangle = 0, \qquad (5.74)$$

where the flux surface average is defined in Eq. (5.30). Next, consider the E variation,

$$\frac{\partial L}{\partial E} - \frac{d}{d\psi}\frac{\partial L}{\partial E_\psi} = 0.$$

Since L depends on E only through R and Z, and on E_ψ only through R_ψ and Z_ψ, this is equivalent to

$$\frac{\partial R}{\partial E}\left[\frac{\partial L}{\partial R} - \frac{d}{d\psi}\frac{\partial L}{\partial R_\psi}\right] + \frac{\partial Z}{\partial E}\left[\frac{\partial L}{\partial Z} - \frac{d}{d\psi}\frac{\partial L}{\partial Z_\psi}\right] = 0,$$

or, from Eqs. (5.70) and (5.71),

$$\left\langle \left(\cos\theta Z_\theta + E^{-2}\sin\theta R_\theta\right)\frac{RG}{J} \right\rangle = 0. \qquad (5.75)$$

By a similar argument, the variation with respect to the R_A^m gives

$$\left\langle \left(-\cos m\theta Z_\theta + \sin m\theta R_\theta\right)\frac{RG}{J} \right\rangle = 0. \qquad (5.76)$$

Suppose that we wish to use this method to solve a problem in which the shape of the outermost flux surface is fixed. This corresponds to prescribing $E(1)$ and $R_A^m(1)$ as well as the free functions $p(\Psi)$ and $g(\Psi)$. We denote by $\mathbf{S}(\psi)$ the geometry vector defined as the $(M+1)$ component vector,

$$\mathbf{S}(\psi) = \left[E(\psi), R_A^0(\psi), R_A^2(\psi), \cdots R_A^m(\psi)\right].$$

Since the second-derivative terms only appear linearly in Eqs. (5.75) and (5.76), they can be written symbolically as

$$\mathbf{A}(\mathbf{S}, \mathbf{S}_\psi, \psi) \cdot \mathbf{S}_{\psi\psi} + \mathbf{D}(\mathbf{S}, \mathbf{S}_\psi, \psi) = 0. \tag{5.77}$$

Here we have introduced the $(M+1) \times (M+1)$ matrix $\mathbf{A}$ and the $(M+1)$ component vector $\mathbf{D}$. Equation (5.77) is a system of $(M+1)$ second-order ordinary differential equations (ODEs) which requires $2(M+1)$ boundary conditions. Half of these are specified at the boundary $\psi = 1$ by specifying the shape of the boundary surface, and half are specified as regularity conditions at the origin. This is solved by a quasi-linearization iteration, each step of which is a boundary value problem. At the kth iteration, we want next to approximate

$$\mathbf{A} \cdot \mathbf{S}_{\psi\psi}^{k+1} + \mathbf{D}^{k+1} = 0. \tag{5.78}$$

To linearize, we compute algebraically the derivatives $\partial D_i / \partial S_j$ and $\partial D_i / \partial S_{\psi j}$. The vector $\mathbf{D}$ is Taylor expanded about its value at iteration k,

$$\mathbf{D}^{k+1} \approx \mathbf{D}^k + \frac{\partial \mathbf{D}}{\partial \mathbf{S}} \cdot \left(\mathbf{S}^{k+1} - \mathbf{S}^k\right) + \frac{\partial \mathbf{D}}{\partial \mathbf{S}_\psi} \cdot \left(\mathbf{S}_\psi^{k+1} - \mathbf{S}_\psi^k\right).$$

Insertion into Eq. (5.78) gives

$$\mathbf{A} \cdot \mathbf{S}_{\psi\psi}^{k+1} + \frac{\partial \mathbf{D}}{\partial \mathbf{S}} \cdot \mathbf{S}^{k+1} + \frac{\partial \mathbf{D}}{\partial \mathbf{S}_\psi} \cdot \mathbf{S}_\psi^{k+1} + \tilde{\mathbf{D}}^k = 0. \tag{5.79}$$

Here we have defined

$$\tilde{\mathbf{D}}^k = \mathbf{D}^k - \frac{\partial \mathbf{D}}{\partial \mathbf{S}} \cdot \mathbf{S}^k - \frac{\partial \mathbf{D}}{\partial \mathbf{S}_\psi} \cdot \mathbf{S}_\psi^k.$$

Equation (5.79) is now linear in the unknown. We finite difference in ψ to obtain the block-tridiagonal system

$$\mathbf{a}_j \cdot \mathbf{S}_{j+1} - \mathbf{b}_j \cdot \mathbf{S}_j + \mathbf{c}_j \cdot \mathbf{S}_{j-1} + \mathbf{d}_j = 0. \tag{5.80}$$

Here, the coefficient blocks are defined as

$$\mathbf{a}_j = \frac{1}{(\Delta\psi)^2} \mathbf{A}_j^k + \frac{1}{(2\Delta\psi)} \left.\frac{\partial \mathbf{D}}{\partial \mathbf{S}_\psi}\right|_j,$$

$$\mathbf{b}_j = \frac{2}{(\Delta\psi)^2} \mathbf{A}_j^k,$$

$$\mathbf{c}_j = \frac{1}{(\Delta\psi)^2} \mathbf{A}_j^k - \frac{1}{(2\Delta\psi)} \left.\frac{\partial \mathbf{D}}{\partial \mathbf{S}_\psi}\right|_j,$$

$$\mathbf{d}_j = \tilde{\mathbf{D}}^k.$$

Equation (5.80) is solved using the standard block-tridiagonal techniques introduced in Section 3.3.3. The regularity conditions at the origin are enforced by noting that at the origin, the following asymptotic forms are valid:

$$\begin{aligned}
E &= E_1 \psi + \cdots, \\
R_A^0 &= R_0^0 + R_1^0 \psi + \cdots, \\
R_A^m &= R_1^m \psi + \cdots, \\
\Psi &= \Psi_0 + \Psi_1 \psi + \Psi_2 \psi^2 + \cdots.
\end{aligned} \tag{5.81}$$

Substituting these forms into Eqs. (5.74), (5.75), and (5.76) gives, after considerable algebra, the relations:

$$\begin{aligned}
R_1^0 &= \frac{-T}{2R_0^0 \left(E_0^2 + 3E_0^{-2}\right)} \left[1 - \frac{p_0' E_0^2 (R_0^0)^2 T}{\Psi_1}\right], \\
E_1 &= \frac{1}{1 + E_0^{-4}} \left[\frac{T}{6E_0 (R_0^0)^2} - \frac{E_0 T^2 p_0'}{24\Psi_1} + \frac{2(R_1^0)^2}{TE_0^5}\right. \\
&\quad + \left.\frac{R_1^0}{6E_0 R_0} \left(E_0^2 + 4E_0^{-2}\right) - \frac{\Psi_2}{3\Psi_1}\left(E_0 - E_0^{-3}\right)\right], \\
\Psi_1 &= \frac{-T}{2\left(E_0^2 + E_0^{-2}\right)} \left[(gg')_0 + (R_0^0)^2 p_0'\right].
\end{aligned}$$

Here, we have defined

$$(gg')_0 \equiv g \frac{dg}{d\Psi}\bigg|_{\Psi = \Psi_0}, \qquad p_0' \equiv \frac{dp}{d\Psi}\bigg|_{\Psi = \Psi_0}.$$

5.5.5 Steepest Descent Method

Another method for calculating inverse equilibrium is based on minimizing the energy functional W defined in Eq. (4.41) while maintaining the ideal MHD constraints [107]. It was shown in Section 5.4.1 that if θ^* is a straight field line poloidal angle, we can write an axisymmetric equilibrium magnetic field as

$$\mathbf{B} = \nabla\phi \times \nabla\Psi + \nabla H \times \nabla\theta^*. \tag{5.82}$$

Here, $2\pi\Psi(\psi)$ is the poloidal magnetic flux, $2\pi H$ is the toroidal magnetic flux, and $q(\psi) \equiv H'/\Psi'$ is the safety factor as defined in Eq. (5.35). A parameter $\lambda(\psi, \theta)$ is introduced to allow flexibility in the angular expansion so that the straight field line coordinate θ^* is related to the expansion angle θ by

$$\theta^* = \theta + \lambda(\psi, \theta).$$

The parameter λ will be determined by the condition that the Jacobian in the (ψ, θ^*, ϕ) coordinate system be proportional to R^2. If the Jacobian in the

(ψ, θ, ϕ) coordinate system is $J \equiv [\nabla\psi \times \nabla\theta \cdot \nabla\phi]^{-1}$, then we have the defining relation

$$\lambda(\psi,\theta) = \int_0^\theta \left[\frac{\alpha(\psi)J}{R^2} - 1\right] d\theta, \qquad \alpha(\psi) \equiv \frac{1}{2\pi}\int_0^{2\pi}\frac{J}{R^2}d\theta. \tag{5.83}$$

This form for the magnetic field manifestly conserves the magnetic flux profiles Ψ' and Φ'. It is shown in [95] that the adiabatic conservation of mass between neighboring flux surfaces can be enforced by expressing the pressure as $p(\psi) = M(\psi)[V']^\gamma$. Here V', defined in Eq. (5.29), is the surface differential volume element, γ is the adiabatic index ($\gamma = 5/3$ for an ideal gas or plasma), and the *mass function*, $M(\psi)$ is held fixed during the variation.

To perform the variation of W, we suppose that the cylindrical coordinates $R(\psi,\theta)$ and $Z(\psi,\theta)$ depend on an artificial time parameter t. Then, for any scalar function $S(R,Z)$, we have

$$\frac{\partial S}{\partial t} = \frac{\partial S}{\partial R}\dot{R} + \frac{\partial S}{\partial Z}\dot{Z}.$$

We can therefore compute the time derivative of W, holding $\Psi'(\psi)$, $\Phi'(\psi)$, and $M(\psi)$ fixed, and holding the boundary shape fixed, from the relation

$$\frac{dW}{dt} = -\int_0^{\psi_{max}} d\psi \int_0^{2\pi} d\theta \left(F_R\dot{R} + F_Z\dot{Z}\right), \tag{5.84}$$

where $F_R = RZ_\theta G$, $F_Z = -RR_\theta G$, with G given by Eq. (5.72).

A solution technique has been implemented whereby the functions $R(\psi,\theta)$ and $Z(\psi,\theta)$ are represented by Fourier series in the angle θ,

$$R(\psi,\theta) = \sum_m [R_c^m(\psi)\cos m\theta + R_s^m(\psi)\sin m\theta],$$
$$Z(\psi,\theta) = \sum_m [Z_c^m(\psi)\cos m\theta + Z_s^m(\psi)\sin m\theta]. \tag{5.85}$$

Substitution into Eq. (5.84) gives

$$\frac{dW}{dt} = -\int_0^{\psi_{max}} \left(F_R^{mc}\dot{R}_c^m + F_R^{ms}\dot{R}_s^m + F_Z^{mc}\dot{Z}_c^m + F_Z^{ms}\dot{Z}_s^m\right) V'd\psi, \tag{5.86}$$

where we have defined

$$F_R^{mc} = \frac{1}{V'}\int_0^{2\pi} F_R\cos m\theta d\theta, \qquad F_R^{ms} = \frac{1}{V'}\int_0^{2\pi} F_R\sin m\theta d\theta, \tag{5.87}$$

and similarly for F_Z^{mc} and F_Z^{ms}.

Since W is bounded from below due to flux and mass conservation and is positive definite, the equilibrium corresponds to a minimum energy state.

Thus, by finding the path along which $\dot{W}$ decreases monotonically, an equilibrium will eventually be reached. Such a path is defined by the steepest descent path

$$\frac{\partial R_c^m}{\partial t} = F_R^{mc}, \qquad \frac{\partial R_s^m}{\partial t} = F_R^{ms} \qquad (5.88)$$

again, with similar equations for Z_c^m and Z_s^m. Substitution into Eq. (5.86) gives for the rate of decrease in W along this descent path

$$\frac{dW}{dt} = -\int_0^{\psi_{max}} \left(|F_R^{mc}|^2 + |F_R^{ms}|^2 + |F_Z^{mc}|^2 + |F_Z^{ms}|^2 \right) V' d\psi. \qquad (5.89)$$

Equations (5.88) and (5.89), together with the prescription for renormalizing the poloidal angle, Eq. (5.83), provide a prescription for relaxing W to its minimum energy state, but for the same reasons given in the discussion of Jacobi relaxation in Section 3.5.1, since the $F_{R,Z}^{mc,s}$ correspond to second-order differential operators, explicit time differencing leads to a prohibitively slow algorithm. The convergence can be greatly accelerated by converting them to damped hyperbolic equations. For example, we replace the $\dot{R}_c^m$ equation in Eq. (5.88) by

$$\frac{\partial^2 R_c^m}{\partial t^2} + \frac{1}{\tau}\frac{\partial R_c^m}{\partial t} = F_R^{mc}, \qquad (5.90)$$

and similarly for R_s^m, Z_c^m, and Z_s^m. This is essentially the dynamic relaxation method introduced in Section 3.5.2. Here the parameter τ has little effect on numerical stability and should be chosen to critically damp the slowest mode. Here we note that by multiplying Eq. (5.90) by $V'\dot{R}_c^m$, inserting into Eq. (5.86), and doing the same for the other Fourier coefficients R_s^m, Z_c^m, and Z_s^m, we obtain a descent equation corresponding to a second-order method,

$$\frac{d}{dt}(W_K + W) = -\frac{2}{\tau}W_K, \qquad (5.91)$$

where

$$W_K = \frac{1}{2}\int_0^{\psi_{max}} \left(|\dot{R}_c^m|^2 + |\dot{Z}_c^m|^2 + |\dot{R}_s^m|^2 + |\dot{Z}_s^m|^2 \right) V' d\psi$$

is the kinetic energy in the system.

Boundary conditions must be supplied for the R_c^m, R_s^m, Z_c^m, and Z_s^m at the origin ($\psi = 0$) and at the boundary, where they prescribe the shape of the last closed flux surface. At the origin, the requirement that these functions remain analytic gives the condition that for $m > 0$ they each vanish, and for $m = 0$ their derivatives with respect to ψ should have the same asymptotic behavior near $\psi = 0$ as does $V'(\psi)$ [107].

5.6 Summary

Many stability and transport calculations are greatly facilitated by the use of coordinates aligned with the magnetic field. For toroidal confinement configurations where the magnetic field lines form surfaces, these surfaces can be used to construct coordinates. If the geometry is axisymmetric, or close to it, there are many advantages to using axisymmetric magnetic coordinates where the toroidal angle ϕ is unchanged from the angle in cylindrical coordinates. The poloidal angle θ can be chosen by specifying the form of the Jacobian. It is often required to solve the equilibrium equation in the flux coordinate system. The quantities to be determined are the cylindrical coordinates in terms of the flux coordinates, $R(\psi, \theta)$ and $Z(\psi, \theta)$. There are many approaches to solving for these quantities. The use of flux coordinates makes it convenient to specify surface functions as one of the free functions in the Grad–Shafranov equation.

Problems

5.1: Show that by applying Ampere's law to a closed constant-ψ contour, you will get the same result for the current interior to the contour as you would by integrating Eq. (5.34) over the contour area.

5.2: A constant ψ magnetic surface has a local normal vector $\hat{n} = (n_R, n_Z)$ and a local tangent vector $\hat{t}$. Let α be the angle that the tangent vector makes with the horizontal. Then the *curvature*, κ, of the boundary curve is defined as the rate of change of the angle α with respect to the arc length s,

$$\kappa = \frac{d\alpha}{ds}.$$

Let $R(\psi, \theta)$ and $Z(\psi, \theta)$ be cylindrical coordinates:
(a) Show that the local curvature is given by:

$$\kappa = \frac{R_\theta Z_{\theta\theta} - Z_\theta R_{\theta\theta}}{(R_\theta^2 + Z_\theta^2)^{3/2}},$$

where subscripts on R and Z denote partial differentiation.
(b) Show that an equivalent expression for the curvature is given by:

$$\kappa = \frac{n_Z^2 \psi_{RR} + n_R^2 \psi_{ZZ} - 2n_R n_Z \psi_{RZ}}{|\nabla \psi|},$$

where subscripts on ψ again denote partial differentiation.

5.3: *Christoffel symbols* arise from differentiating the basis vectors and projecting the results back onto the original basis. The first two Christoffel symbols are defined by:

$$\frac{\partial}{\partial \psi} \nabla \psi = \Gamma^1_{11} \nabla \psi + \Gamma^2_{11} \nabla \theta.$$

(a) Show that

$$\Gamma^1_{11} = \frac{R}{J} \left(R_{\psi\psi} Z_\theta - Z_{\psi\psi} R_\theta \right).$$

(b) Calculate Γ^2_{11}.

5.4: Use the methods in Section 5.2.3 to calculate formulas for curl, div, and grad in the standard spherical coordinate system (r, θ, ϕ).

5.5: Starting from the first line in Eq. (5.43), derive the next two lines using relations in this chapter.

5.6: Derive the q-solver form of the equilibrium equation, Eq. (5.56).

5.7: Derive the expression given in Eq. (5.62) that the function G obeys in the J-solver form of the equilibrium equation.

5.8: *Helical Grad–Shafranov equation.* Let (r, θ, z) be a standard cylindrical coordinate system. Define a new helical coordinate, $\tau \equiv \theta + \epsilon z$, where ϵ is a constant, and work in the (r, τ, z) non-orthogonal coordinate system. Assume helical symmetry so that all scalar quantities are functions only of r and τ.
(a) Show that the most general expression for the magnetic field $\mathbf{B}$, a vector field that satisfies $\nabla \cdot \mathbf{B} = 0$ and $\mathbf{B} \cdot \nabla \Psi(r, \tau) = 0$, is

$$\mathbf{B} = \nabla z \times \nabla \Psi + F(r, \tau) \nabla r \times \nabla \tau,$$

where $F(r, \tau)$ is an arbitrary function.
(b) Show that in equilibrium, when $\mathbf{J} \times \mathbf{B} - \nabla p = 0$, where $\mathbf{J} = \mu_0^{-1} \nabla \times \mathbf{B}$ is the current density, F and and scalar pressure p must satisfy

$$F = rg^2 \left[f(\Psi) + \epsilon r \frac{\partial \Psi}{\partial r} \right],$$

$$p = p(\psi),$$

where $g \equiv \left(1 + \epsilon^2 r^2 \right)^{-1/2}$ and $f(\Psi)$ and $p(\Psi)$ are functions only of Ψ.
(c) Derive the helical Grad–Shafranov equation [108, 109]:

$$\frac{1}{r} \frac{\partial}{\partial r} \left(g^2 r \frac{\partial \Psi}{\partial r} \right) + \frac{1}{r^2} \frac{\partial^2 \Psi}{\partial \tau^2} - 2\epsilon f g^4 + g^2 f f' + \mu_0 p' = 0.$$

Chapter 6

Diffusion and Transport in Axisymmetric Geometry

6.1 Introduction

In this chapter we consider the time evolution of magnetically confined plasmas over time scales that are very long compared to the Alfvén transit time, and are thus characterized by resistive diffusion and particle and heat transport. This will lead into and provide motivation for a discussion of finite difference methods for parabolic equations in Chapter 7. Because the electron and ion heat fluxes $\mathbf{q}_e$ and $\mathbf{q}_i$ are extremely anisotropic in a highly magnetized plasma, it becomes essential to work in a coordinate system that is aligned with the magnetic field. The derivation of an appropriate set of transport equations to deal with this is presented in Section 6.2 and its subsections. These transport equations need to be supplemented by an equilibrium constraint obtained by solving a particular form of the equilibrium equation as described in Section 6.3. Together, this system of equations provides an accurate description of the long time scale evolution of a MHD stable plasma.

6.2 Basic Equations and Orderings

Here we consider the scalar pressure two-fluid MHD equations of Section 1.2.1, which for our purposes can be written:

$$nm_i \left(\frac{\partial \mathbf{u}}{\partial t} + \mathbf{u} \cdot \nabla \mathbf{u} \right) + \nabla p = \mathbf{J} \times \mathbf{B}, \tag{6.1}$$

$$\frac{\partial n}{\partial t} + \nabla \cdot (n\mathbf{u}) = S_n, \tag{6.2}$$

$$\frac{\partial \mathbf{B}}{\partial t} = -\nabla \times \mathbf{E}, \tag{6.3}$$

$$\mathbf{E} + \mathbf{u} \times \mathbf{B} = \mathbf{R}, \tag{6.4}$$

$$\frac{3}{2}\frac{\partial p}{\partial t} + \nabla \cdot \left[\mathbf{q} + \frac{3}{2}p\mathbf{u}\right] = -p\nabla \cdot \mathbf{u} + \mathbf{J} \cdot \mathbf{R} + S_e, \tag{6.5}$$

$$\frac{3}{2}\frac{\partial p_e}{\partial t} + \nabla \cdot \left[\mathbf{q}_e + \frac{3}{2}p_e\mathbf{u}\right] = -p_e\nabla \cdot \mathbf{u} + \mathbf{J} \cdot \mathbf{R} + Q_{\Delta ei} + S_{ee}. \tag{6.6}$$

We have denoted by $\mathbf{R}$ the inhomogeneous term in the generalized Ohm's law, Eq. (6.4), by $\mathbf{q} = \mathbf{q}_i + \mathbf{q}_e$ the random heat flux vector due to ions and electrons, by $Q_{\Delta ei}$ the electron-ion equipartition term, and by S_n and $S_e = S_{ei} + S_{ee}$ sources of particles and energy. Note that in comparing Eq. (6.4) with Eq. (1.32), we have the relation

$$\mathbf{R} = \frac{1}{ne}[\mathbf{R}_e + \mathbf{J} \times \mathbf{B} - \nabla p_e - \nabla \cdot \boldsymbol{\pi}_e]. \tag{6.7}$$

We are also neglecting (assumed small) terms involving the ion and electron stress tensors $\boldsymbol{\pi}_e$, $\boldsymbol{\pi}_i$, and the heating due to $\nabla T_e \cdot \mathbf{J}$.

We now apply a *resistive time scale ordering* [110, 111, 112] to these equations to isolate the long time scale behavior. This consists of ordering all the source and transport terms to be the order of the inverse magnetic Lundquist number, Eq. (1.59), $S^{-1} = \epsilon \ll 1$, where ϵ is now some small dimensionless measure of the dissipation. Thus

$$\eta \sim \mathbf{R} \sim S_n \sim S_e \sim \mathbf{q} \sim \epsilon \ll 1. \tag{6.8}$$

We look for solutions in which all time derivatives and velocities are also small, of order ϵ ,

$$\frac{\partial}{\partial t} \sim \mathbf{u} \sim \epsilon \ll 1, \tag{6.9}$$

as is the electric field, $\mathbf{E} \sim \epsilon$.

Applying this ordering to Eqs. (6.1)–(6.6), we find that the last five equations remain unchanged, merely picking up the factor ϵ in every term, which can then be canceled. However, the momentum equation, Eq. (6.1), does change, with a factor of ϵ^2 multiplying only the inertial terms,

$$\epsilon^2 nm_i \left(\frac{\partial \mathbf{u}}{\partial t} + \mathbf{u} \cdot \nabla \mathbf{u}\right) + \nabla p = \mathbf{J} \times \mathbf{B}. \tag{6.10}$$

Thus in the limit $\epsilon \to 0$ we can neglect the inertial terms, replacing the momentum equation with the equilibrium condition

$$\nabla p = \mathbf{J} \times \mathbf{B}. \tag{6.11}$$

Equation (6.11) is correct to second order in ϵ, and provides significant simplifications since replacing Eq. (6.10) by this removes all the wave propagation characteristics from the system.

We note here that the system of equations given by Eqs. (6.11) and (6.2) through (6.6) involve the plasma velocity $\mathbf{u}$, but there is no longer a time advancement equation for $\mathbf{u}$. We will derive a method to solve this system asymptotically in spite of this apparent difficulty. We restrict consideration here to axisymmetric geometry. The most general form for an axisymmetric magnetic field consistent with Eq. (6.11) was shown in Section 4.2 to be given by

$$\mathbf{B} = \nabla\phi \times \nabla\Psi + g(\Psi)\nabla\phi. \tag{6.12}$$

Let us first consider the poloidal part of the magnetic field evolution equation, Eq. (6.3). Insertion of Eq. (6.12) into Eq. (6.3) and taking the $(\hat{R}, \hat{Z})$ projections gives an equation to evolve the poloidal flux function

$$\frac{\partial\Psi}{\partial t} = R^2\mathbf{E} \cdot \nabla\phi + C(t). \tag{6.13}$$

The integration constant $C(t)$ can be set to zero by adopting the convention that Ψ be proportional to the actual poloidal flux that must vanish at $R = 0$. Setting $C(t) = 0$ and using Eq. (6.4) to eliminate the electric field, we have

$$\frac{\partial\Psi}{\partial t} + \mathbf{u} \cdot \nabla\Psi = R^2\nabla\phi \cdot \mathbf{R}. \tag{6.14}$$

(Note that the symbol R is being used to represent the cylindrical coordinate, while $\mathbf{R}$, defined in Eq. (6.7), is the vector inhomogeneous term in the generalized Ohm's law.) The $\nabla\phi$ projection of Eq. (6.3) gives

$$\frac{\partial g}{\partial t} = R^2\nabla \cdot [\nabla\phi \times \mathbf{E}],$$

or, upon substituting from Eq. (6.4),

$$\frac{\partial g}{\partial t} + R^2\nabla \cdot \left[\frac{g}{R^2}\mathbf{u} - (\nabla\phi \cdot \mathbf{u})\nabla\phi \times \nabla\Psi - \nabla\phi \times \mathbf{R}\right] = 0. \tag{6.15}$$

The system of time evolution equations that we are solving is thus reduced to the five scalar equations (6.2), (6.5), (6.6), (6.14), and (6.15) as well as the equilibrium equation (6.11).

6.2.1 Time-Dependent Coordinate Transformation

We adopt here the axisymmetric magnetic flux coordinate system developed in Chapter 5. Since the magnetic field and flux surfaces evolve and change in time, the coordinate transformation being considered will be a time-dependent one. At any given time we have the flux coordinates (ψ, θ, ϕ) and

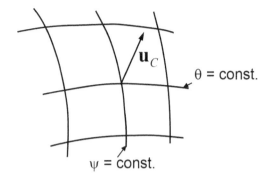

FIGURE 6.1: $\mathbf{u}_C$ is the velocity of a fluid element with a given ψ,θ value relative to a fixed Cartesian frame.

the inverse representation $\mathbf{x}(\psi, \theta, \phi)$, where $\mathbf{x}$ are Cartesian coordinates. We define the *coordinate velocity* at a particular (ψ, θ, ϕ) location as the time rate of change of the Cartesian coordinate at a fixed value of the flux coordinate as shown in Figure 6.1,

$$\mathbf{u}_C = \left. \frac{\partial \mathbf{x}}{\partial t} \right|_{\psi,\theta,\phi}. \tag{6.16}$$

Consider now a scalar function, α, that may be thought of either as a function of Cartesian coordinates and time, $\alpha(\mathbf{x}, t)$, or of flux coordinates and time, $\alpha(\psi, \theta, \phi, t)$. Time derivatives at a fixed spatial location $\mathbf{x}$ and at fixed coordinates (ψ, θ, ϕ) are related by the chain rule of partial differentiation,

$$\left. \frac{\partial \alpha}{\partial t} \right|_{\psi,\theta,\phi} = \left. \frac{\partial \alpha}{\partial t} \right|_{\mathbf{x}} + \frac{\partial \alpha}{\partial \mathbf{x}} \cdot \left. \frac{\partial \mathbf{x}}{\partial t} \right|_{\psi,\theta,\phi}.$$

Using Eq. (6.16), we therefore have the relation

$$\left. \frac{\partial}{\partial t} \right|_{\mathbf{x}} = \left. \frac{\partial}{\partial t} \right|_{\psi,\theta,\phi} - \mathbf{u}_C \cdot \nabla. \tag{6.17}$$

We will also make use of the relation for the time derivative of the Jacobian, defined in Eqs. (5.1) and (5.19),

$$\left. \frac{\partial J}{\partial t} \right|_{\psi,\theta,\phi} = J \nabla \cdot \mathbf{u}_C. \tag{6.18}$$

This may be verified directly.

With the introduction of the coordinate velocity in Eq. (6.16), it follows that the fluid velocity appearing in the MHD equations can be thought of as consisting of two parts,

$$\mathbf{u} = \mathbf{u}_C + \mathbf{u}_R, \tag{6.19}$$

where $\mathbf{u}_C$ is the coordinate velocity already discussed, and $\mathbf{u}_R$ is the velocity of the fluid relative to the coordinates. Since the total velocity $\mathbf{u}$ is a physical quantity which must be determined by the MHD equations, constraining either $\mathbf{u}_C$ or $\mathbf{u}_R$ to be of a particular form will determine the other.

The two parts of the velocity field can be thought of as a *Lagrangian* part and an *Eulerian* part. If the total velocity were represented with $\mathbf{u}_C$ only and there were no dissipation, the coordinates would be frozen into the fluid as it moves and distorts. If the total velocity were represented with $\mathbf{u}_R$, the coordinates would be fixed in space, and the fluid would move through them. We will see that an attractive choice is to split the velocity between these two parts so that the coordinates can move just enough to stay flux coordinates, but we allow the fluid to diffuse relative to them.

6.2.2   Evolution Equations in a Moving Frame

We now transform each of the scalar evolution equations, Eqs. (6.14), (6.15), (6.2), (6.5), and (6.6), into the moving flux coordinate frame by using the identities in Eqs. (6.17), (6.18), and (6.19). After some manipulations, we obtain the scalar equations

$$\frac{\partial \Psi}{\partial t} + \mathbf{u}_R \cdot \nabla \Psi = R^2 \nabla \phi \cdot \mathbf{R}, \tag{6.20}$$

$$\frac{\partial}{\partial t}\left(g \frac{J}{R^2}\right) + J\nabla \cdot \left[\frac{g}{R^2}\mathbf{u}_R - (\nabla \phi \cdot \mathbf{u}_R)\nabla \phi \times \nabla \Psi - \nabla \phi \times \mathbf{R}\right] = 0, \tag{6.21}$$

$$\frac{\partial}{\partial t}(nJ) + J\nabla \cdot (n\mathbf{u}_R) = JS_n, \tag{6.22}$$

$$\frac{\partial}{\partial t}\left(p^{3/5}J\right) + J\nabla \cdot \left[p^{3/5}\mathbf{u}_R\right] + \frac{2}{5}Jp^{-2/5}[\nabla \cdot q - \mathbf{J} \cdot \mathbf{R} - S_e] = 0, \tag{6.23}$$

$$\frac{\partial}{\partial t}\left(p_e^{3/5}J\right) + J\nabla \cdot \left[p_e^{3/5}\mathbf{u}_R\right] \quad + \quad \frac{2}{5}Jp_e^{-2/5}[\nabla \cdot q_e - \mathbf{J} \cdot \mathbf{R} - Q_{\Delta ei} - S_{ee}] = 0. \tag{6.24}$$

Here, and in what follows, the time derivatives are with ψ and θ held fixed, and are thus in a moving flux coordinate frame. We note that the coordinate velocity $\mathbf{u}_C$ does not appear in Eqs. (6.20)–(6.23), but only the velocity of the fluid relative to the moving coordinate, $\mathbf{u}_R = \mathbf{u} - \mathbf{u}_C$, is present. This is because the equations are of conservation form, and thus valid in a moving frame. We will use this fact to define a coordinate transformation in which the velocity vanishes altogether.

We first derive a reduced set of equations by integrating Eqs. (6.21)–(6.24) over the angle θ at fixed values of the coordinate ψ. This *flux surface averaging* leads to a set of one-dimensional evolution equations that depend only on the coordinate ψ and time t. Furthermore, by using the fact that $g = g(\psi, t)$ and $p = p(\psi, t)$ from the equilibrium constraint, and that $n \simeq n(\psi, t)$ and $p_e \simeq p_e(\psi, t)$ from the fact that the temperatures are nearly constant on flux surfaces because the parallel conductivities are large compared to their perpendicular values, we can obtain a closed set of equations that depend only on these surface averages.

From Eqs. (5.8) and (5.11), and the definition of differential volume and surface average in Eqs. (5.29) and (5.30), it follows that for any vector $\mathbf{A}$, we have the identity

$$2\pi \int_0^{2\pi} J\nabla \cdot \mathbf{A}\, d\theta = \frac{\partial}{\partial \psi} \left[V' \langle \mathbf{A} \cdot \nabla \psi \rangle \right].$$

Using this, the following set of equations are obtained by integrating Eqs. (6.21)–(6.24) over the angle θ at fixed value of the flux coordinate ψ:

$$\frac{\partial}{\partial t} \left[gV' \langle R^{-2} \rangle \right] + \frac{\partial}{\partial \psi} \left[gV' \langle R^{-2}\mathbf{u}_R \cdot \nabla \psi \rangle - V' \langle \nabla \phi \times \mathbf{R} \cdot \nabla \psi \rangle \right] = 0, \quad (6.25)$$

$$\frac{\partial}{\partial t} \left[nV' \right] + \frac{\partial}{\partial \psi} \left[nV' \langle \nabla \psi \cdot \mathbf{u}_R \rangle \right] = V' \langle S_n \rangle, \quad (6.26)$$

$$\frac{\partial}{\partial t} \left[p^{\frac{3}{5}}V' \right] + \frac{\partial}{\partial \psi} \left[p^{\frac{3}{5}}V' \langle \mathbf{u}_R \cdot \nabla \psi \rangle \right] \quad (6.27)$$
$$+ \frac{2}{5} p^{-\frac{2}{5}} \left[\frac{\partial}{\partial \psi} \left(V' \langle \mathbf{q} \cdot \nabla \psi \rangle \right) - V' \langle \mathbf{J} \cdot \mathbf{R} \rangle - V' \langle S_e \rangle \right] = 0.$$

$$\frac{\partial}{\partial t} \left[p_e^{\frac{3}{5}}V' \right] + \frac{\partial}{\partial \psi} \left[p_e^{\frac{3}{5}}V' \langle \mathbf{u}_R \cdot \nabla \psi \rangle \right] \quad (6.28)$$
$$+ \frac{2}{5} p_e^{-\frac{2}{5}} \left[\frac{\partial}{\partial \psi} \left(V' \langle \mathbf{q}_e \cdot \nabla \psi \rangle \right) - V' \left(\langle \mathbf{J} \cdot \mathbf{R} \rangle + \langle Q_{\Delta ei} \rangle + \langle S_{ee} \rangle \right) \right] = 0.$$

As discussed above, an important constraint that must be incorporated is that the coordinates (ψ, θ, ϕ) remain *flux coordinates* as they evolve in time. To this end, we require that in the moving frame, the flux function Ψ evolve in time in such a way that the coordinate ψ remain a flux coordinate, i.e.,

$$\nabla \phi \times \nabla \psi \cdot \frac{\partial \Psi}{\partial t} = 0. \quad (6.29)$$

From Eq. (6.20), this implies

$$\nabla\Psi \cdot \mathbf{u}_R - R^2\mathbf{R} \cdot \nabla\phi = f(\psi). \tag{6.30}$$

Here $f(\psi)$ is a presently undetermined function only of ψ. Equation (6.30) puts an important constraint on the relative velocity $\mathbf{u}_R$ and hence on the coordinate velocity $\mathbf{u}_C$ through Eq. (6.19), but it also leaves some freedom in that we are free to prescribe the function $f(\psi)$. This freedom will be used to identify the flux coordinate ψ with a particular surface function.

The system of equations given by Eqs. (6.20), (6.25)–(6.28), and (6.30) still depend upon the relative velocity $\mathbf{u}_R \cdot \nabla\psi$. This is determined up to a function only of ψ by Eq. (6.30), which follows from the constraint that constant Ψ surfaces align with constant ψ surfaces as they both evolve. We thus have a freedom in the velocity decomposition that we can use to simplify the problem. The remaining function of ψ is determined by specifying which flux function ψ is. Three common choices are the following.

(i)   Constant poloidal flux:

$$\mathbf{u}_R \cdot \nabla\Psi = R^2\nabla\phi \cdot \mathbf{R}. \tag{6.31}$$

(ii)   Constant toroidal flux:

$$-g\left\langle R^{-2}\mathbf{u}_R \cdot \nabla\psi \right\rangle = \left\langle \nabla\phi \times \nabla\psi \cdot \mathbf{R} \right\rangle. \tag{6.32}$$

(iii)   Constant mass:

$$\left\langle \nabla\psi \cdot \mathbf{u}_R \right\rangle = 0. \tag{6.33}$$

Here, we choose number (ii), the toroidal magnetic flux, as it is most appropriate for most magnetic fusion applications, particularly for describing tokamaks. The toroidal field in the tokamak is primarily produced by the external field magnets, and is generally much stronger than the poloidal field. This makes it the most immobile, and thus most suitable for use as a coordinate. Also, unlike the poloidal magnetic flux, the toroidal flux at the magnetic axis does not change in time, always remaining zero.

6.2.3   Evolution in Toroidal Flux Coordinates

By combining Eq. (6.32) and the constraint Eq. (6.30), we can eliminate the free function $f(\psi)$ and solve explicitly for the normal relative velocity. This gives

$$f(\psi) = -\frac{\langle \mathbf{B} \cdot \mathbf{R} \rangle}{\langle \mathbf{B} \cdot \nabla\phi \rangle},$$

which when inserted into Eq. (6.30) yields

$$\mathbf{u}_R \cdot \nabla\psi = \frac{1}{\Psi'}\left[R^2\mathbf{R} \cdot \nabla\phi - \frac{\langle \mathbf{B} \cdot \mathbf{R} \rangle}{\langle \mathbf{B} \cdot \nabla\phi \rangle} \right]. \tag{6.34}$$

Using Eq. (6.34) to eliminate the relative velocity $\mathbf{u}_R$ from the surface averaged transport equations allows us to identify the flux coordinate ψ with the toroidal magnetic flux inside a constant flux surface, Φ. This can be verified by calculating directly from the definition of Φ in Eq. (5.31),

$$\left.\frac{\partial \Phi}{\partial t}\right|_{\psi} = \frac{1}{2\pi} \int_0^{\psi} [gV' \langle R^{-2} \rangle]_t \, d\psi = 0,$$

where we used Eqs. (6.25) and (6.34). It is seen that the use of Eq. (6.34) causes Φ and ψ to be stationary with respect to each other, and we may therefore adopt Φ as the flux surface label. Equation (6.25) need no longer be solved as it is intrinsically satisfied by our adoption of Φ as the flux coordinate. Since ψ and Φ are the same, we obtain a useful identity, valid for toroidal flux coordinates, by differentiating each side of Eq. (5.31) by $\psi \equiv \Phi$,

$$V' = \frac{2\pi}{g \langle R^{-2} \rangle}. \tag{6.35}$$

Using Eq. (6.34) to eliminate the relative velocity from Eqs. (6.20), (6.26), (6.27), and (6.28) then yields the surface-averaged transport equations relative to surfaces of constant toroidal flux. In deriving Eqs. (6.46) and (6.47) that follow, we make use of the equilibrium conditions in Eqs. (4.4)–(4.8) to express the equilibrium current density as

$$\mathbf{J} = -R^2 \frac{dp}{d\Psi} \nabla\phi - \frac{1}{\mu_0} \frac{dg}{d\Psi} \mathbf{B}.$$

We also use Eq. (6.34) and the surface average of the inverse equilibrium equation, Eq. (5.53), to obtain the intermediate result

$$V' \langle \mathbf{J} \cdot \mathbf{R} \rangle = -p'V' \langle \mathbf{u}_R \cdot \nabla\Phi \rangle + \frac{\langle \mathbf{B} \cdot \mathbf{R} \rangle}{\langle \mathbf{B} \cdot \nabla\phi \rangle} \frac{d}{d\Phi} \left[\frac{V'}{2\pi\mu_0 q} \left\langle \frac{|\nabla\Phi|^2}{R^2} \right\rangle \right]. \tag{6.36}$$

To express the final form of the surface-averaged transport equations, we first define the rotational transform

$$\iota \equiv \frac{1}{q} = 2\pi \frac{d\Psi}{d\Phi}, \tag{6.37}$$

the loop voltage

$$V_L \equiv 2\pi \frac{\langle \mathbf{B} \cdot \mathbf{R} \rangle}{\langle \mathbf{B} \cdot \nabla\phi \rangle}, \tag{6.38}$$

the differential particle number, or number of particles in the differential volume between surfaces Φ and $\Phi + d\Phi$,

$$N' \equiv nV', \tag{6.39}$$

the particle flux

$$\Gamma \equiv 2\pi q n \left[\langle R^2 \mathbf{R} \cdot \nabla \phi \rangle - \frac{\langle \mathbf{B} \cdot \mathbf{R} \rangle}{\langle \mathbf{B} \cdot \nabla \phi \rangle} \right], \tag{6.40}$$

the differential total and electron entropy densities

$$\sigma = p V'^{\frac{5}{3}}, \qquad \sigma_e = p_e V'^{\frac{5}{3}}, \tag{6.41}$$

the surface integrated current density

$$K \equiv \frac{V'}{(2\pi)^2 \mu_0 q} \left\langle \frac{|\nabla \Phi|^2}{R^2} \right\rangle, \tag{6.42}$$

and the electron and ion heat fluxes

$$Q_e \equiv V' \left[\langle \mathbf{q}_e \cdot \nabla \Phi \rangle + \frac{5}{2} \frac{p_e}{n} \Gamma \right],$$

$$Q_i \equiv V' \left[\langle \mathbf{q}_i \cdot \nabla \Phi \rangle + \frac{5}{2} \frac{p_i}{n} \Gamma \right]. \tag{6.43}$$

With these definitions, the basic transport equations take on the compact form:

$$\frac{\partial \Psi}{\partial t} = \frac{1}{2\pi} V_L, \tag{6.44}$$

$$\frac{\partial N'}{\partial t} + \frac{\partial}{\partial \Phi} V' \Gamma = V' \langle S_n \rangle, \tag{6.45}$$

$$\frac{3}{2} (V')^{-\frac{2}{3}} \frac{\partial \sigma}{\partial t} + \frac{\partial}{\partial \Phi} (Q_e + Q_i) = V_L \frac{\partial K}{\partial \Phi} + V' \langle S_e \rangle, \tag{6.46}$$

$$\frac{3}{2} (V')^{-\frac{2}{3}} \frac{\partial \sigma_e}{\partial t} + \frac{\partial Q_e}{\partial \Phi} + V' \left(\frac{\Gamma}{n} \frac{\partial p_i}{\partial \Phi} - Q_{\Delta ei} \right) = V_L \frac{\partial K}{\partial \Phi} + V' \langle S_{ee} \rangle. \tag{6.47}$$

By differentiating Eq. (6.44) with respect to the toroidal flux Φ, we obtain an evolution equation for the rotational transform, defined in Eq. (6.37),

$$\frac{\partial \iota}{\partial t} = \frac{\partial}{\partial \Phi} V_L. \tag{6.48}$$

In a similar manner, we can obtain an evolution equation for the *toroidal angular momentum density*, $\Omega(\Phi) = m_i N' \langle R^2 \rangle \omega(\Phi)$, where ω is the toroidal angular velocity as defined in Eq. (4.11):

$$\frac{\partial \Omega}{\partial t} + \frac{\partial}{\partial \Phi} V' \Gamma_\Omega = V' \langle S_\Omega \rangle. \tag{6.49}$$

The physical reason we were able to eliminate the fluid velocity entirely from the system of transport equations derived here is that transport coefficients only determine the *relative transport* of the magnetic field and plasma densities and energies with respect to one another. We cast these equations in a frame moving with the toroidal magnetic flux, and so only the relative motion remains. The absolute motion of the toroidal magnetic surfaces is determined by the equilibrium constraint and the interaction with externally applied magnetic fields. This is addressed in Section 6.3.

6.2.4   Specifying a Transport Model

Specifying a transport model consists of providing the transport fluxes, Γ, $\langle \mathbf{q}_i \cdot \nabla \Phi \rangle$, $\langle \mathbf{q}_e \cdot \nabla \Phi \rangle$, V_L, and Γ_Ω and the equipartition term $Q_{\Delta ei}$ as functions of the thermodynamic and magnetic field variables and the metric quantities. Also required are the source functions for mass, energy, electron energy, and angular momentum, $\langle S_n \rangle$, $\langle S_e \rangle$, $\langle S_{ee} \rangle$, $\langle S_\Omega \rangle$. The radiative loss function $\langle S_{RAD} \rangle$ must be subtracted from the energy and electron energy source functions.

At a given time, when the geometry is fixed, the plasma pressures, densities, and toroidal angular velocities are obtained from the adiabatic variables through the relations, valid for a two-component plasma with charge $Z = 1$ and with $n_e = n_i = n$,

$$
\begin{aligned}
p_e(\Phi) &= \sigma_e / V'^{5/3}, \\
p_i(\Phi) &= p - p_e = (\sigma - \sigma_e) / V'^{5/3}, \\
n(\Phi) &= N'/V', \\
k_B T_e(\Phi) &= p_e/n, \\
k_B T_i(\Phi) &= p_i/n, \\
\omega(\Phi) &= \Omega / \left(m_i N' \langle R^2 \rangle \right).
\end{aligned}
$$

The electron-ion equipartition term, $Q_{\Delta ei}$, is normally taken to be the classical value given by Eq. (1.45).

It is convenient to define a five-component force vector given by Φ derivatives of the density, total and electron pressures, current density, and toroidal angular velocity relative to the toroidal magnetic flux,

$$
\mathbf{F} = \left[n'(\Phi), p'(\Phi), p_e'(\Phi), (A(\Phi)\iota(\Phi))', \omega'(\Phi) \right]. \tag{6.50}
$$

Here we have defined the geometrical quantity

$$
A(\Phi) \equiv \frac{V'}{g} \left\langle \frac{|\nabla \Phi|^2}{R^2} \right\rangle.
$$

This is related to the surface-averaged parallel current density by

$$
\frac{\langle \mathbf{J} \cdot \mathbf{B} \rangle}{\langle \mathbf{B} \cdot \nabla \phi \rangle} = \frac{g^2}{(2\pi)^2 \mu_0} \left[\frac{V'}{g} \left\langle \frac{|\nabla \Phi|^2}{R^2} \right\rangle \iota \right]' = \frac{g^2}{(2\pi)^2 \mu_0} \left(A(\Phi)\iota(\Phi) \right)'.
$$

The transport fluxes can now be expressed as a matrix of functions multiplying the force vector:

$$
\Gamma = \sum_{j=1}^{5} \Gamma^j F_j, \qquad \langle \mathbf{q}_i \cdot \nabla \Phi \rangle = \sum_{j=1}^{5} q_i^j F_j, \qquad \langle \mathbf{q}_e \cdot \nabla \Phi \rangle = \sum_{j=1}^{5} q_e^j F_j,
$$

$$
V_L = \sum_{j=1}^{5} V_L^j F_j + V_L^6, \qquad \Gamma_\Omega = \sum_{j=1}^{5} \Gamma_\Omega^j F_j.
$$

With this convention, specification of the 25 scalar functions $\Gamma^1, \Gamma^2, \cdots \Gamma_\Omega^5$ will specify the transport model. The additional term V_L^6 will be used to incorporate a source term for current drive.

A. Pfirsch–Schlüter regime plasma

The non-zero scalar coefficients needed to evaluate the transport fluxes for an electron-ion plasma in the collision-dominated regime are as follows [113, 114]:

$$
\begin{aligned}
\Gamma^1 &= -L_{12}p_e, & \Gamma^2 &= -L_{11}n, & \Gamma^3 &= L_{12}n, \\
\Gamma^4 &= -\frac{\eta_\| q g^2 n}{2\pi\mu_0} \left\langle \frac{|\nabla\psi|^2}{R^2} \right\rangle \langle B^2 \rangle^{-1}, & q_i^1 &= L_i p_i^2/n, \\
q_i^2 &= -L_i p_i, & q_i^3 &= L_i p_i, & q_e^1 &= L_{22} p_e^2/n, \\
q_e^2 &= L_{12} p_e, & q_e^3 &= -L_{22} p_e, & V_L^4 &= \frac{\eta_\| g^2}{2\pi\mu_0}.
\end{aligned}
$$

The transport coefficients are:

$$
\begin{aligned}
L_{11} &= L_0 \left[1 + 2.65(\eta_\|/\eta_\perp) q_*^2 \right], \\
L_{12} &= (3/2)L_0 \left[1 + 1.47(\eta_\|/\eta_\perp) q_*^2 \right], \\
L_{22} &= 4.66 L_0 \left[1 + 1.67(\eta_\|/\eta_\perp) q_*^2 \right], \\
L_i &= \sqrt{2} L_0 (m_i/m_e)^{1/2} (T_e/T_i)^{3/2} \left[1 + 1.60 q_*^2 \right], \\
L_0 &= \frac{\eta_\perp}{\mu_0} \langle |\nabla\Phi|^2/B^2 \rangle.
\end{aligned}
$$

Here, we have introduced the function

$$
q_*^2 = \frac{1}{2} \left[\langle B^{-2} \rangle - \langle B^2 \rangle^{-1} \right] \frac{g^2}{\langle |\nabla\Psi|^2/B^2 \rangle}, \tag{6.51}
$$

which reduces to the square of the safety factor, q^2, in the low beta, large aspect ratio limit. The resistivity functions $\eta_\|$ and $\eta_\perp$ are given by Eqs. (1.43) and (1.44).

If an external source of current drive is present, we can incorporate it into this model with the V_L^6 coefficient by defining

$$
V_L^6 = -2\pi\eta_\| J_\|^{CD}, \tag{6.52}
$$

where

$$J_\parallel^{CD} \equiv \frac{\langle \mathbf{J}^{CD} \cdot \mathbf{B} \rangle}{\langle \mathbf{B} \cdot \nabla\phi \rangle}, \tag{6.53}$$

with $\mathbf{J}^{CD}$ being the external current drive vector.

B. Banana Regime Plasma

The transport coefficients for a low-collisionality plasma including trapped and circulating particles have been computed for an arbitrary aspect ratio and cross section shape [114] with the restriction that $|\mathbf{B}|$ have only a single maximum on each flux surface [112, 114, 115, 116]. The fraction of trapped particles on a flux surface is given by [112]

$$f_t = 1 - (3/4)\langle B^2 \rangle \int_0^{B_c^{-1}} \lambda d\lambda / \langle (1 - \lambda B)^{1/2} \rangle, \tag{6.54}$$

where B_c is the maximum value of $|B|$ on a flux surface. Using this, we can express the banana regime transport model as follows:

$$\Gamma^1 = p_i L_{11}^{bp} y - p_e \left(L_{12}^{bp} + \tilde{L}_{12} + L_E L_{13} L_{23} \right),$$
$$\Gamma^2 = n(1+y)\left[-L_{11}^{bp} - L_E(L_{13})^2 \right] - n\tilde{L}_{11},$$
$$\Gamma^3 = n\left[L_{11}^{bp} y + L_{12}^{bp} + \tilde{L}_{12} + L_E L_{13}(L_{23} + yL_{13}) \right],$$
$$\Gamma^4 = -L_E L_{13}(2\pi)^{-2} g^3 n \langle R^{-2} \rangle,$$

$$q_i^1 = L_i^{nc} p_i^2/n^2,$$
$$q_i^2 = -L_i^{nc} p_i/n,$$
$$q_i^3 = L_i^{nc} p_i/n,$$

$$q_e^1 = \left[-L_{12}^{bp} y p_i + L_{22}^{nc} p_e + L_E L_{23}(-L_{13} y p_i + L_{23} p_e) \right] p_e/n,$$
$$q_e^2 = \left[\left(L_{12}^{bp} + L_E L_{23} L_{13} \right)(1+y) + \tilde{L}_{12} \right] p_e,$$
$$q_e^3 = \left[-L_{22}^{nc} - L_{12}^{bp} y - L_E L_{23}(L_{13} y + L_{23}) \right] p_e,$$
$$q_e^4 = L_E L_{23}(2\pi)^{-2} g^3 \langle R^{-2} \rangle,$$

$$V_L^1 = L_E^* L_{13} p/n,$$
$$V_L^2 = L_E^* L_{13} n(y+1),$$
$$V_L^3 = L_E^* (L_{13} - L_{23}) n,$$
$$V_L^4 = L_E g^2 \langle B^2 \rangle /2\pi.$$

Here, the transport coefficients are:

$$
\begin{aligned}
L_{11}^{bp} &= L_*(1.53 - 0.53 f_t), \\
L_{12}^{bp} &= L_*(2.13 - 0.63 f_t), \\
L_{13} &= (2\pi q) g f_t (1.68 - 0.68 f_t), \\
L_{23} &= 1.25 (2\pi q) g f_t (1 - f_t),
\end{aligned}
$$

$$
\begin{aligned}
\tilde{L}_{11} &= L_0(1 + 2q_*^2), \\
\tilde{L}_{12} &= (3/2) L_0 (1 + 2q_*^2),
\end{aligned}
$$

$$
\begin{aligned}
L_{22}^{nc} &= 4.66 \left[L_0(1 + 2q_*^2) + L_* \right], \\
L_i^{nc} &= \sqrt{2}(m_i/m_e)^{1/2}(T_e/T_i)^{3/2} \left[L_0(1 + 2q_*^2) + 0.46 L_*(1 - 0.54 f_t)^{-1} \right], \\
y &= -1.17(1 - f_t)(1 - 0.54 f_t)^{-1},
\end{aligned}
$$

$$
\begin{aligned}
L_{33} &= -1.26 f_t (1 - 0.18 f_t), \\
L_* &= f_t (2\pi q g)^2 \frac{\eta_\perp}{\mu_0} / \langle B^2 \rangle, \\
L_E &= \frac{\eta_\|(1 + L_{33})^{-1}}{\mu_0 \langle B^2 \rangle}, \\
L_E^* &= L_E \frac{2\pi \langle B^2 \rangle}{g \langle R^{-2} \rangle}
\end{aligned}
$$

Note that in steady state, when the loop voltage V_L is a spatial constant, the banana regime model implies a current driven by the temperature and density gradients,

$$
\left\langle \frac{1}{R^2} \right\rangle \frac{\langle \mathbf{J} \cdot \mathbf{B} \rangle}{\langle \mathbf{B} \cdot \nabla \phi \rangle} = -f_t p \left[\frac{L_{13}^*}{n} \frac{dn}{d\Psi} + \frac{L_{13}^*(1 + y)}{T_e + T_i} \frac{dT_i}{d\Psi} + \frac{L_{13}^* - L_{23}^*}{T_e + T_i} \frac{dT_e}{d\Psi} \right].
$$
(6.55)

Here $L_{13}^* \equiv L_{13}/(2\pi q g f_t)$, $L_{23}^* \equiv L_{23}/(2\pi q g f_t)$, and y are all dimensionless and of order unity. In deriving Eq. (6.55) we have used the relation $2\pi q d/d\Phi = d/d\Psi$, where Ψ is the poloidal flux function. This current is called the *bootstrap current*. These relations have been extended to multi-charged ions [117] and to arbitrary collisionality [118]. An excellent overview of the results for the transport coefficients in all of the collisionality regimes is given in Helander and Sigmar [119].

As in the collisional regime case, an external source of current drive can be included by defining the equivalent of Eq. (6.52). In the presence of external current drive, we would add the coefficient

$$
V_L^6 = -2\pi \eta_\| (1 + L_{33})^{-1} J_\|^{CD},
$$

where $J_\parallel^{CD}$ is defined in Eq. (6.53).

C. Anomalous Transport Model

Although this is still an area of active research, there are a number of anomalous transport models available that purport to calculate local values of the surface averaged transport coefficients based on local values of the surface averaged profiles of density, the temperatures, angular velocity, and current profile, and their gradients. These profiles are often obtained from a fit to a subsidiary micro-instability calculation [120, 121, 122, 123].

A typical model would return the particle diffusivity D, the electron thermal diffusivity χ_e, the ion thermal diffusivity χ_i, and the toroidal angular velocity diffusivity, χ_ω, all with dimension m^2/s, and a parallel resistivity function $\eta_\parallel$ with units $\Omega - m$. A relatively simple diagonal model would fit into the above formalism as follows [124]:

$$\Gamma^1 = -D \left\langle |\nabla \Phi|^2 \right\rangle,$$

$$q_i^1 = \chi_i \frac{p_i}{n} \left\langle |\nabla \Phi|^2 \right\rangle, \qquad q_i^2 = -\chi_i \left\langle |\nabla \Phi|^2 \right\rangle, \qquad q_i^3 = \chi_i \left\langle |\nabla \Phi|^2 \right\rangle,$$

$$q_e^1 = \chi_e \frac{p_e}{n} \left\langle |\nabla \Phi|^2 \right\rangle, \qquad q_e^3 = -\chi_e \left\langle |\nabla \Phi|^2 \right\rangle,$$

$$V_L^4 = \frac{\eta_\parallel g^2}{2\pi \mu_0},$$

$$\Gamma_\Omega^5 = -\chi_\omega \left\langle |\nabla \Phi|^2 \right\rangle m_i n \left\langle R^2 \right\rangle.$$

Bootstrap and current drive terms are included in these models by defining the additional V_L^i coefficients as discussed above.

The micro-instability-based models tend to return transport coefficients that have strong dependences on the gradients of the corresponding surface averaged profiles. We discuss in the next chapter special computational techniques for dealing with these.

6.3 Equilibrium Constraint

The variables N', σ, ι, σ_e, and Ω introduced in Section 6.2 are called *adiabatic variables*. If there is no dissipation and no explicit sources of mass or energy, then the time derivatives of these quantities are zero in the toroidal flux coordinate system being used here.

In the presence of dissipation, the surface-averaged transport equations of the last section describe how these adiabatic variables evolve relative to equilibrium magnetic surfaces with fixed values of toroidal magnetic flux. To complete the description, we need to solve a global equation to describe how these surfaces evolve relative to a fixed laboratory frame in which the toroidal field and poloidal field magnetic coils are located. This is the associated equilibrium problem. In the next subsection, we describe the circuit equations that describe how the nearby coil currents evolve in time due to applied and induced voltages. Then, we describe two approaches for incorporating the equilibrium constraint, the Grad–Hogan method and an accelerated form of the Taylor method.

6.3.1 Circuit Equations

The toroidal plasma is coupled electromagnetically to its surroundings; both passive structures and poloidal field coils that are connected to power supplies. We assume here that the external conductors are all axisymmetric, and that their currents and applied voltages are in the toroidal (ϕ) direction. Although the passive structures are continuous, it is normally adequate to subdivide them into discrete elements, each of which obeys a discrete circuit equation. Thus, each of the poloidal field coils and passive structure elements obeys a circuit equation of the form

$$\frac{d}{dt}\Psi_{Pi} + R_i I_i = V_i, \qquad (6.56)$$

where the poloidal flux at each coil i is defined by

$$\Psi_{Pi} = L_i I_i + \sum_{i \neq j} M_{ij} I_j + 2\pi \int_P J_\phi(\mathbf{R}')G(\mathbf{R}_i, \mathbf{R}')d\mathbf{R}'. \qquad (6.57)$$

Here, R_i and V_i are the resistance and applied voltage at coil i, L_i is the self-inductance of coil i, and M_{ij} is the mutual inductance of coil i with coil j. For a passive conductor, the corresponding $V_i = 0$.

The last term in Eq. (6.57) is an integral over the plasma volume. This represents the mutual inductance between the distributed plasma current and the conductor with index i.

6.3.2 Grad–Hogan Method

The Grad–Hogan method [110, 125] splits every time step into two parts. In the first part, the adiabatic variables, including the poloidal flux at the conductors, are advanced from time t to time $t + \delta t$ by solving Eqs. (6.45)–(6.49) and (6.56) using techniques discussed in Chapter 7. In the second part, these adiabatic variables are held fixed while we solve the appropriate form of the equilibrium equation, where the "free functions" $p'(\Psi)$ and $gg'(\Psi)$ have been expressed in terms of the adiabatic variables $\sigma(\Phi)$ and $\iota(\Phi)$. The individual PF coil and conductor currents will change during this part of the time step in order to keep the poloidal flux fixed at each coil location and at the plasma magnetic axis. This part of the time step effectively determines the absolute motion of the toroidal flux surfaces relative to a fixed frame.

The adiabatic variables may also be expressed in terms of the poloidal flux function Ψ. This is most convenient when solving the Grad–Shafranov equation in the Grad–Hogan method as the poloidal flux function is what is being solved for. Thus, if we define $V_\Psi \equiv dV/d\Psi$ and $\sigma_\Psi \equiv pV_\Psi^{5/3}$, the form of the equilibrium equation that needs to be solved is:

$$\Delta^*\Psi + \mu_0 R^2 \frac{d}{d\Psi}\left[\frac{\sigma_\Psi(\Psi)}{V_\Psi^{5/3}}\right] + \frac{(2\pi)^4 q(\Psi)}{V_\Psi \langle R^{-2}\rangle}\frac{d}{d\Psi}\left[\frac{q(\Psi)}{V_\Psi \langle R^{-2}\rangle}\right] = 0. \qquad (6.58)$$

The functions $\sigma_\Psi(\Psi)$ and $q(\Psi)$ must be held fixed while finding the equilibrium solution.

Equation (6.58) for Ψ can be solved using a free boundary analogue to the techniques discussed in Section 5.5.1. Note that the poloidal flux at the magnetic axis, Ψ_{MA}, must be held fixed during the equilibrium solution as well. Since $\nabla\Psi = 0$ at the magnetic axis, from Eq. (4.42), the evolution equation for Ψ at the axis is simply

$$\frac{\partial \Psi_{MA}}{\partial t} = R^2 \nabla\phi \cdot \mathbf{R}. \qquad (6.59)$$

(Note that this is equivalent to applying Eq. (6.44) to Ψ at the magnetic axis.) The value of Ψ_{MA} is evolved during the transport part of the time step using Eq. (6.59) and Ψ_{MA} and the adiabatic variables are then held fixed during the equilibrium solution part of the time step.

This completes the formalism needed to describe the transport in an axisymmetric system. The one-dimensional evolution equations, Eqs. (6.45)–(6.49), advance the adiabatic variables in time on the resistive time scale. Equation (6.59) is used to advance the value of Ψ at the magnetic axis, and Eq. (6.56) is used to advance the value of Ψ at the nearby conductors. The equilibrium equation, Eq. (6.58), defines the flux surface geometry consistent with these adiabatic variables and the boundary conditions. It does not introduce any new time scales into the equations.

6.3.3   Taylor Method (Accelerated)

J. B. Taylor [126] suggested an alternative to the Grad–Hogan method that does not require solving the equilibrium equation with the adiabatic constraints, Eq. (6.58). His approach involves solving for the velocity field $\mathbf{u}$, which when inserted into the field and pressure evolution equations, Eqs. (6.3), (6.4), and (6.5), will result in the equilibrium equation, Eq. (6.11), continuing to be satisfied as time evolves. It was shown by several authors [111, 112] that an elliptic equation determining this velocity field could be obtained by time differentiating the equilibrium equation and substituting in from the time evolution equations. Taking the time derivative of Eq. (6.11) gives

$$\nabla\dot{p} = \dot{\mathbf{J}} \times \mathbf{B} + \mathbf{J} \times \dot{\mathbf{B}}, \qquad (6.60)$$

or, by substituting in from Eqs. (6.3), (6.4), and (6.5),

$$\frac{2}{3}\nabla\left[-\nabla\cdot\left(\mathbf{q}+\frac{5}{2}p\mathbf{u}\right)+\mathbf{J}\cdot\left(-\mathbf{u}\times\mathbf{B}+\mathbf{R}\right)+S_e\right]$$
$$+\mu_0^{-1}\nabla\times[\nabla\times(\mathbf{u}\times\mathbf{B}-\mathbf{R})]\times\mathbf{B}+\mathbf{J}\times[\nabla\times(\mathbf{u}\times\mathbf{B}-\mathbf{R})].\qquad(6.61)$$

This equation can, in principle, be solved for $\mathbf{u}$, and that velocity field $\mathbf{u}$ can be used to keep the system in equilibrium without repeatedly solving the equilibrium equation. While this approach has been shown to be viable [127], it suffers from "drifting" away from an exact solution of the equilibrium equation. A preferred approach [128, 129] to obtaining this velocity is to use the accelerated steepest descent algorithm which involves obtaining the velocity from the residual equation,

$$\dot{\mathbf{u}}+\frac{1}{\tau}\mathbf{u}=D\left[\mathbf{J}\times\mathbf{B}-\nabla p\right].\qquad(6.62)$$

By choosing the proportionality and damping factors, D and τ, appropriately, the system can be kept arbitrarily close to an equilibrium state as it evolves. This is equivalent to applying the dynamic relaxation method of Section 3.5.2 to the plasma equilibrium problem.

In using this approach, the magnetic field variables Ψ and g must be evolved in time as two-dimensional functions from Eqs. (6.14) and (6.15) in order to preserve the magnetic flux constraints. Equation (6.48) for ι is then redundant but is useful as a check. The total entropy constraint is preserved by representing the pressure as $p(\psi)=\sigma(\psi)/V'^{5/3}$. The other adiabatic variables, including the poloidal flux at the conductors, are still evolved by solving Eqs. (6.45), (6.46), (6.47), (6.49), and (6.56).

6.4   Time Scales

Here we discuss some of the time scales associated with diffusion of the particles, energy and fields in a collisional regime tokamak plasma. In doing so, we show the relation of the variables and equations introduced here to more common variables used in simpler geometries and in more approximate descriptions.

A. Flux Diffusion

Recall that the rotational transform evolves according to Eq. (6.48) where, for an arbitrary inhomogeneous term $\mathbf{R}$ in Ohm's law, the loop voltage is defined by Eq. (6.38). Here, for simplicity in discussing the basic time scales, we consider the Ohm's law appropriate for a single fluid resistive MHD plasma,

$$\mathbf{R}=\eta_\parallel\mathbf{J}_\parallel+\eta_\perp\mathbf{J}_\perp.\qquad(6.63)$$

When inserted into Eq. (6.38), this gives

$$V_L = 2\pi\eta_\| \frac{\langle \mathbf{B} \cdot \mathbf{J} \rangle}{\langle \mathbf{B} \cdot \nabla\phi \rangle}. \tag{6.64}$$

Evaluating Eq. (6.64) and inserting into Eq. (6.48) yields a diffusion-like equation for the rotational transform $\iota(\Phi, t)$,

$$\frac{\partial}{\partial t}\iota = \frac{d}{d\Phi}\left[\frac{\eta_\| g}{V' \langle R^{-2} \rangle \mu_o} \frac{d}{d\Phi}\left(\frac{V'}{g} \left\langle \frac{|\nabla\Phi|^2}{R^2} \right\rangle \iota \right) \right]. \tag{6.65}$$

We can rewrite this equation in the circular cross section *large aspect ratio limit* in which the major radius and toroidal field strength can be assumed constants and the cylindrical coordinate r is the only independent variable. Thus, we have

$$
\begin{aligned}
R &\to R_o(constant), & g &\to R_o B_z^0, \\
B_z &\to B_z^o \gg B_\theta, & V' &\to 2\pi R_0 / B_z^0, \\
\Phi &\to \pi r^2 B_z^o, & \iota = q^{-1} &\to \frac{R_0 B_\theta}{r B_z^0}.
\end{aligned}
$$

In this limit, Eq. (6.65) becomes

$$\frac{\partial}{\partial t} B_\theta = \frac{d}{dr}\left[\frac{\eta_\|}{\mu_o} \frac{1}{r} \frac{d}{dr}(r B_\theta) \right]. \tag{6.66}$$

From this, we can evaluate the characteristic time for the solution to decay,

$$t_o = \frac{a^2 \mu_o}{\eta_\|},$$

where a is the minor radius. This is known as the *skin time* of the device.

B. Particle Diffusion

The evolution equation for the differential particle number N' is given by Eq. (6.45). In this resistive MHD model, the particle flux is completely defined in terms of the inhomogeneous term $\mathbf{R}$ in Eq. (6.40). By making use of the identity

$$R^2\nabla\phi = \frac{1}{B^2}\left[g\mathbf{B} - \mathbf{B} \times \nabla\Psi \right], \tag{6.67}$$

it can be seen that Γ can be written in the following form

$$\Gamma = 2\pi q n \left[g\left(\left\langle \frac{\mathbf{R} \cdot \mathbf{B}}{B^2} \right\rangle - \frac{\langle \mathbf{R} \cdot \mathbf{B} \rangle}{\langle B^2 \rangle} \right) - \left\langle \frac{\mathbf{R} \times \mathbf{B} \cdot \nabla\Psi}{B^2} \right\rangle - \frac{V_L}{2\pi}\frac{\langle B_p^2 \rangle}{\langle B^2 \rangle} \right]. \tag{6.68}$$

Here $\langle B_p^2 \rangle = \langle |\nabla\Psi|^2 / R^2 \rangle$. The first term in brackets, which only depends on the part of $\mathbf{R}$ parallel to the magnetic field $\mathbf{B}$, corresponds to neo-classical Pfirsch–Schlüter diffusion, the second term to classical diffusion, and the third term to the classical inward pinch. We evaluate this for the simple Ohm's law in Eq. (6.63) to find

$$\frac{\partial N'}{\partial t} = \frac{d}{d\Phi} N' \left[\eta_\perp \left\langle \frac{|\nabla\Phi|^2}{B^2} \right\rangle [1 + 2q_*^2 (\eta_\parallel/\eta_\perp)] \frac{dp}{d\Phi} + qV_L \frac{\langle B_p^2 \rangle}{\langle B_T^2 \rangle} \right]. \qquad (6.69)$$

Again, we can evaluate Eq. (6.69) in the large aspect ratio, circular limit to find

$$\frac{\partial}{\partial t}(nr) = \frac{d}{dr} n \left[\frac{r}{B^2}(\eta_\perp + 2q^2\eta_\parallel) \frac{dp}{dr} + r\frac{E_\phi B_\theta}{B^2} \right].$$

The first term in brackets corresponds to diffusion, and the second term to an inward $\mathbf{E} \times \mathbf{B}$ drift. Noting that $p = 2nk_BT$, we can estimate the classical diffusion time in this limit,

$$t_n = \frac{a^2\mu_o}{\eta(1 + 2q^2)\beta}, \qquad (6.70)$$

where $\beta \equiv \mu_0 p/B^2$. We note that $t_n \gg t_o$ if $\beta \ll 1$.

C. Heat Conduction Equation

The evolution of the differential entropy density is given by Eq. (6.46). Since the inhomogeneous vector $\mathbf{R}$ does not enter into the heat flux relation, we must rely on an additional kinetic calculation to define this. The heat flux in a collisional plasma has been shown to be [130]

$$Q = -p\eta_\perp \left(\frac{2m_i}{m_e} \right)^{\frac{1}{2}} (1 + 2q_*^2) \left\langle \frac{|\nabla\Phi|^2}{B^2} \right\rangle nV' \frac{dT}{d\Phi}.$$

By balancing the time derivative term with the heat conduction term, and evaluating in the large aspect ratio limit, we obtain the heat conduction time

$$t_h = \frac{a^2\mu_o}{\hat{\eta}(1 + 2q^2)\beta \left(\frac{2m_i}{m_e} \right)^{\frac{1}{2}}}.$$

Normally, for tokamak parameters, we have

$$t_n > t_o > t_h \gg t_a,$$

where t_a is the Alfvén wave transit time of the device.

D. Equilibration

For times longer than each of the characteristic times computed above, one or more of the evolution equations will be in steady state. Thus for $t > t_A$ the system will be in force balance equilibrium with

$$\nabla p = \mathbf{J} \times \mathbf{B},$$

or, in particular, the form of the Grad–Shafranov equation given by (6.58) will be satisfied. For $t > t_h$ the ion heat conduction will balance Joule heating in an ohmically heated tokamak, thus

$$V_L \frac{d}{d\Phi}\left[\frac{V'}{(2\pi)^2 q}\left\langle \frac{|\nabla\Phi|^2}{R^2} \right\rangle \right]$$
$$+ \frac{d}{d\Phi}\left[p\eta_\perp \left(\frac{2m_i}{m_e} \right)^{\frac{1}{2}} (1+2q^2)\left\langle \frac{|\nabla\Phi|^2}{B^2} \right\rangle N'\frac{dT}{d\Phi} \right] = 0. \quad (6.71)$$

This will determine the temperature profile. For longer times $t > t_o$, the toroidal current, or ι, evolves diffusively to obtain the steady-state condition

$$V_L = \text{const.} = \frac{\eta_\parallel g}{\mu_o V' \langle R^{-2} \rangle} \frac{d}{d\Phi}\left[\frac{V'}{g}\left\langle \frac{|\nabla\Phi|^2}{R^2} \right\rangle \iota \right].$$

Note that this allows the energy equation, Eq. (6.46), to be integrated so that we have the steady-state relation

$$Q_e + Q_i = V_L K + c, \qquad (6.72)$$

where c is an integration constant. Evaluating this for our collisional transport model gives

$$\frac{V_L}{(2\pi)^2 q}\left\langle \frac{|\nabla\Phi|^2}{R^2} \right\rangle + p\eta_\perp \left(\frac{2m_i}{m_e} \right)^{\frac{1}{2}} (1+2q^2)\left\langle \frac{|\nabla\Phi|^2}{B^2} \right\rangle n\frac{dT}{d\Phi} = c.$$

6.5 Summary

In a magnetic confinement device such as a tokamak, the magnetic field lines form nested surfaces. Because the thermal conductivities and particle diffusivities are so much greater parallel to the field than across it, we can assume that the temperatures and densities are constant on these surfaces when solving for the relatively slow diffusion of particles, energy, magnetic field, and angular momentum across the surfaces. Because this cross-field diffusion is slow compared to the Alfvén time, we can assume that the plasma is always in force balance equilibrium. The toroidal magnetic flux is a good coordinate because it is relatively immobile and is always zero at the magnetic axis. We surface average the evolution equation to get a set of 1D evolution equations for the adiabatic invariants. These are given by Eqs. (6.45), (6.46), (6.48), (6.47), and (6.49). These need to be closed by supplying transport coefficients from a subsidiary transport or kinetic calculation. These 1D evolutions need to be solved together with the 2D equilibrium constraint, which defines the geometry and relates the magnetic fluxes to the coil currents and other boundary conditions. The time scales present in the equations were estimated for a collisional regime plasma and it was found that the the heating time is fastest, followed by the magnetic field penetration time (or the skin time), followed by the particle diffusion time. These times are all much longer than the Alfvén transit time.

Problems

6.1: Derive Eq. (6.18) using the definition of $\mathbf{u}_C$ in Eq. (6.16) and the expression for the Jacobian in Eq. (5.19).

6.2: Derive the evolution equations in a moving frame, Eqs. (6.20)–(6.24).

6.3: Derive the intermediate result, Eq. (6.36).

6.4: Derive a set of evolution equations analogous to Eqs. (6.45), (6.46), and (6.48) using the poloidal flux velocity, defined by Eq. (6.31) as the reference velocity (instead of the toroidal flux velocity).

6.5: Show that the expression for q_* in Eq. (6.51) reduces to the normal definition for the safety factor q in the large aspect ratio, low β limit.

6.6: Calculate the time scales as done in Section 6.4, but for the banana regime plasma model introduced in Section 6.2.4.

Chapter 7

Numerical Methods for Parabolic Equations

7.1 Introduction

In this chapter we consider numerical methods for solving parabolic equations. The canonical equation of this type is given by

$$\frac{\partial u}{\partial t} = \nabla \cdot \mathbf{D} \cdot \nabla u. \tag{7.1}$$

This is sometimes called the *heat equation* as it describes the conduction of heat throughout a material. It is also prototypical of many types of diffusive processes that occur in a magnetized plasma, including the transport of heat, particles, angular momentum, and magnetic flux in real space, and the evolution of the distribution function in velocity space due to particle collisions. By allowing the diffusion coefficient $\mathbf{D}$ in Eq. (7.1) to be a tensor, we can describe anisotropic processes such as the diffusion of heat and particles when a magnetic field is present. In the next section we discuss the basic numerical algorithms for solving equations of the parabolic type in one dimension, such as the surface-averaged equations derived in the last chapter. We then discuss extension of these methods to multiple dimensions.

7.2 One-Dimensional Diffusion Equations

We begin by analyzing common methods for solving scalar linear diffusion equations in one dimension. We next consider treatment of non-linear diffusion equations such as those that often arise in practice. We further discuss considerations when solving diffusion equations in cylindrical-like geometries that possess a singularity at the origin. We then generalize to vector forms and illustrate how the surface-averaged transport equations derived in the last chapter can be put into this form.

7.2.1   Scalar Methods

Consider the basic one-dimensional scalar diffusion equation in slab geometry

$$\frac{\partial u}{\partial t} = \sigma \frac{\partial^2 u}{\partial x^2},\tag{7.2}$$

where σ is a positive constant. We discretize time and space with the notation $t^n = n\delta t$ and $x_j = j\delta x$, and denote $u(x_j, t^n)$ by u_j^n. Consider the following methods for solving Eq. (7.2).

A. Forward-Time Centered Space (FTCS)

The forward-time centered-space algorithm is an explicit method defined by

$$u_j^{n+1} = u_j^n + \frac{\delta t\, \sigma}{\delta x^2} \left[u_{j+1}^n - 2u_j^n + u_{j-1}^n\right].$$

We perform von Neumann stability analysis as described in Section 2.5 by examining the amplification factor for an individual wavenumber. Letting

$$u_j^n \rightarrow \tilde{u}_k r^n e^{ij\theta_k},$$

and defining $s = \delta t \sigma / \delta x^2$, we find for the amplification factor

$$r = 1 + 2s\left[\cos\theta_k - 1\right].$$

The stability condition, $|r| \le 1$ for all θ_k, gives $s \le \frac{1}{2}$, or

$$\delta t \le \frac{\delta x^2}{2\sigma}.\tag{7.3}$$

The leading order truncation error for this method is found by Taylor series analysis to be

$$T_\Delta = -\frac{\delta t}{2}\frac{\partial^2 u}{\partial t^2} + \frac{\sigma\,\delta x^2}{12}\frac{\partial^4 u}{\partial x^4} + \cdots.\tag{7.4}$$

If we always operate at the stability limit, we can use the equality in Eq. (7.3) to eliminate δt from Eq. (7.4) to give an expression for the truncation error at the stability limit

$$T_\Delta^{SL} = -\frac{\delta x^2}{4\sigma}\left[\frac{\partial^2 u}{\partial t^2} - \frac{\sigma^2}{3}\frac{\partial^4 u}{\partial x^4}\right] + \cdots.\tag{7.5}$$

This illustrates that the leading order error terms arising from the time and spatial discretization are of the same order.

The disadvantage of this method, of course, is that the condition Eq. (7.3) on the time step is very restrictive. If the spatial increment δx is reduced by a factor of 2, then the time step δt must be reduced by a factor of 4 to remain stable, resulting in a factor of 8 more space-time points that need to be computed to get to the same integration time T.

B. DuFort–Frankel Method

The method of DuFort and Frankel [132] is defined by

$$u_j^{n+1} = u_j^{n-1} + \frac{2\delta t \sigma}{\delta x^2} \left[u_{j+1}^n - \left(u_j^{n+1} + u_j^{n-1} \right) + u_{j-1}^n \right]. \tag{7.6}$$

This method looks implicit since advanced time variables, u^{n+1}, appear on the right side of the equal sign. However, it can be rewritten in the form

$$u_j^{n+1} = \left[\frac{1 - 2s}{1 + 2s} \right] u_j^{n-1} + \frac{2s}{1 + 2s} \left(u_{j+1}^n + u_{j-1}^n \right),$$

showing that it is actually explicit. Von Neumann stability analysis yields for the amplification factor

$$r = \frac{2s \cos \theta_k \pm \left[1 - 4s^2 \sin^2 \theta_k \right]^{\frac{1}{2}}}{1 + 2s},$$

which is less than or equal to 1 in magnitude for all values of θ_k, making this method *unconditionally stable*. This is an example of an explicit method which is stable for any value of the time step.

Let us examine the form of the leading order truncation error for Eq. (7.6). Taylor expanding about u_j^n gives

$$T_\Delta = -\sigma \left(\frac{\delta t}{\delta x} \right)^2 \frac{\partial^2 u}{\partial t^2} + \mathcal{O} \left(\delta t^2 \right) + \mathcal{O} \left(\delta x^2 \right) + \mathcal{O} \left(\frac{\delta t^4}{\delta x^2} \right).$$

Consistency requires that when a convergence study is performed, it must be done such that $(\delta t / \delta x) \to 0$ as $\delta t \to 0$. That is, the DuFort–Frankel finite difference equation, Eq. (7.6), is consistent with the differential equation, Eq. (7.2), if and only if we take the zero time step zero gridsize limit in such a way that δt goes to zero faster than δx. If, on the other hand, the ratio $\delta t / \delta x$ is kept fixed, say $(\delta t / \delta x) = \beta$, then Eq. (7.6) is consistent not with the diffusion equation, Eq. (7.2), but with the damped wave equation

$$\sigma \beta^2 \frac{\partial^2 u}{\partial t^2} + \frac{\partial u}{\partial t} = \sigma \frac{\partial^2 u}{\partial x^2}.$$

Nevertheless, Eq. (7.6) is still a useful method since it is numerically stable and it interfaces naturally when using the leap frog method for convective or hyperbolic terms as we will see in Chapter 9. To perform convergence studies, we must let δt decrease more rapidly than proportional to δx, i.e.,

$$\delta t \sim \delta x^{1+\alpha}$$

for some $\alpha > 0$.

C. Backward-Time Centered-Space (BTCS)

Consider the implicit backward-time centered-space method

$$u_j^{n+1} = u_j^n + \frac{\delta t \sigma}{\delta x^2} \left[u_{j+1}^{n+1} - 2u_j^{n+1} + u_{j-1}^{n+1} \right]. \tag{7.7}$$

The amplification factor for BTCS is given by

$$r = \left[1 - 2s \left(\cos \theta_k - 1 \right) \right]^{-1},$$

which has amplitude $|r| \leq 1$ for all θ_k. This method is *unconditionally stable* for constant coefficient σ, which is typical for implicit methods. The truncation error can be found by expanding about u_j^{n+1}. We find

$$T_\Delta = \frac{\delta t}{2} \frac{\partial^2 u}{\partial t^2} + \frac{\sigma \delta x^2}{12} \frac{\partial^4 u}{\partial x^4} + \cdots,$$

which will be dominated by the first-order δt term, unless $\delta t \sim \delta x^2$ as required by the stability condition Eq. (7.3) for the explicit FTCS method. Thus, for accuracy reasons, we would normally be required to choose the time step such that $\delta t \sim \delta x^2$, even though stability allows a larger time step. A discussion of the generalization of this method to the case where the coefficient σ is a strong non-linear function of the solution u is given in Section 7.2.2.

D. θ-Implicit and Crank–Nicolson

Consider now the θ-implicit method. We introduce the implicit parameter θ and consider the method

$$
\begin{aligned}
u_j^{n+1} = u_j^n \ &+\ \theta \frac{\delta t \sigma}{\delta x^2} \left[u_{j+1}^{n+1} - 2u_j^{n+1} + u_{j-1}^{n+1} \right] \\
&+\ (1 - \theta) \frac{\delta t \sigma}{\delta x^2} \left[u_{j+1}^n - 2u_j^n + u_{j-1}^n \right].
\end{aligned}
\tag{7.8}
$$

The amplification factor is given by

$$r = \frac{1 - s^*(1 - \theta)}{1 + s^* \theta}, \tag{7.9}$$

where $s^* = 2s \left[1 - \cos \theta_k \right] \geq 0$. We see that for $\theta \geq \frac{1}{2}$, we have $|r| \leq 1$ which implies the algorithm given by Eq. (7.8) is stable for all values of δt. For $\theta < \frac{1}{2}$, we have the stability condition $s^* < 2/(1 - 2\theta)$, or

$$\delta t < \frac{\delta x^2}{2\sigma(1 - 2\theta)}.$$

Choosing $\theta = \frac{1}{2}$ in Eq. (7.8) gives the Crank–Nicolson method. This is seen to be time centered as well as unconditionally stable. Because it is time centered, the truncation error for $\theta = \frac{1}{2}$ is of the form

$$T_\Delta = \mathcal{O}\left(\delta t^2\right) + \mathcal{O}\left(\delta x^2\right). \tag{7.10}$$

The truncation error term arising from the time differencing is therefore of the same order as that arising from the spatial differencing for $\delta t \sim \delta x$. This implies that for the Crank–Nicolson method with $\theta = \frac{1}{2}$, unlike for either the DuFort–Frankel method or the BTCS method, it is clearly advantageous to use a value of δt substantially larger than the explicit limit as given by Eq. (7.3) because there is no degradation in accuracy for δt up to the size of δx. This is a substantial advantage in practice, as it implies that if δx is reduced by 2, then we need only reduce δt by 2 to reduce the truncation error by 4. This results in only a factor of 4 more space-time points to get to the same integration time T instead of the factor of 8 required for the fully explicit FTCS method.

E. Modified Crank–Nicolson

The Crank–Nicolson method is desirable because it is unconditionally stable and second-order accurate in both space and time. However, it is seen from Eq. (7.9) that the absolute value of the amplification factor approaches unity for large values of s^*, i.e.,

$$|r| \to 1 \quad \text{for} \quad s^* \to \infty \quad \text{at} \quad \theta = \frac{1}{2}. \tag{7.11}$$

This can lead to difficulty for large values of the time step because the short wavelength perturbations are not damped.

Let $\delta_x^2 u_j \equiv u_{j+1} - 2u_j + u_{j-1}$ be the centered second difference operator. Consider now the *modified* Crank–Nicolson method [131] that takes a weighted sum of the spatial derivatives at three time levels:

$$u_j^{n+1} = u_j^n + \frac{\delta t \sigma}{\delta x^2}\left[\frac{9}{16}\delta_x^2 u_j^{n+1} + \frac{3}{8}\delta_x^2 u_j^n + \frac{1}{16}\delta_x^2 u_j^{n-1}\right]. \tag{7.12}$$

This can be shown to be second-order accurate in space and time and unconditionally stable, and the amplification factor has the property that

$$|r| \to \frac{1}{3} \quad \text{for} \quad s^* \to \infty. \tag{7.13}$$

It therefore exhibits strong damping of the short, gridscale wavelength modes, in the limit of very large time steps, and for that reason may be preferred over the standard Crank–Nicolson method for some applications.

7.2.2 Non-Linear Implicit Methods

When micro-instability-based anomalous transport models are used to calculate the thermal conductivity in the flux-surface-averaged equations described in Chapter 6, the thermal conductivity functions can have a strong non-linear dependence on the local temperature gradient [120]. If a method such as BTCS or Crank–Nicolson is used without accounting for this non-linearity, large oscillations or numerical instability can occur. Here we describe a modification to the standard Crank–Nicolson algorithm to account for this non-linearity [133].

Consider the one-dimensional diffusion equation for the temperature T in cylindrical geometry,

$$\frac{\partial T}{\partial t} = \frac{1}{r}\frac{\partial}{\partial r}\left[r\chi\frac{\partial T}{\partial r}\right]. \tag{7.14}$$

Here r is the radial coordinate. We consider the case where the diffusion coefficient χ is a strong function of the spatial gradient of the solution. It is convenient to introduce as the dependent variable the quantity $\Phi \equiv \frac{1}{4}r^2$, which can be thought of as a toroidal flux coordinate. Then, Eq. (7.14) can be written in the form

$$\frac{\partial T}{\partial t} = \frac{\partial}{\partial \Phi}\left[\Phi\,\chi\,(T')\frac{\partial T}{\partial \Phi}\right]. \tag{7.15}$$

Here we have emphasized the dependence of the diffusion coefficient χ on the gradient of the solution $T' \equiv \partial T/\partial \Phi$ by writing $\chi\,(T')$.

It is shown in [134] that if the implicit BTCS method of Eq. (7.7) is applied to this equation with the diffusion coefficient $\chi\,(T')$ evaluated at the old time level n when advancing to the new time level $n + 1$, instability can develop. They considered a particular analytic form where the thermal conductivity was proportional to the temperature gradient raised to some power p, $\chi = \chi_0\,(T')^p$. In this case, a linearized analysis shows that the BTCS method will be numerically unstable for $|p| > 1$. It is not possible to evaluate the numerical stability of the general case where χ is an unknown, possibly non-analytic, function of T', but we must assume that it can be unstable.

Consider now the θ-implicit method applied to Eq. (7.15), but with the diffusion coefficient being evaluated at the same time level as the derivatives of the solution. Using the convention $\Phi_j \equiv \left(j - \frac{1}{2}\right)\delta\Phi$ (so that $\Phi_{\frac{1}{2}} = 0$) and introducing $s_1 \equiv \delta t/\delta\Phi^2$, conservative differencing gives

$$T_j^{n+1} = T_j^n + s_1\theta\left[\Phi\chi|_{j+\frac{1}{2}}^{n+1}\left(T_{j+1}^{n+1} - T_j^{n+1}\right) - \Phi\chi|_{j-\frac{1}{2}}^{n+1}\left(T_j^{n+1} - T_{j-1}^{n+1}\right)\right]$$
$$+ s_1(1-\theta)\left[\Phi\chi|_{j+\frac{1}{2}}^n\left(T_{j+1}^n - T_j^n\right) - \Phi\chi|_{j-\frac{1}{2}}^n\left(T_j^n - T_{j-1}^n\right)\right]. \tag{7.16}$$

To approximate the diffusion coefficient at the advanced time, we Taylor expand in time as follows:

$$\chi|_{j+\frac{1}{2}}^{n+1} = \chi|_{j+\frac{1}{2}}^n + \frac{\partial\chi}{\partial T'}\bigg|_{j+\frac{1}{2}}^n\left(T_{j+\frac{1}{2}}'^{n+1} - T_{j+\frac{1}{2}}'^n\right).$$

It is convenient to define the change in temperature in going from time step

n to $n+1$ at grid point j as

$$\delta T_j \equiv T_j^{n+1} - T_j^n.$$

In terms of this, we have the relations

$$T_j^{n+1} = T_j^n + \delta T_j,$$

and

$$T'^{n+1}_{j+\frac{1}{2}} - T'^n_{j+\frac{1}{2}} = \frac{1}{\delta\Phi}\left(\delta T_{j+1} - \delta T_j\right).$$

Substituting into Eq. (7.16), and keeping terms to first order in $\delta T \sim \delta t$, we get a tridiagonal system for δT_j,

$$A_j \delta T_{j+1} - B_j \delta T_j + C_j \delta T_{j-1} = D_j. \tag{7.17}$$

Here,

$$A_j = s_1\theta\Phi_{j+\frac{1}{2}}\left[\chi|^n_{j+\frac{1}{2}} + \left.\frac{\partial\chi}{\partial T'}\right|^n_{j+\frac{1}{2}} T'^n_{j+\frac{1}{2}}\right],$$

$$C_j = s_1\theta\Phi_{j-\frac{1}{2}}\left[\chi|^n_{j-\frac{1}{2}} + \left.\frac{\partial\chi}{\partial T'}\right|^n_{j-\frac{1}{2}} T'^n_{j-\frac{1}{2}}\right],$$

$$B_j = 1 + A_j + C_j,$$

$$D_j = -s_1\left[\Phi_{j+\frac{1}{2}}\,\chi|^n_{j+\frac{1}{2}}\left(T^n_{j+1} - T^n_j\right) - \Phi_{j-\frac{1}{2}}\,\chi|^n_{j-\frac{1}{2}}\left(T^n_j - T^n_{j-1}\right)\right].$$

Equation (7.17) is equivalent to the first iteration of a multivariable Newton's method iteration applied to Eq. (7.16). If the function $\chi(T')$ is well behaved, this will normally give an accurate solution. However, if $\chi(T')$ exhibits strong non-linearities, Newton's method can be applied recursively to solve for the value of T_j^{n+1} that solves this equation exactly. Keeping the non-linear terms, Eq. (7.16) can be written as

$$F_j(T_k^{n+1}) = -T_j^{n+1} + T_j^n + s_1\theta\left[\Phi\,\chi|^{n+1}_{j+\frac{1}{2}}\left(T_{j+1}^{n+1} - T_j^{n+1}\right)\right.$$

$$\left. - \Phi\,\chi|^{n+1}_{j-\frac{1}{2}}\left(T_j^{n+1} - T_{j-1}^{n+1}\right)\right] \tag{7.18}$$

$$+ s_1(1-\theta)\left[\Phi\,\chi|^n_{j+\frac{1}{2}}\left(T_{j+1}^n - T_j^n\right) - \Phi\,\chi|^n_{j-\frac{1}{2}}\left(T_j^n - T_{j-1}^n\right)\right] = 0.$$

Multivariable Newton's method applied to Eq. (7.18) corresponds to the iteration (with iteration index i out of N)

$$\left.\frac{\partial F_j}{\partial\left(T_k^{n+1}\right)}\right|^{n+(i-1)/N}\left(T_k^{n+i/N} - T_k^{n+(i-1)/N}\right) = -F_j^{n+(i-1)/N}, \tag{7.19}$$

where the sum over index k is implied. Written without implied summation but keeping only the non-zero terms, this takes the form

$$A_j^* \, T_{j+1}^{n+i/N} - B_j^* \, T_j^{n+i/N} + C_j^* \, T_{j-1}^{n+i/N} = D_j^*. \qquad (7.20)$$

Here, we have defined

$$A_j^* = s_1 \theta \Phi_{j+\frac{1}{2}} \left[\chi|_{j+\frac{1}{2}}^{n+(i-1)/N} + \left.\frac{\partial \chi}{\partial T'}\right|_{j+\frac{1}{2}}^{n+(i-1)/N} T'_{j+\frac{1}{2}}^{n+(i-1)/N} \right],$$

$$C_j^* = s_1 \theta \Phi_{j-\frac{1}{2}} \left[\chi|_{j-\frac{1}{2}}^{n+(i-1)/N} + \left.\frac{\partial \chi}{\partial T'}\right|_{j-\frac{1}{2}}^{n+(i-1)/N} T'_{j-\frac{1}{2}}^{n+(i-1)/N} \right],$$

$$B_j^* = 1 + A_j^* + C_j^*,$$

$$\begin{aligned} D_j^* = \ & -T_j^n \\ & - s_1(1-\theta) \left[\Phi_{j+\frac{1}{2}} \, \chi|_{j+\frac{1}{2}}^n \left(T_{j+1}^n - T_j^n\right) - \Phi_{j-\frac{1}{2}} \, \chi|_{j-\frac{1}{2}}^n \left(T_j^n - T_{j-1}^n\right) \right] \\ & - s_1\theta \left[\Phi_{j+\frac{1}{2}} \left.\frac{\partial \chi}{\partial T'}\right|_{j+\frac{1}{2}}^{n+(i-1)/N} T'_{j+\frac{1}{2}}^{n+(i-1)/N} \left(T_j^{n+(i-1)/N} - T_{j+1}^{n+(i-1)/N}\right) \right. \\ & \left. \qquad + \Phi_{j-\frac{1}{2}} \left.\frac{\partial \chi}{\partial T'}\right|_{j-\frac{1}{2}}^{n+(i-1)/N} T'_{j-\frac{1}{2}}^{n+(i-1)/N} \left(T_j^{n+(i-1)/N} - T_{j-1}^{n+(i-1)/N}\right) \right] \end{aligned}$$

Equation (7.20) is iterated from $i = 1, \cdots, N$, where N is chosen according to some tolerance. It is seen that Eqs. (7.17) and (7.20) are equivalent if $N = 1$.

As an illustration, we present comparison results of solving Eq. (7.15) with a model thermal conductivity function that mimics the critical gradient thermal diffusivity model GLF23 [120], which is in wide use. The functional specification is:

$$\chi(T') = \begin{cases} k \left(|T'| - T_c'\right)^\alpha + \chi_0 & \text{for } |T'| > T_c', \\ \chi_0 & \text{for } |T'| \le T_c'. \end{cases} \qquad (7.21)$$

This is a non-analytic function with discontinuous derivatives in the vicinity of $|T'| = T_c'$.

In this example we set $\chi_0 = 1.0$, $\alpha = 0.5$, $k = 10$, and $T_c' = 0.5$. We define a mesh going from 0 to 1 with 100 mesh points. We initialize the solution to $T = 1 - \Phi$, set the source term to $S = 1$, apply the boundary condition of $T = 0$ at $\Phi = 1$, and integrate forward in time. In Figure 7.1 we plot the computed solution at a particular space-time point, $(\Phi, t) = (.10, 0.16)$ as a function of the value of the time step δt for a series of calculations to compare the convergence properties of two algorithms applied to the same equation. In the curve labeled "Backward Euler," we used the BTCS method of Eq. (7.7), where the thermal conductivity function $\chi(T')$ was evaluated at the old time level. In the curve labeled "Newton Method," we used the single iteration non-linear implicit method, Eq. (7.17). It is seen that for this highly non-linear problem, the non-linear implicit method is far superior to the BTCS method, allowing a time step several orders of magnitude larger for the same accuracy.

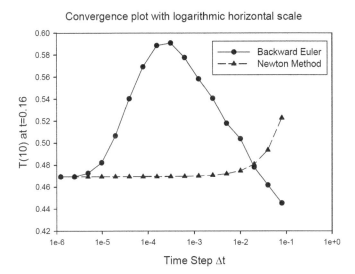

FIGURE 7.1: Comparison of the convergence properties of the Backward Euler method (BTCS) and the non-linear implicit Newton iterative method.

7.2.3 Boundary Conditions in One Dimension

One-dimensional scalar diffusion equations such as Eq. (7.2) would normally require two boundary conditions, one at each physical boundary. These can either be of the Dirichlet (fixed value) or Neumann (fixed derivative) type, or some linear combination such as $u + a\partial u/\partial x = b$ for some constants a and b.

Coordinate systems such as magnetic flux coordinates or cylindrical coordinates have a singularity at the origin. For one-dimensional systems, such as the surface-averaged transport equations or theta-independent solutions in cylindrical coordinates such as in Eq. (7.14) or Eq. (7.15), one of the coordinate boundaries can be the coordinate origin. The origin is not a physical boundary, but it can be a computational boundary. The "boundary condition" applied at the coordinate origin is thus not a true boundary condition but rather a condition that the solution be regular there. These regularity conditions can all be derived from performing a Taylor series expansion near the origin and imposing symmetry constraints to eliminate terms.

For example, if we were solving Eq. (7.15), we would note that the temperature T must have a Taylor series expansion about the origin. In Cartesian coordinates, this takes the form

$$T(x, y) = T_0 + T_x x + T_y y + \frac{1}{2}T_{xx}x^2 + T_{xy}xy + \frac{1}{2}T_{yy}y^2 + \dots \qquad (7.22)$$

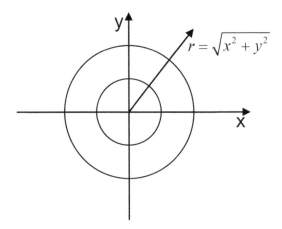

FIGURE 7.2: Cylindrical symmetry restricts the allowable terms in a Taylor series expansion in x and y about the coordinate origin.

However, we see from Figure 7.2 that by symmetry, all terms in the series that are odd in x or y must vanish, and the even terms must only be functions of $r^2 = x^2 + y^2$, or equivalently $\Phi - \frac{1}{4}r^2$. We therefore have the expansion, valid near the origin,

$$T(\Phi) = T_0 + C_1\Phi + \mathcal{O}(\delta\Phi^2), \qquad (7.23)$$

where T_0 and C_1 are constants obtained by fitting to the values at the first two grid values. One can often center the definition of variables so that it is the heat flux that needs to be evaluated at the origin, where it vanishes by symmetry. As in the previous section, if we define $T_j = T(\Phi_j)$ with the index j offset by $\frac{1}{2}$ so that $\Phi_j \equiv \left(j - \frac{1}{2}\right)\delta\Phi$ for $j \geq 1$, then $\Phi_{j-\frac{1}{2}}$ in Eq. (7.16) vanishes for $j = 1$ and no other boundary condition needs to be supplied. We can then fit the truncated series expansion, Eq. (7.23), to the first two zone values, T_1 and T_2, shown in Figure 7.3, to calculate to second-order in $\delta\Phi$

$$T_0 = \frac{3}{2}T_1 - \frac{1}{2}T_2. \qquad (7.24)$$

Equation (7.24) looks like an extrapolation formula, but it is actually performing a second-order interpolation, since the center point is surrounded on all sides by the first and second zones as can be seen in Figure 7.2.

7.2.4 Vector Forms

We saw in Section 6.2.2 that by choosing a suitable set of coordinates and dependent variables, the transport equations describing the resistive time scale evolution of an axisymmetric high-temperature toroidal plasma can be

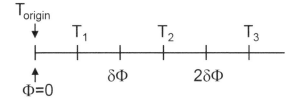

FIGURE 7.3: Temperatures are defined at equal intervals in $\delta\Phi$, offset $\frac{1}{2}\delta\Phi$ from the origin so that the condition of zero flux can be imposed there.

written as a set of non-linear one-dimensional evolution equations. These can be written in the general form:

$$\mathbf{D}^{(1)} \cdot \frac{\partial}{\partial t}\mathbf{Y} = \frac{\partial}{\partial \Phi}\left[\mathbf{K}^{(1)} \cdot \frac{\partial}{\partial \Phi}\left(\mathbf{D}^{(2)} \cdot \mathbf{Y}\right)\right]$$
$$+ \mathbf{K}^{(2)} \cdot \frac{\partial}{\partial \Phi}\left(\mathbf{D}^{(2)} \cdot \mathbf{Y}\right) + \mathbf{K}^{(3)} \cdot \mathbf{Y} + \mathbf{S}. \qquad (7.25)$$

For the model described in Chapter 6, the vector of unknowns consists of the five surface functions

$$\mathbf{Y}(\Phi, t) = \begin{bmatrix} N'(\Phi, t) \\ \sigma(\Phi, t) \\ \sigma_e(\Phi, t) \\ \iota(\Phi, t) \\ \Omega(\Phi, t) \end{bmatrix}.$$

In Eq. (7.25), $\mathbf{D}^{(1)}$ and $\mathbf{D}^{(2)}$ are diagonal matrices, and $\mathbf{K}^{(1)}$, $\mathbf{K}^{(2)}$, and $\mathbf{K}^{(3)}$ are in general full matrices with the rank of $\mathbf{Y}$. The general transport model defined in Chapter 6 could be put in this form by defining the two diagonal matrices:

$$D^{(1)} = \mathrm{diag}\begin{bmatrix} 1 \\ \frac{3}{2}(V')^{-2/3} \\ \frac{3}{2}(V')^{-2/3} \\ 1 \\ 1 \end{bmatrix}, \qquad D^{(2)} = \mathrm{diag}\begin{bmatrix} (V')^{-1} \\ (V')^{-5/3} \\ (V')^{-5/3} \\ A \\ [m_i \langle R^2 \rangle N']^{-1} \end{bmatrix},$$

and the non-zero elements of the matrices $\mathbf{K}^{(1)}$ - $\mathbf{K}^{(3)}$, which contain the transport coefficients, are given by, for $j = 1, \cdots, 5$:

$$
\begin{aligned}
K_{1j}^{(1)} &= -V'\Gamma^j, & K_{32}^{(2)} &= -V'\Gamma/n, \\
K_{2j}^{(1)} &= -V'\left(q_e^j + q_i^j + \tfrac{5}{2}\tfrac{p}{n}\Gamma^j\right), & K_{33}^{(2)} &= V'\Gamma/n, \\
K_{3j}^{(1)} &= -V'\left(q_e^j + \tfrac{5}{2}\tfrac{p_e}{n}\Gamma^j\right), & K_{32}^{(3)} &= (V')^{-2/3}\tau_{ei}^{-1}, \\
K_{4j}^{(1)} &= V_L^j, & K_{33}^{(3)} &= -2(V')^{-2/3}\tau_{ei}^{-1}, \\
K_{5j}^{(1)} &= \Gamma_\Omega^j & \tau_{ei}^{-1} &= \tfrac{3e^2 n}{m_i}\eta_\perp.
\end{aligned}
$$

The vector $\mathbf{S}$ contains the source terms:

$$
S = \begin{bmatrix} V'\langle S_n \rangle \\ V_L \frac{dK}{d\Phi} + V'\langle S_e \rangle \\ V_L \frac{dK}{d\Phi} + V'\langle S_{ee} \rangle \\ \frac{\partial}{\partial\Phi} V_L^6 \\ V'\langle S_\Omega \rangle \end{bmatrix}.
$$

Equation (7.25) is a non-linear one-dimensional *vector diffusion equation*. Note that it is not purely parabolic since it contains both a first derivative term and a zero derivative term. In many applications the non-parabolic terms involving $\mathbf{K}^{(2)}$ and $\mathbf{K}^{(3)}$ may be dominant. Because the θ-implicit method is unconditionally stable for linear parabolic and hyperbolic equations, it is normally the method of choice for numerical solution of this general form for the equations. The θ-implicit method applied to the vector diffusion equation, Eq. (7.25), using conservative centered differencing, leads to a block tridiagonal system of the form

$$
\mathbf{A}_j \cdot \mathbf{Y}_{j+1}^{n+1} - \mathbf{B}_j \cdot \mathbf{Y}_j^{n+1} + \mathbf{C}_j \cdot \mathbf{Y}_{j-1}^{n+1} = \mathbf{D}_j. \tag{7.26}
$$

This is solved for the solution at the advanced time level $\mathbf{Y}_j^{n+1}$ for $j = 1, \cdots, N-1$. Here $\mathbf{A}_j, \mathbf{B}_j$ and $\mathbf{C}_j$ are $N \times N$ matrices and $\mathbf{D}_j$ is an N vector. These are defined at each mesh point j as follows:

$$
\begin{aligned}
\mathbf{A}_j &= \theta\delta t \left[\frac{1}{\delta\Phi^2} \mathbf{K}_{j+\frac{1}{2}}^{(1)} + \frac{1}{2\delta\Phi} \mathbf{K}_j^{(2)} \right] \cdot \mathbf{D}_{j+1}^{(2)} \\
\mathbf{C}_j &= \theta\delta t \left[\frac{1}{\delta\Phi^2} \mathbf{K}_{j-\frac{1}{2}}^{(1)} - \frac{1}{2\delta\Phi} \mathbf{K}_j^{(2)} \right] \cdot \mathbf{D}_{j-1}^{(2)} \\
\mathbf{B}_j &= \mathbf{D}_j^{(1)} + \theta\frac{\delta t}{\delta\Phi^2} \left[\mathbf{K}_{j+\frac{1}{2}}^{(1)} + \mathbf{K}_{j-\frac{1}{2}}^{(1)} \right] \cdot \mathbf{D}_j^{(2)} - \theta\delta t \mathbf{K}_j^{(3)} \\
\mathbf{D}_j &= -\mathbf{D}_j^{(1)} \cdot \mathbf{Y}_j^n - \delta t \mathbf{S} \\
&\quad - (1-\theta)\,\delta t \left\{ \frac{1}{\delta\Phi^2} \left[\mathbf{K}_{j+\frac{1}{2}}^{(1)} \left(\mathbf{D}_{j+1}^{(2)} \cdot \mathbf{Y}_{j+1}^n - \mathbf{D}_j^{(2)} \cdot \mathbf{Y}_j^n \right) \right.\right. \\
&\qquad\qquad \left. - \mathbf{K}_{j-\frac{1}{2}}^{(1)} \left(\mathbf{D}_j^{(2)} \cdot \mathbf{Y}_j^n - \mathbf{D}_{j-1}^{(2)} \cdot \mathbf{Y}_{j-1}^n \right) \right] \\
&\qquad \left. + \frac{1}{2\delta\Phi} \mathbf{K}_j^{(2)} \left(\mathbf{D}_{j+1}^{(2)} \cdot \mathbf{Y}_{j+1}^n - \mathbf{D}_{j-1}^{(2)} \cdot \mathbf{Y}_{j-1}^n \right) + \mathbf{K}_j^{(3)} \cdot \mathbf{Y}_j^n \right\}
\end{aligned}
$$

Equation (7.26) is solved by introducing the $N \times N$ matrices $\mathbf{E}_j$ and the N vectors $\mathbf{F}_j$ and using the block tridiagonal method introduced in Section 3.3.3.

It should be noted that θ-implicit method, even for $\theta = 1$, does not guarantee that non-linear instabilities will not occur or that the solution of the non-linear equations will be accurate. Depending on the nature of the non-linearities, techniques such as presented in Section 7.2.2 or some "predictor-corrector" treatment may be needed for the non-linear terms. It has also been found by several authors that if the convective terms, such as those involving $\mathbf{K}^{(\mathbf{2})}$, are dominant, it may be better to treat those terms using "upwind differencing," such as described in Section 9.3. These modifications are still compatible with the block tridiagonal structure of the equations.

It is normally also necessary to monitor the change in the solution from time step to time step, and to adjust the time step so that the solution changes only to within some tolerance from step to step. Repeating a step with a smaller time step may sometimes be required.

7.3 Multiple Dimensions

There are some applications in which we desire to solve a parabolic equation in more than one spatial dimension. This would be the case, for example, if one were solving the magnetic flux diffusion equations in real space, rather than magnetic flux coordinates. The methods to be discussed here are also applicable to the treatment of viscous terms in the momentum equation.

Let us then consider a model equation of the form

$$\frac{\partial u}{\partial t} = \sigma \nabla^2 u, \tag{7.27}$$

where ∇^2 is the Laplacian operator in either two or three Cartesian coordinates. The one-dimensional form of Eq. (7.27) is just Eq. (7.2). In one dimension, implicit difference equations for parabolic equations lead to tridiagonal matrix equations that are readily solved. In two or more dimensions, these matrix equations are no longer tridiagonal and, as is the case for elliptic equations in two or more dimensions, there are a variety of techniques for solving them.

7.3.1 Explicit Methods

The forward-time centered-space (FTCS) and DuFort–Frankel explicit methods discussed in Section 7.2.1 for one-dimensional equations have straightforward generalization to 2D or 3D. For the (FTCS) method, we find that the stability condition in M dimensions becomes

$$\delta t \le \frac{\delta x^2}{2\sigma M}, \tag{7.28}$$

where we have assumed $\delta x = \delta y = \delta z$. The DuFort–Frankel method is easily generalized and remains unconditionally stable. In 2D, with $u_{j,k} = u(j\delta x, k\delta y)$, it is given by

$$
\begin{aligned}
u_{j,k}^{n+1} = u_{j,k}^{n-1} \; &+ \; \frac{2\sigma\delta t}{\delta x^2} \left[u_{j+1,k}^n - \left(u_{j,k}^{n+1} + u_{j,k}^{n-1} \right) + u_{j-1,k}^n \right] \\
&+ \; \frac{2\sigma\delta t}{\delta y^2} \left[u_{j,k+1}^n - \left(u_{j,k}^{n+1} + u_{j,k}^{n-1} \right) + u_{j,k-1}^n \right].
\end{aligned}
$$

7.3.2 Fully Implicit Methods

The θ-implicit (or Crank–Nicolson for $\theta = \frac{1}{2}$) method can be defined in two dimensions as a straightforward generalization of Eq. (7.8). If $\delta x = \delta y$, we have

$$
\begin{aligned}
u_{j,k}^{n+1} \; = \; & u_{j,k}^n + \theta \frac{\sigma\delta t}{\delta x^2} \left[u_{j+1,k}^{n+1} + u_{j-1,k}^{n+1} + u_{j,k+1}^{n+1} + u_{j,k-1}^{n+1} - 4u_{j,k}^{n+1} \right] \\
& + \; (1-\theta) \frac{\sigma\delta t}{\delta x^2} \left[u_{j+1,k}^n + u_{j-1,k}^n + u_{j,k+1}^n + u_{j,k-1}^n - 4u_{j,k}^n \right]. \tag{7.29}
\end{aligned}
$$

It is readily shown that this is unconditionally stable for $\theta \ge \frac{1}{2}$ and second-order accurate in space and time for $\theta = \frac{1}{2}$, however, it requires solving a sparse matrix equation each time step. If the matrix is not of a special form, as considered in the next section, then one of the iterative methods discussed in Sections 3.6 or 3.7 can be used. It should be kept in mind that the resulting implicit method is only worthwhile if the number of iterations required to solve the matrix equation to sufficient tolerance is less than the ratio of the time step used to the explicit time step given by Eq. (7.28).

7.3.3 Semi-Implicit Method

The semi-implicit method is a technique for the implicit differencing of Eq. (7.27) so that the difference equations can be solved using the fast Fourier transform techniques of Section 3.8.2 even when σ is a function of position. To implement the semi-implicit method, we first define a constant σ_o, which is greater than or equal to $\sigma/2$ in Eq. (7.27) every place in the domain,

$$\sigma_0 \ge \frac{1}{2}\sigma(\mathbf{x}). \tag{7.30}$$

We then rewrite Eq. (7.27) in the form

$$\frac{\partial u}{\partial t} = \sigma_o \nabla^2 u + (\sigma - \sigma_o) \nabla^2 u. \tag{7.31}$$

The first term on the right side is spatially differenced at the advanced time $(n+1)$, while the second term is differenced using the old time level n. In two dimensions, the finite difference equation being considered is

$$
\begin{aligned}
u_{j,k}^{n+1} = {} & u_{j,k}^{n} + \frac{\delta t \sigma_o}{\delta x^2} \left[u_{j+1,k}^{n+1} + u_{j-1,k}^{n+1} + u_{j,k+1}^{n+1} + u_{j,k-1}^{n+1} - 4u_{j,k}^{n+1} \right] \\
& + \frac{\delta t (\sigma - \sigma_o)}{\delta x^2} \left[u_{j+1,k}^{n} + u_{j-1,k}^{n} + u_{j,k+1}^{n} + u_{j,k-1}^{n} - 4u_{j,k}^{n} \right]. \quad (7.32)
\end{aligned}
$$

Equation (7.32) can be solved for the $u_{j,k}^{n+1}$ using transform methods since $S_o = \delta t \sigma_o / (\delta x)^2$ is a spatial constant. Von Neumann stability analysis shows the method to be stable for $\sigma_o \geq \sigma/2$, which is just the condition given in Eq. (7.30).

The truncation error associated with the finite difference equation (7.32) is given by

$$
T_\Delta = \frac{1}{2} \delta t \left(2\sigma_o - \sigma \right) \nabla^2 \frac{\partial u}{\partial t} + \mathcal{O}\left(\delta t^2\right) + \mathcal{O}\left(\delta x^2\right) \cdots . \quad (7.33)
$$

We note that the first term, which is absent from the Crank–Nicolson method, will dominate for either $\delta t \gg \delta x^2/\sigma$ or for $\sigma_o \gg \sigma/2$, so that one cannot make σ_0 arbitrarily large.

7.3.4 Fractional Steps or Splitting

There is an interesting class of methods known as *alternating direction*, *splitting*, and *fractional steps* [135] that replace the solution of a complicated multidimensional problem by a succession of simpler one-dimensional problems. To illustrate the fractional step method, let us consider the time-centered ($\theta = \frac{1}{2}$) Crank–Nicolson method in multiple dimensions. It is convenient to define some new notation so that the Crank–Nicolson method applied to Eq. (7.27) is given by

$$
\begin{aligned}
\frac{u^{n+1} - u^n}{\delta t} &= \sigma \frac{\left(\delta_x^2 + \delta_y^2 + \cdots + \delta_q^2\right)\left(u^{n+1} + u^n\right)}{2\delta x^2} \\
&= -\left(A_1 + \cdots + A_q\right)\left(\frac{u^{n+1} + u^n}{2}\right). \quad (7.34)
\end{aligned}
$$

Here $\delta_x^2 = u\left(x + \delta x\right) - 2u\left(x\right) + u\left(x - \delta x\right)$, etc. are second difference operators, and $A_i = -\sigma/\left(\delta x\right)^2 \delta_{x_i}^2$. The fractional step method replaces Eq. (7.34) by a sequence of difference equations defined by

$$
\frac{u^{n+i/q} - u^{n+(i-1)/q}}{\delta t} = -A_i \frac{u^{n+i/q} + u^{n+(i-1)/q}}{2} \quad (7.35)
$$

for $i = 1, ..., q$. The time step between n and $n+1$ is broken up into fractional values of size $\delta t/q$, where q is the number of spatial dimensions. To advance from one fractional time to the next, one need only invert a one-dimensional implicit operator of the tridiagonal form.

The von Neumann stability amplification factor r for going from time level n to level $n + 1$ is seen to be just the product of the amplification factors of each of the 1D difference equations defined by Eq. (7.35) for fixed i. Since each fractional operator is the same as the Crank–Nicolson operator evaluated in Section 7.2.1, each amplification factor is always less than or equal to unity, making the combined method unconditionally stable.

To examine the accuracy of Eq. (7.35), we rewrite it in the form

$$\left(I + \frac{\delta t}{2} A_i\right) u^{n+i/q} = \left(I - \frac{\delta t}{2} A_i\right) u^{n+(i-1)/q}; \qquad i = 1, 2, \cdots, q. \quad (7.36)$$

Now *if* the operators A_i commute, as they do in our example, Eq. (7.36) is equivalent to

$$\left(I + \frac{\delta t}{2} A_1\right) \cdots \left(I + \frac{\delta t}{2} A_q\right) u^{n+1} = \left(I - \frac{\delta t}{2} A_1\right) \cdots \left(I - \frac{\delta t}{2} A_q\right) u^n.$$

Multiplying out gives

$$u^{n+1} - u^n + \delta t\, A \frac{u^{n+1} + u^n}{2}$$
$$= -\frac{(\delta t)^2}{4} \left(A_1 A_2 + \cdots + A_{q-1} A_q\right) \left(u^{n-1} - u^n\right) + \mathcal{O}\left(\delta t^3\right).$$

Since the left side is just the Crank–Nicolson operator, and since $u^{n+1} - u^n \sim \delta t$, the truncation error is $\mathcal{O}\left(\delta t^2\right) + \mathcal{O}\left(\delta x^2\right)$. The split difference scheme will therefore be second-order accurate if the split operators are commutable. Otherwise, it is only first order accurate.

If the split operators A_1 did not commute, the split scheme of only first order accuracy can give second-order results in two cycles *if the cycle is repeated in the opposite direction*. For every even n value, we replace Eq. (7.36) by the equivalent method, but with i running backwards from q to 1. Examination of the truncation error shows that the non-commutative terms would then cancel.

7.3.5 Alternating Direction Implicit (ADI)

The fractional step method outlined in the last subsection was not meant to be applied to steady-state problems because each of the fractional time solutions $u^{n+1/q}$ satisfy different equations, and none of them approximates the steady-state equations. Even if the exact steady-state solution of a given problem is substituted and used as the initial data, the fractional time step method will generate solutions for different fractional steps which will not quite settle down to a steady-state limit.

The situation can be remedied by retaining at each fractional step the derivative terms in the other directions, but evaluated at the previous or otherwise known fractional time level. For example, for the two-dimensional approximation to Eq. (7.27) we consider

$$u^{n+\frac{1}{2}} - u^n = \frac{1}{2}\sigma \frac{\delta t}{(\delta x)^2} \left[\delta_x^2 u^{n+\frac{1}{2}} + \delta_y^2 u^n\right], \qquad (7.37)$$

$$u^{n+1} - u^{n+\frac{1}{2}} = \frac{1}{2}\sigma \frac{\delta t}{(\delta x)^2} \left[\delta_x^2 u^{n+\frac{1}{2}} + \delta_y^2 u^{n+1}\right]. \qquad (7.38)$$

Each step in Eqs. (7.37) and (7.38) only involves inversion of a tridiagonal matrix. The two fractional time steps constitute a cycle of the calculation so that the amplification factor r is a product $r = r'r''$ of two factors, one coming from each equation. If θ_1 and θ_2 are the angles appearing from Fourier transforming in the x and y directions and $s = \sigma \delta t/\delta x^2$ then the amplification factors are given by

$$[1 + s(1 - \cos\theta_1)]\, r' = 1 - s(1 - \cos\theta_2)\,,$$

$$[1 + s(1 - \cos\theta_2)]r'' = 1 - s(1 - \cos\theta_1),$$

and thus

$$r = r'r'' = \frac{[1 - s(1 - \cos\theta_1)] \cdot [1 - s(1 - \cos\theta_2)]}{[1 + s(1 - \cos\theta_1)] \cdot [1 + s(1 - \cos\theta_2)]}.$$

Since $s(1 - \cos\theta_{1,2}) \geq 0$, r lies between 0 and 1, and unconditional stability follows.

To find the order of accuracy of Eqs. (7.37) and (7.38), we subtract Eq. (7.38) from Eq. (7.37) to give an expression for $u^{n+\frac{1}{2}}$ which is then substituted into the sum of the two equations to give

$$\frac{u^{n+1} - u^n}{\delta t} = \sigma \frac{(\delta_x^2 + \delta_y^2)(u^{n+1} + u^n)}{2\delta x^2} - \sigma^2 \delta t \frac{\delta_x^2 \delta_y^2 (u^{n+1} - u^n)}{4\delta x^4}.$$

The last term is $\mathcal{O}(\delta t^2)$ and the other terms are centered as in the Crank–Nicolson equation with $\theta = \frac{1}{2}$. The truncation error is seen to be $T_\Delta = \mathcal{O}(\delta t^2) + \mathcal{O}(\delta x^2)$.

The ADI method is also useful for hyperbolic problems, especially for equation sets of mixed type. This is discussed in Section 9.5.2.

7.3.6 Douglas–Gunn Method

Unfortunately, the method described in the last section does not generalize to three or more dimensions in a way that retains unconditional stability and second-order accuracy. However, a different method exists [136] which does have these properties. This procedure is best described as a succession of approximate solutions u^*, u^{**} to the Crank–Nicolson difference equations. The last of the sequence, u^{***} for three dimensions, is simply renamed u^{n+1} for use in the next cycle. The equations are

$$u^* - u^n = s\left[\frac{1}{2}\delta_x^2\left(u^* + u^n\right) + \delta_y^2 u^n + \delta_z^2 u^n\right],$$

$$u^{**} - u^n = s\left[\frac{1}{2}\delta_x^2\left(u^* + u^n\right) + \frac{1}{2}\delta_y^2\left(u^{**} + u^n\right) + \delta_z^2 u^n\right],$$

$$u^{n+1} - u^n = s\left[\frac{1}{2}\delta_x^2\left(u^* + u^n\right) + \frac{1}{2}\delta_y^2\left(u^{**} + u^n\right) + \frac{1}{2}\delta_z^2\left(u^{n+1} + u^n\right)\right].$$

$$(7.39)$$

This method can be generalized in an obvious way to any number of space variables. In the case of two variables x and y, it can be shown to be equivalent to Eqs. (7.37) and (7.38), but in a modified form. Again, evaluation of u^{n+1} consists only of solving a succession of tridiagonal linear systems.

The amplification factor for a complete cycle can be calculated to give

$$r = \frac{1 - a - b - c + ab + ac + bc + abc}{(1 + a)(1 + b)(1 + c)},$$

where a, b, c stand for $s(1 - \cos\theta_i)$, for $i = 1, 2, 3$. The denominator, when multiplied out, contains exactly the same terms as the numerator, except that the signs are all $+$ so that again $0 \le r \le 1$ and the method is unconditionally stable. The exact prescription given by Eq. (7.39) must be followed to guarantee unconditional stability. For example, if u^* is replaced by u^{**} in the x-derivative term in the third line of Eq. (7.39) the method can be unstable.

The accuracy of Eq. (7.39) can be found by eliminating u^* and u^{**}. The result is

$$\frac{u^{n+1} - u^n}{\delta t} = \sigma\frac{\left(\delta_x^2 + \delta_y^2 + \delta_z^2\right)\left(u^{n+1} + u^n\right)}{2\delta x^2}$$
$$- \sigma^2\delta t\frac{\left(\delta_x^2\delta_y^2 + \delta_x^2\delta_z^2 + \delta_y^2\delta_z^2\right)\left(u^{n+1} - u^n\right)}{4\delta x^4}$$
$$+ \sigma^3\delta t^2\frac{\delta_x^2\delta_y^2\delta_z^2\left(u^{n+1} - u^n\right)}{8\delta x^6}.$$

The total truncation error is therefore

$$T_\Delta = \mathcal{O}(\delta t^2) + \mathcal{O}(\delta x^2).$$

7.3.7   Anisotropic Diffusion

As has been previously discussed, the presence of a strong magnetic field results in the heat conductivity being highly anisotropic. Chapter 6 discussed the method of constructing magnetic flux coordinates and performing flux surface averaging, which is valid when the anisotropy ratio approaches infinity and closed magnetic surfaces are present. Here we consider the fully two- or three-dimensional treatment of strongly anisotropic heat conductivity when one of the coordinates is not aligned with the magnetic field.

The method described here is due to Günter, et al. [137]. It is based on the observation that symmetrically defining all components of the parallel temperature gradient at the same points on a staggered grid with respect to the temperatures dramatically improved the convergence behavior for large anisotropy ratios. Consider now the temperature evolution equation in the presence of anisotropic thermal conduction. Neglecting terms arising from convection and compression heating, we have

$$\frac{3}{2} n \frac{\partial}{\partial t} T = -\nabla \cdot \mathbf{q} + S, \tag{7.40}$$

where the heat flux vector is of the form

$$\mathbf{q} = -n \left[\chi_\| \mathbf{bb} + \chi_\perp (\mathbf{I} - \mathbf{bb}) \right] \cdot \nabla T. \tag{7.41}$$

Here, $\chi_\|$ and $\chi_\perp$ are the thermal conductivities parallel and perpendicular to the direction of the magnetic field, $\mathbf{b}$ is a unit vector in the direction of the magnetic field, and S represents a volumetric source term.

The temperature is defined on integer grid indices, $T_{i,j}$. The finite difference scheme involves first defining temperature gradients at staggered locations,

$$\left. \frac{\partial T}{\partial x} \right|_{i+\frac{1}{2}, j+\frac{1}{2}} = \frac{1}{2\delta x} \left[(T_{i+1,j+1} + T_{i+1,j}) - (T_{i,j+1} + T_{i,j}) \right], \tag{7.42}$$

$$\left. \frac{\partial T}{\partial y} \right|_{i+\frac{1}{2}, j+\frac{1}{2}} = \frac{1}{2\delta y} \left[(T_{i+1,j+1} + T_{i,j+1}) - (T_{i+1,j} + T_{i,j}) \right], \tag{7.43}$$

and defining the parallel heat flux vector at these same locations as

$$q_{\|,x,i+\frac{1}{2},j+\frac{1}{2}} = -\left(n \chi_\| b_x \right)_{i+\frac{1}{2},j+\frac{1}{2}}$$
$$\times \left(b_{x,i+\frac{1}{2},j+\frac{1}{2}} \cdot \left. \frac{\partial T}{\partial x} \right|_{i+\frac{1}{2},j+\frac{1}{2}} + b_{y,i+\frac{1}{2},j+\frac{1}{2}} \cdot \left. \frac{\partial T}{\partial y} \right|_{i+\frac{1}{2},j+\frac{1}{2}} \right),$$

$$q_{\|,y,i+\frac{1}{2},j+\frac{1}{2}} = -\left(n \chi_\| b_y \right)_{i+\frac{1}{2},j+\frac{1}{2}}$$
$$\times \left(b_{x,i+\frac{1}{2},j+\frac{1}{2}} \cdot \left. \frac{\partial T}{\partial x} \right|_{i+\frac{1}{2},j+\frac{1}{2}} + b_{y,i+\frac{1}{2},j+\frac{1}{2}} \cdot \left. \frac{\partial T}{\partial y} \right|_{i+\frac{1}{2},j+\frac{1}{2}} \right).$$

The divergence of the parallel heat flux is then calculated as

$$
\begin{aligned}
\nabla \cdot \mathbf{q}_\parallel |_{i,j} &= \frac{1}{2\delta x} \left[\left(q_{\parallel,x,i+\frac{1}{2},j+\frac{1}{2}} + q_{\parallel,x,i+\frac{1}{2},j-\frac{1}{2}} \right) \right. \\
&\quad - \left. \left(q_{\parallel,x,i-\frac{1}{2},j+\frac{1}{2}} + q_{\parallel,x,i-\frac{1}{2},j-\frac{1}{2}} \right) \right] \\
&\quad + \frac{1}{2\delta y} \left[\left(q_{\parallel,y,i+\frac{1}{2},j+\frac{1}{2}} + q_{\parallel,y,i-\frac{1}{2},j+\frac{1}{2}} \right) \right. \\
&\quad - \left. \left(q_{\parallel,y,i+\frac{1}{2},j-\frac{1}{2}} + q_{\parallel,y,i-\frac{1}{2},j-\frac{1}{2}} \right) \right].
\end{aligned}
\tag{7.44}
$$

This method of differencing is seen to be conservative and also to maintain the self-adjointness of the operator $\nabla \cdot \chi_\parallel \mathbf{bb} \cdot \nabla T$. It is shown in [137] that this method can provide accurate solutions for extreme anisotropic ratios of $\chi_\parallel/\chi_\perp = 10^9$ or higher.

While this technique, referred to as "the symmetric method," is able to handle large anisotropy ratios, it is shown in [138] that it does not strictly preserve monotonicity and can lead to negative temperatures in some extreme applications where very large temperature gradients suddenly appear. Modifications to this method have been proposed using slope limiters [138] that cure the monotonicity problem, but which also destroy the desirable property that the perpendicular numerical diffusion is independent of the anisotropy ratio $\chi_\parallel/\chi_\perp$.

For large values of $\chi_\parallel/\chi_\perp$, the spatial difference scheme defined by Eqs. (7.40)–(7.44) is normally solved with a fully implicit method as discussed in Section 7.3.2 due to the time scale discrepancy implied by this ratio. However, the next section discusses an alternative solution method that is less computationally demanding.

7.3.8 Hybrid DuFort–Frankel/Implicit Method

A noteworthy hybrid time advance for the anisotropic diffusion equation has recently been proposed [139, 140] and demonstrated. It is motivated by the desire to reduce the size of the matrices compared to those of a fully implicit solution of the anisotropic diffusion equations. The computational domain is decomposed into N_x subdomains by introducing domain boundaries at various coordinate values $X_k, k = 1, 2, \cdots, N_x$ as shown in Figure 7.4. For the grid points that lie inside each of these boundaries, we apply a fully implicit BTCS method as defined in Section 7.3.7, with the temperatures being defined at the advanced level. Equations (7.42) and (7.43) become

$$
\left. \frac{\partial T}{\partial x} \right|_{i+\frac{1}{2},j+\frac{1}{2}} = \frac{1}{2\delta x} \left[\left(T^{n+1}_{i+1,j+1} + T^{n+1}_{i+1,j} \right) - \left(T^{n+1}_{i,j+1} + T^{n+1}_{i,j} \right) \right],
$$

$$
\left. \frac{\partial T}{\partial y} \right|_{i+\frac{1}{2},j+\frac{1}{2}} = \frac{1}{2\delta y} \left[\left(T^{n+1}_{i+1,j+1} + T^{n+1}_{i,j+1} \right) - \left(T^{n+1}_{i+1,j} + T^{n+1}_{i,j} \right) \right].
\tag{7.45}
$$

However, at the subdomain boundaries, the heat fluxes are calculated using a

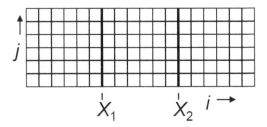

FIGURE 7.4: In the hybrid DuFort–Frankel/implicit method, BTCS differencing, Eq. (7.45), is used in the subdomain interiors, but DuFort–Frankel differencing, Eqs. (7.46) and (7.47), are used on the subdomain boundaries. The matrix equation for the new time level quantities then decomposes into separate smaller matrices for each subdomain.

generalization of the DuFort–Frankel method which uses advanced time values only along the domain boundaries. For example, if the index i is on a domain boundary, we calculate the temperature gradients to the right of the boundary as:

$$
\begin{aligned}
\left.\frac{\partial T}{\partial x}\right|_{i+\frac{1}{2},j+\frac{1}{2}} &= \frac{1}{2\delta x}\left[\left(T^n_{i+1,j+1} + T^n_{i+1,j}\right)\right.\\
&\left.- \frac{1}{2}\left(T^{n+1}_{i,j+1} + T^{n+1}_{i,j} + T^{n-1}_{i,j+1} + T^{n-1}_{i,j}\right)\right],\\
\left.\frac{\partial T}{\partial y}\right|_{i+\frac{1}{2},j+\frac{1}{2}} &= \frac{1}{2\delta y}\left[\left(T^n_{i+1,j+1} - T^n_{i+1,j}\right)\right.\\
&\left.+ \frac{1}{2}\left(T^{n+1}_{i,j+1} - T^{n+1}_{i,j} + T^{n-1}_{i,j+1} - T^{n-1}_{i,j}\right)\right], \qquad (7.46)
\end{aligned}
$$

and those to the left of the boundary as:

$$
\begin{aligned}
\left.\frac{\partial T}{\partial x}\right|_{i-\frac{1}{2},j+\frac{1}{2}} &= \frac{1}{2\delta x}\left[\frac{1}{2}\left(T^{n+1}_{i,j+1} + T^{n+1}_{i,j} + T^{n-1}_{i,j+1} + T^{n-1}_{i,j}\right)\right.\\
&\left.- \left(T^n_{i-1,j+1} + T^n_{i-1,j}\right)\right],\\
\left.\frac{\partial T}{\partial y}\right|_{i-\frac{1}{2},j+\frac{1}{2}} &= \frac{1}{2\delta y}\left[\frac{1}{2}\left(T^{n+1}_{i,j+1} - T^{n+1}_{i,j} + T^{n-1}_{i,j+1} - T^{n-1}_{i,j}\right)\right.\\
&\left.+ \left(T^n_{i-1,j+1} - T^n_{i-1,j}\right)\right], \qquad (7.47)
\end{aligned}
$$

etc. Once all the temperature gradients are defined adjacent to each domain boundary using the generalized DuFort–Frankel differencing as per Eqs. (7.46) and (7.47), we can compute the divergence of the heat flux using Eq. (7.44) (and the corresponding expression for $\nabla\cdot\mathbf{q}_\perp$). The temperatures on the domain boundaries can then be advanced to time level $n + 1$ using only time level n values of the temperatures within the domains. The difference equations for each of the domain boundaries will therefore be of tridiagonal form, coupling together only the new time temperatures on the same domain boundary. Once these temperatures are advanced to the new time levels, the solutions at the subdomain boundaries will give the Dirichlet boundary conditions which are applied for the implicit scheme in each of the the subdomain interiors. This leads to smaller, decoupled matrix equations for the new time temperatures in each of the subdomains.

It is shown in [140] that this hybrid method is unconditionally stable. However, as for the standard DuFort–Frankel method, there is a time-step constraint to avoid spurious oscillations of the solution. It is given by

$$\delta t < \frac{(\delta x)l}{\chi_{\text{eff}}} \sqrt{N_{\text{DF}}}. \tag{7.48}$$

Here, δx is the zone size, l is the spatial scale of the spurious oscillation that is being excited, χ_{eff} is an effective heat conductivity perpendicular to the boundaries of the subdomains, and N_{DF} is the number of rows of zones within each subdomain. (For a fixed total number of zones, the number of subdomains scales like N_{DF}^{-1}.) It is clear from this scaling that if one of the coordinates (the x coordinate and i index in this example) lies nearly parallel to the magnetic field, the method will provide the most benefit.

7.4   Summary

In one dimension, parabolic equations can be solved using the θ-implicit algorithm without any significant performance penalty since the difference equations are in the tridiagonal form. Strongly non-linear equations may require special treatment to obtain accurate solutions with large time steps, but even iterative methods for converging the non-linear terms remain in tridiagonal form and are thus not computationally expensive. In multiple dimensions, implicit methods lead to an "elliptic-like" equation which must be solved each time step. Implicit methods in 2D or 3D are worthwhile as long as the ratio of the work required in solving the implicit equations compared to that required in solving the explicit equations is less than the ratio of time step used in the implicit method to that used in the explicit method (for the same accuracy). For Cartesian (or other orthogonal) coordinates, some type of splitting method may be used. Anisotropic transport can be computed accurately by first defining the "heat flux" vector in staggered zone locations. The anisotropic equations normally need to be solved implicitly, but a hybrid DuFort–Frankel/implicit method may offer some advantages.

Problems

7.1: Consider the explicit second-order "leapfrog" centered-time centered-space algorithm applied to the 1D diffusion problem.

$$u_j^{n+1} = u_j^{n-1} + \frac{2\delta t \sigma}{\delta x^2} \left[u_{j+1}^n - 2u_j^n + u_{j-1}^n \right].$$

Use the von Neumann stability analysis method to show that the method is *unconditionally unstable*.

7.2: The Lax–Friedrichs method applied to the 1D diffusion problem is

$$u_j^{n+1} = \frac{1}{2} \left(u_{j+1}^n + u_{j-1}^n \right) + \frac{\delta t \sigma}{\delta x^2} \left[u_{j+1}^n - 2u_j^n + u_{j-1}^n \right].$$

Recall that this method was described in Section 2.5.2 as a means of stabilizing the FTCS method for a hyperbolic problem. Use the von Neumann stability analysis to show that the method is *unconditionally unstable* for the diffusion equation.

7.3: Verify that the modified Crank–Nicolson method as defined by Eq. (7.12) is second-order accurate in time, and that the amplification factor obeys Eq. (7.13) in that limit.

7.4: Consider the non-linear diffusion equation in one-dimensional slab geometry:

$$\frac{\partial T}{\partial t} = \frac{\partial}{\partial x} \left[\chi \frac{\partial}{\partial x} T \right],$$

where the thermal conductivity is proportional to the temperature gradient raised to the power p, $\chi = \chi_0 \left(\frac{\partial T}{\partial x} \right)^p$. Perform a linear analysis of the stability of the BTCS method where the thermal conductivity term is evaluated at the old time level, i.e.,

$$T^{n+1} = T^n + \delta t \frac{\partial}{\partial x} \left[\chi^n \frac{\partial}{\partial x} T^{n+1} \right]. \tag{7.49}$$

Assume the temperature is slightly perturbed from its equilibrium value, $T^n = T_0 + \tilde{T}^n$, where T_0 satisfies the steady-state equation $\frac{\partial}{\partial x} \left[\chi_0 \frac{\partial}{\partial x} T_0 \right] = 0$ and $\tilde{T} \ll T_0$. Let $\tilde{T}^n = r^n e^{ikx}$. Show that for short wavelength perturbations, the criteria for instability is

$$(p+1) \left[-2 + (p-1)k^2 \chi \delta t \right] > 0. \tag{7.50}$$

7.5 Consider the anisotropic diffusion equation in two dimensions with a coordinate system that is aligned with the anisotropy:

$$\frac{\partial T}{\partial t} = \kappa_x \frac{\partial^2 T}{\partial x^2} + \kappa_y \frac{\partial^2 T}{\partial y^2}.$$

Assume that $\kappa_x \gg \kappa_y$. (a) Write down a finite difference equation for this that is implicit (BTCS) in the x direction and explicit (FTCS) in the y direction. (b) Use von Neumann stability analysis to calculate the stability criteria. (c) What are the leading order truncation errors?

Chapter 8

Methods of Ideal MHD Stability Analysis

8.1 Introduction

In this and the following chapters we discuss methods for determining the stability and subsequent evolution of configurations that deviate from the equilibrium solutions discussed in Chapters 4, 5, and 6. We begin in this chapter by discussing the theoretical basis for linear methods which are suitable for computing the stability of small deviations from equilibrium using the ideal MHD description of the plasma given in Section 1.2.3. It is shown that a variational form is especially convenient for this. We then discuss special theoretical techniques that have been developed for cylindrical and toroidal geometry to enable efficient computation. In the following chapters we present different computational techniques for solving the equations presented here in both the linear and the non-linear regimes, and for extending the solutions to include non-ideal effects.

8.2 Basic Equations

Here we are working with the ideal MHD equations as presented in Section 1.2.3. We divide the configuration into two regions as illustrated in Figure 8.1. A plasma region is surrounded by either a vacuum region or by a pressureless, currentless plasma region, which is in turn surrounded by a perfectly conducting wall. We employ an Eulerian description, for which we summarize the relevant equations.

8.2.1 Linearized Equations about Static Equilibrium

The linearized ideal MHD equations are found by looking for solutions which deviate only infinitesimally from the equilibrium, and by keeping terms only to first order in this small deviation. We take as equilibrium quantities

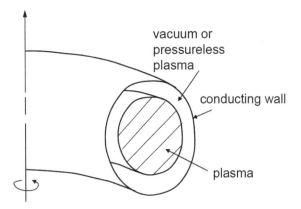

FIGURE 8.1: In ideal MHD, plasma is surrounded by either a vacuum region or a pressureless plasma, which is in turn surrounded by a conducting wall.

the current density $\mathbf{J}_0(\mathbf{x})$, the magnetic field $\mathbf{B}_0(\mathbf{x})$, the pressure $p_0(\mathbf{x})$, and the fluid velocity $\mathbf{V}_0(\mathbf{x})$. These satisfy

$$\mathbf{J}_0 \times \mathbf{B}_0 = \nabla p_0, \tag{8.1}$$

$$\nabla \times \mathbf{B}_0 = \mu_0 \mathbf{J}_0, \tag{8.2}$$

$$\nabla \cdot \mathbf{B}_0 = 0, \tag{8.3}$$

$$\mathbf{V}_0 = 0. \tag{8.4}$$

Equations (8.1) through (8.3) are equivalent to the Grad–Shafranov equation, Eq. (4.8), in axisymmetric toroidal geometry. Note that Eq. (8.4) is an important restriction on the equilibrium being considered. We remove this constraint in Section 8.2.5.

The quantities $\mathbf{B}, \mathbf{J}, p$, and $\mathbf{v}$ are decomposed into equilibrium and perturbed parts,

$$\mathbf{B}(\mathbf{x}, t) = \mathbf{B}_0(\mathbf{x}) + \mathbf{B}_1(\mathbf{x}, t),$$

$$\mathbf{J}(\mathbf{x}, t) = \mathbf{J}_0(\mathbf{x}) + \mathbf{J}_1(\mathbf{x}, t),$$

$$p(\mathbf{x}, t) = p_0(\mathbf{x}) + p_1(\mathbf{x}, t),$$

$$\mathbf{v}(\mathbf{x}, t) = \mathbf{v}_1(\mathbf{x}, t),$$

where the perturbed variables $\mathbf{B}_1, \mathbf{J}_1$, and p_1 are much smaller than their equilibrium parts. (From here on we drop the subscripts "0" from the equilibrium quantities.) It is customary to introduce a displacement field $\boldsymbol{\xi}(\mathbf{x}, t)$ as shown in Figure 8.2, which is just the displacement of a fluid element originally located at position $\mathbf{x}$. If we take as initial conditions that the perturbed quantities vanish at time zero,

$$\boldsymbol{\xi}(\mathbf{x}, 0) = \mathbf{B}_1(\mathbf{x}, 0) = p_1(\mathbf{x}, 0) = 0,$$

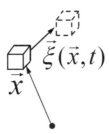

FIGURE 8.2: $\boldsymbol{\xi}(\mathbf{x}, t)$ is the displacement field.

then the linearized ideal MHD equations for the pressure and magnetic field can be integrated once in time to take the form

$$p_1 = -\boldsymbol{\xi} \cdot \nabla p - \gamma p \nabla \cdot \boldsymbol{\xi}, \tag{8.5}$$
$$\mathbf{B}_1 = \nabla \times (\boldsymbol{\xi} \times \mathbf{B}). \tag{8.6}$$

The perturbed current density follows from Eq. (1.26),

$$\mathbf{J}_1 = \frac{1}{\mu_0} \nabla \times \mathbf{B}_1. \tag{8.7}$$

The linearized momentum equation then becomes

$$\rho_0 \cdot \frac{\partial^2 \boldsymbol{\xi}}{\partial t^2} = \mathbf{J} \times \mathbf{B}_1 + \mathbf{J}_1 \times \mathbf{B} - \nabla p_1,$$

or

$$\rho_0 \frac{\partial^2 \boldsymbol{\xi}}{\partial t^2} = \mathbf{F}\{\boldsymbol{\xi}\}. \tag{8.8}$$

Here, we have introduced the ideal MHD linear force operator, given by [141]

$$\mathbf{F}(\boldsymbol{\xi}) = \frac{1}{\mu_0} \left[(\nabla \times \mathbf{B}) \times \mathbf{Q} + (\nabla \times \mathbf{Q}) \times \mathbf{B} \right] + \nabla(\boldsymbol{\xi} \cdot \nabla p + \gamma p \nabla \cdot \boldsymbol{\xi}), \tag{8.9}$$

and the perturbed magnetic field in the plasma

$$\mathbf{Q}(\mathbf{x}, t) \equiv \mathbf{B}_1(\mathbf{x}, t) = \nabla \times (\boldsymbol{\xi} \times \mathbf{B}). \tag{8.10}$$

Equation (8.8) is known as the linear *equation of motion* for an ideal MHD plasma. Note that only the equilibrium density, ρ_0, appears so there is no need to include an evolution equation for the linearized density in the absence of equilibrium flow.

The equations for the magnetic field in the vacuum are already linear. If we denote the perturbed field in the vacuum as $\hat{\mathbf{Q}}$, then these become

$$\nabla \times \hat{\mathbf{Q}} = 0,$$
$$\nabla \cdot \hat{\mathbf{Q}} = 0.$$

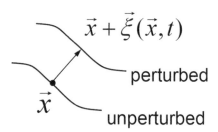

$$\vec{x} + \vec{\xi}(\vec{x}, t)$$

perturbed

$$\vec{x}$$

unperturbed

FIGURE 8.3: The physical boundary conditions are applied at the perturbed boundary, which is related to the unperturbed boundary through the displacement field $\boldsymbol{\xi}(\mathbf{x}, t)$.

The pressure balance boundary condition, Eq. (1.77), simplifies to

$$\left[\left[p + \frac{1}{2\mu_0}B^2\right]\right] = 0.$$

This is applied to the perturbed values at the perturbed plasma-vacuum interface as shown in Figure 8.3. If we denote the vacuum equilibrium field by $\hat{\mathbf{B}}$, we have the matching condition at the plasma-vacuum boundary:

$$p_1\left(\mathbf{x} + \boldsymbol{\xi}, t\right) + \frac{1}{\mu_0}\mathbf{B} \cdot \mathbf{Q}\left(\mathbf{x} + \boldsymbol{\xi}, t\right) = \frac{1}{\mu_0}\hat{\mathbf{B}} \cdot \hat{\mathbf{Q}}\left(\mathbf{x} + \boldsymbol{\xi}, t\right).$$

Letting $p_1\left(\mathbf{x} + \boldsymbol{\xi}, t\right) = p_1 + \boldsymbol{\xi} \cdot \nabla p$, etc. and substituting from Eq. (8.5), we have

$$-\gamma p \nabla \cdot \boldsymbol{\xi} + \frac{1}{\mu_0}\mathbf{B} \cdot \mathbf{Q} + \frac{1}{2\mu_0}\boldsymbol{\xi} \cdot \nabla B^2 = \frac{1}{\mu_0}\hat{\mathbf{B}} \cdot \hat{\mathbf{Q}} + \frac{1}{2\mu_0}\boldsymbol{\xi} \cdot \nabla \hat{B}^2. \qquad (8.11)$$

Equation (8.11) is to be applied at the unperturbed boundary.

Note that we have denoted the boundary equilibrium magnetic field in the plasma as $\mathbf{B}$ and that in the vacuum as $\hat{\mathbf{B}}$. If these are not equal, it implies that there is an infinitesimally thin "surface current" present in the equilibrium. While an equilibrium with surface currents is mathematically allowable and sometimes useful for analytic studies, it is not physical and normally not allowed in computational studies. For the remainder of this chapter we will assume that there are no equilibrium surface currents. This implies that the equilibrium plasma and vacuum fields are equal on the plasma/vacuum boundary, $\mathbf{B} = \hat{\mathbf{B}}$, and furthermore that the equilibrium pressure $p = 0$ at the boundary. With these assumptions and using the pressure balance condition on the equilibrium fields, together with Eq. (8.5) for the perturbed pressure, Eq. (8.11) reduces to a boundary condition on the total perturbed pressure at the unperturbed boundary,

$$p_1 + \frac{1}{\mu_0}\mathbf{B} \cdot \mathbf{Q} = \frac{1}{\mu_0}\mathbf{B} \cdot \hat{\mathbf{Q}}. \tag{8.12}$$

The remaining boundary condition at the plasma-vacuum interface is the continuity of the normal component of the magnetic field at the perturbed boundary. Let $\hat{\mathbf{n}}$ be a unit vector oriented so as to be perpendicular to the tangent of the equilibrium boundary curve. The requirement that the normal field be continuous can be shown to be equivalent to the following condition evaluated at the unperturbed boundary:

$$\hat{\mathbf{n}} \cdot \hat{\mathbf{Q}} = \hat{\mathbf{n}} \cdot \nabla \times \left(\boldsymbol{\xi} \times \hat{\mathbf{B}}\right).$$

Again, if there are no equilibrium surface currents, this becomes simply

$$\hat{\mathbf{n}} \cdot \hat{\mathbf{Q}} = \hat{\mathbf{n}} \cdot \nabla \times (\boldsymbol{\xi} \times \mathbf{B}). \tag{8.13}$$

8.2.2 Methods of Stability Analysis

There are two distinctly different methods for studying the stability of the system described by Eq. (8.8). These are the initial value approach and the normal modes approach.

(i) Initial Value Approach
The most straightforward method of analyzing Eq. (8.8) is to specify initial conditions $\boldsymbol{\xi}(\mathbf{x}, 0), \partial\boldsymbol{\xi}(\mathbf{x}, 0)/\partial t$ and the boundary conditions on $\boldsymbol{\xi}$, and to integrate Eq. (8.8) forward in time as an initial value problem. This has the advantage of being physical and closely modeling the evolution of the physical system for small departures from equilibrium. It also has the advantage that it easily extends to non-linear equations and non-ideal effects and is, in fact, the method overwhelmingly used in non-linear analysis. The disadvantage of this approach is that it can only find the most unstable mode as this will grow up exponentially and quickly dominate the solution.

(ii) Normal Modes Approach
In this approach, we look for separable eigenmode solutions. If we assume time dependence

$$\boldsymbol{\xi}(\mathbf{x}, t) = \boldsymbol{\xi}(\mathbf{x})e^{-i\omega t},$$

then Eq. (8.8) takes the time-independent form

$$-\rho_0\omega^2\boldsymbol{\xi}(\mathbf{x}) = \mathbf{F}(\boldsymbol{\xi}). \tag{8.14}$$

This is an eigenvalue equation for the frequency ω^2. Eq. (8.14) needs only boundary conditions to define a problem. Its advantages are that it is efficient numerically and can be amenable to analysis. The disadvantages of this approach is that it is strictly linear, and that it requires the eigenvalues to be discrete and distinguishable.

Some stability studies start directly from Eq. (8.14), which, like Eq. (8.8), is often referred to as "the equation of motion." However, since the operator $\mathbf{F}$ in ideal MHD without equilibrium flow is self-adjoint, a variational statement of Eq. (8.14) is possible and offers many advantages over a direct solution. This is discussed in the following sections.

8.2.3   Self-Adjointness of F

It can be proven by explicit calculation [142] that for any two vector fields $\boldsymbol{\eta}(\mathbf{x})$ and $\boldsymbol{\xi}(\mathbf{x})$ that satisfy the boundary conditions,

$$\int d\tau \boldsymbol{\eta} \cdot \mathbf{F}(\boldsymbol{\xi}) = \int d\tau \boldsymbol{\xi} \cdot \mathbf{F}(\boldsymbol{\eta}), \qquad (8.15)$$

where the integral is over the plasma volume. This property is called self-adjointness. It has several consequences which we now discuss.

(i) The eigenvalues of F are real
To prove this we subtract the inner product of $\boldsymbol{\xi}^*(\mathbf{x})$, the complex conjugate of $\boldsymbol{\xi}(\mathbf{x})$, with Eq. (8.14) from the inner product of $\boldsymbol{\xi}(\mathbf{x})$ with the complex conjugate of Eq. (8.14) to obtain

$$(\omega^2 - \omega^{*2})\rho_0 \int |\boldsymbol{\xi}|^2 d\tau = -\int d\tau \left[\boldsymbol{\xi}^* \cdot \mathbf{F}(\boldsymbol{\xi}) - \boldsymbol{\xi} \cdot \mathbf{F}(\boldsymbol{\xi}^*) \right] = 0,$$

where we used Eq. (8.15). It follows that

$$\omega^2 = \omega^{*2}$$

or, in words, that ω^2 is purely real. This is an important feature which we make much use of. It greatly simplifies the process of finding roots or searching for eigenvalues since it allows one to search only on the real axis, and it tells us that mode crossings from stable to unstable behavior always occur at the origin in the ω^2 plane.

(ii) Orthogonality
Suppose $\boldsymbol{\xi}_n$ is an eigenvector of the operator $\mathbf{F}$ with eigenvalue ω_n^2 and $\boldsymbol{\xi}_m$ is an eigenvector with eigenvalue of ω_m^2. We subtract

$$\int d\tau \boldsymbol{\xi}_m^* \cdot \left[-\rho_0 \omega_n^2 \boldsymbol{\xi}_n = \mathbf{F}(\boldsymbol{\xi}_n) \right]$$

from

$$\int d\tau \boldsymbol{\xi}_n \cdot \left[-\rho_0 \omega_m^2 \boldsymbol{\xi}_m^* = \mathbf{F}(\boldsymbol{\xi}_m^*) \right]$$

to obtain

$$\rho_0 \left(\omega_n^2 - \omega_m^2 \right) \int d\tau \boldsymbol{\xi}_m^* \cdot \boldsymbol{\xi}_n = 0.$$

This implies that the inner product of $\boldsymbol{\xi}_m$ and $\boldsymbol{\xi}_n$ is zero, or that the eigenfunctions are orthogonal, if ω_n^2 and ω_m^2 are different.

(iii) Completeness

If we restrict ourselves to physical solutions $\boldsymbol{\xi}$ that are square integrable, that is $\int d\tau |\boldsymbol{\xi}|^2$ is finite, then it can be shown that $\mathbf{F}$ has a complete set of eigenfunctions $\boldsymbol{\xi}_n$ with eigenvalues ω_n^2 [143, 144, 145]. It follows that any square integrable displacement field $\boldsymbol{\xi}$ can be represented as a linear combination of the $\boldsymbol{\xi}_n$,

$$\boldsymbol{\xi} = \sum_{n=0}^{\infty} a_n \boldsymbol{\xi}_n.$$

However, there is a complication due to the existence of the *continuous spectrum*. These are bands of "improper eigenvalues" for which the eigenvalue equation is solved, but not by square integrable functions. We can understand these further by classifying solutions according to the spectral properties of the operator $\mathbf{F}$.

8.2.4 Spectral Properties of F

Suppose there is a time periodic forcing term S so that the equation of motion, Eq. (8.14), becomes

$$\left(-\frac{\mathbf{F}}{\rho_0} - \omega^2 \right) \boldsymbol{\xi} = S.$$

We look for solutions which are formally denoted by

$$\boldsymbol{\xi} = \left(-\frac{\mathbf{F}}{\rho_0} - \omega^2 \right)^{-1} S.$$

There are three possibilities which we consider [146, 147, 148].

(i) The operator $(-\mathbf{F}/\rho_0 - \omega^2)^{-1}$ does not exist because

$$\left(-\frac{\mathbf{F}}{\rho_0} - \omega^2 \right) \boldsymbol{\xi} = 0.$$

In this case, ω^2 belongs to the *discrete spectrum* of $\mathbf{F}$ and the displacement vector $\boldsymbol{\xi}$ is a square integrable eigenfunction of the operator.

(ii) The operator $(-\mathbf{F}/\rho_0 - \omega^2)^{-1}$ exists, but is unbounded. This implies that

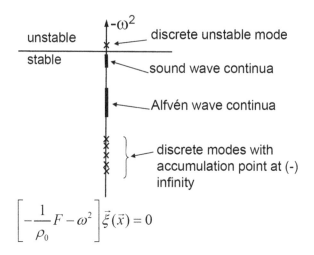

FIGURE 8.4: A typical ideal MHD spectrum.

the displacement vector $\boldsymbol{\xi}$ is not square integrable, even though the forcing function S is. Since this normally occurs for a range of ω^2 values for which a resonance condition is met, in this case we say that ω^2 belongs to the *continuous spectrum* of $\mathbf{F}$.

(iii) The operator $(-\mathbf{F}/\rho_0 - \omega^2)^{-1}$ exists and is bounded. Here, ω^2 belongs to the *resolvent set* of $\mathbf{F}$.

We illustrate a typical, although simplified, ideal MHD spectrum in Figure 8.4. In a low-β magnetized plasma, the spectrum largely separates into subsets associated with the three wave types identified in Chapter 1. Both the sound (or slow magnetoacoustic) wave and the shear Alfvén wave have continuous bands associated with them. This is explored more in Section 8.4.1 for cylindrical geometry and in Section 8.5.1 for toroidal geometry. In cylindrical geometry, there may also be a small number of discrete stable or unstable modes that are shear Alfvén oscillations modified by gradients in the background pressure and current profiles.

In toroidal geometry, the shear Alfvén continuous spectrum can break up into several bands, and discrete, global, toroidicity-induced shear Alfvén eigenmodes (TAE modes) can exist with frequencies that lie in the gaps between the bands [149]. The spectrum becomes even more complex in fully three-dimensional equilibrium such as in stellarators [150].

The fast magnetoacoustic wave is associated with the high frequency discrete band of stable oscillations as shown in the figure. The exact details will, of course, depend on the configuration being analyzed.

The continuum modes often present a practical problem when trying to solve numerically for the discrete spectrum. A truncated numerical basis may make these appear to be finite, when in reality they are not. The discrete TAE and other modes that appear in the gaps between the bands of the the shear Alfvén continuous spectrum have special physical importance because instabilities can occur when these modes resonate with an energetic particle population that may be present [151, 152, 153].

8.2.5 Linearized Equations with Equilibrium Flow

In the presence of equilibrium plasma flow it is still possible to formulate a linear stability problem by linearizing about this equilibrium. In analogy with Eqs. (8.1)–(8.4) we take as equilibrium quantities $\mathbf{J}_0(\mathbf{x})$, $\mathbf{B}_0(\mathbf{x})$, $p_0(\mathbf{x})$, $\mathbf{V}_0(\mathbf{x})$, which now must satisfy

$$
\begin{aligned}
\rho_0 \mathbf{V}_0 \cdot \nabla \mathbf{V}_0 &= -\nabla p_0 + \mathbf{J}_0 \times \mathbf{B}_0, \\
\nabla \times \mathbf{B}_0 &= \mu_0 \mathbf{J}_0, \\
\nabla \cdot \mathbf{B}_0 &= 0, \\
\nabla \cdot (\rho_0 \mathbf{V}_0) &= 0, \\
\nabla \times (\mathbf{V}_0 \times \mathbf{B}_0) &= 0, \\
\mathbf{V}_0 \cdot \nabla \left(p_0 \rho_0^{-\gamma} \right) &= 0.
\end{aligned}
$$

We again linearize the ideal MHD equations, Eqs. (1.60)–(1.63), now in the presence of an equilibrium flow field $\mathbf{V}_0$. Again, dropping the subscript "0" from the equilibrium quantities, and introducing the linear perturbed velocity field $\mathbf{v}_1(\mathbf{x})$ and linear perturbed density $\rho_1(\mathbf{x})$, we obtain [154]:

$$
\frac{\partial \rho_1}{\partial t} + \nabla \cdot \rho_1 \mathbf{V} + \nabla \cdot \rho \mathbf{v}_1 = 0, \tag{8.16}
$$

$$
\frac{\partial \mathbf{B}_1}{\partial t} = \nabla \times (\mathbf{v}_1 \times \mathbf{B} + \mathbf{V} \times \mathbf{B}_1), \tag{8.17}
$$

$$
\frac{\partial p_1}{\partial t} + \mathbf{v}_1 \cdot \nabla p + \mathbf{V} \cdot \nabla p_1 + \gamma p_1 \nabla \cdot \mathbf{V} + \gamma p \nabla \cdot \mathbf{v}_1 = 0, \tag{8.18}
$$

$$
\rho \frac{\partial \mathbf{v}_1}{\partial t} + \rho_1 \mathbf{V} \cdot \nabla \mathbf{V} + \rho \mathbf{v}_1 \cdot \nabla \mathbf{V} + \rho \mathbf{V} \cdot \nabla \mathbf{v}_1
$$
$$
+ \nabla p_1 = \mathbf{J}_1 \times \mathbf{B} + \mathbf{J} \times \mathbf{B}_1. \tag{8.19}
$$

The linear perturbed current density is again given by $\mathbf{J}_1 \equiv \mu_o^{-1} \nabla \times \mathbf{B}_1$.

We introduce the displacement field $\boldsymbol{\xi}(\mathbf{x}, t)$ that represents the infinitesimal displacement relative to a location that is moving with the equilibrium velocity $\mathbf{V}(\mathbf{x})$. By equating the velocity of an element displaced a distance $\boldsymbol{\xi}$ from a point moving with the equilibrium velocity, as viewed in the fixed Eulerian frame, with the velocity of the fluid element as viewed in a Lagrangian frame moving with the element, and then transforming the time derivative back to the fixed Eulerian frame, we obtain the relation between $\mathbf{v}_1$ and $\boldsymbol{\xi}$ [155, 156]:

$$\mathbf{v}_1 + \boldsymbol{\xi} \cdot \nabla \mathbf{V} = \frac{\partial \boldsymbol{\xi}}{\partial t} + \mathbf{V} \cdot \nabla \boldsymbol{\xi}. \tag{8.20}$$

Taking the time dependence as $\exp(-i\omega t)$ and insertion of Eq. (8.20) into Eqs. (8.16)–(8.18) gives expressions for the perturbed density, magnetic field, and pressure:

$$\rho_1 = -\nabla \cdot (\rho \boldsymbol{\xi}), \tag{8.21}$$
$$\mathbf{B}_1 = \nabla \times (\boldsymbol{\xi} \times \mathbf{B}). \tag{8.22}$$
$$p_1 = -\boldsymbol{\xi} \cdot \nabla p - \gamma p \nabla \cdot \boldsymbol{\xi}, \tag{8.23}$$

These are seen to be the same as the relations that hold without flow. Using Eqs. (8.20) and (8.21) in Eq. (8.19) yields the generalization of the equation of motion

$$-\omega^2 \rho \boldsymbol{\xi} - 2i\omega\rho \left(\mathbf{V} \cdot \nabla \boldsymbol{\xi} \right) = \mathbf{F}(\boldsymbol{\xi}) + \nabla \cdot (\boldsymbol{\xi} \mathbf{K}) - \rho \left(\mathbf{V} \cdot \nabla \right)^2 \boldsymbol{\xi}, \tag{8.24}$$

where $\mathbf{K} = \rho \mathbf{V} \cdot \nabla \mathbf{V}$ and $\mathbf{F}$ is defined in Eq. (8.9). Since the eigenvalue ω enters non-linearly in Eq. (8.24), it is convenient to break it into two equations, in each of which ω appears only linearly. We can introduce a new variable $\mathbf{u}$, and write [156]

$$\omega \rho \boldsymbol{\xi} = -i\rho \mathbf{V} \cdot \nabla \boldsymbol{\xi} + \rho \mathbf{u},$$
$$\omega \rho \mathbf{u} = -\mathbf{F}(\boldsymbol{\xi}) - \nabla \cdot (\boldsymbol{\xi} \mathbf{K}) - i\rho \mathbf{V} \cdot \nabla \mathbf{u}. \tag{8.25}$$

This system is no longer self-adjoint and cannot be solved by the variational methods described in the next section, but the normal mode equations can be solved directly [157].

8.3 Variational Forms

Variational forms play a central role in the evaluation of linear ideal MHD stability. We first discuss the basic concepts, and then go on to use the properties of the variational form to simplify the evaluation of ideal MHD stability in both cylindrical and toroidal geometry.

8.3.1 Rayleigh Variational Principle

Let us define the quantity δW which is the change in potential energy due to a perturbation

$$\delta W(\boldsymbol{\xi}^*, \boldsymbol{\xi}) = -\frac{1}{2} \int d\tau \boldsymbol{\xi}^* \cdot \mathbf{F}(\boldsymbol{\xi}). \tag{8.26}$$

Let us also define the quantity K such that $\omega^2 K$ is the kinetic energy

$$K(\boldsymbol{\xi}^*, \boldsymbol{\xi}) = \frac{1}{2} \int d\tau \rho |\boldsymbol{\xi}|^2. \tag{8.27}$$

Consider now the functional Ω^2, known as the Rayleigh quotient,

$$\Omega^2(\boldsymbol{\xi}^*, \boldsymbol{\xi}) \equiv \frac{\delta W(\boldsymbol{\xi}^*, \boldsymbol{\xi})}{K(\boldsymbol{\xi}^*, \boldsymbol{\xi})}. \tag{8.28}$$

The Rayleigh variational principle says that any allowable function $\boldsymbol{\xi}$ (and $\boldsymbol{\xi}^*$) for which Ω^2 becomes stationary is an eigenfunction of the ideal MHD normal mode equations with eigenvalue

$$\omega^2 = \Omega^2(\boldsymbol{\xi}^*, \boldsymbol{\xi}). \tag{8.29}$$

To prove this, consider the variation

$$
\begin{aligned}
\delta\left[\Omega^2\right] &= \frac{\delta\left[\delta W\right]}{K} - \frac{\delta W \delta[K]}{K^2} \\
&= \frac{1}{K}\left[\delta[\delta W] - \Omega^2 \delta[K]\right] \\
&= -\frac{1}{2K} \int d\tau \left\{ \delta\boldsymbol{\xi}^* \cdot \left[\mathbf{F}(\boldsymbol{\xi}) + \rho\Omega^2\boldsymbol{\xi}\right] + \delta\boldsymbol{\xi} \cdot \left[\mathbf{F}(\boldsymbol{\xi}^*) + \rho\Omega^2\boldsymbol{\xi}^*\right] \right\} \\
&= 0.
\end{aligned}
\tag{8.30}
$$

Here use was made of the self-adjointness property of $\mathbf{F}$. Since the functions $\delta\boldsymbol{\xi}^*$ and $\delta\boldsymbol{\xi}$ are arbitrary variations, for the variation of Ω^2 to vanish, we must have

$$\mathbf{F}(\boldsymbol{\xi}) = -\rho\Omega^2\boldsymbol{\xi}. \tag{8.31}$$

This completes the proof that $\Omega^2(\boldsymbol{\xi}, \boldsymbol{\xi}^*)$ is stationary when $\boldsymbol{\xi}$ is an eigenfunction of $\mathbf{F}(\boldsymbol{\xi})$, at which point it is equal to the frequency ω^2. It is also possible to prove that the *most negative* stationary value of Ω^2 corresponds to a minimum in δW [158].

8.3.2 Energy Principle

The energy principle [141, 159] states that there is an instability if, and only if, there exists a vector field $\boldsymbol{\eta}(\mathbf{x})$ that satisfies the boundary conditions, and such that

$$\delta W(\boldsymbol{\eta}^*, \boldsymbol{\eta}) < 0. \tag{8.32}$$

It follows that for a given vector field $\boldsymbol{\eta}$, Eq. (8.32) is sufficient to show instability.

One can thus investigate the stability of a system by looking at the sign of δW for various trial functions $\boldsymbol{\eta}$ without ever solving for the eigenfunctions $\boldsymbol{\xi}(\mathbf{x})$. Noting that since K is positive definite, it follows from Rayleigh's principle that an actual eigenvalue ω^2 must exist such that

$$\omega^2 \leq \frac{\delta W(\boldsymbol{\eta}^*, \boldsymbol{\eta})}{K(\boldsymbol{\eta}^*, \boldsymbol{\eta})},$$

so that the actual system will always be more unstable than that found with the trial function.

The existence of the energy principle provides insight into the significance of the different terms in δW. In particular, positive definite terms will always be stabilizing. Terms which *can* be negative will in general be destabilizing. The minimizing vector field $\boldsymbol{\eta}$ then tries to make the destabilizing terms negative without making the positive definite terms more positive to offset. Note that it follows from Rayleigh's principle that if the trial function $\boldsymbol{\eta}$ is within some small distance ϵ of the actual eigenfunction, $\boldsymbol{\xi}$, so that $(\boldsymbol{\eta} - \boldsymbol{\xi}) \sim \epsilon$, then since we are near a stationary point, the eigenvalue ω^2 will be correct to order ϵ^2.

The energy principle forms the basis for most linear ideal MHD numerical stability codes such as PEST [160] and ERATO [161]. Suppose we have a truncated set of basis functions $(\boldsymbol{\xi}_n; n = 0, 1, \cdots, N)$. We can form trial functions $\boldsymbol{\xi}$ by taking linear combinations of these

$$\boldsymbol{\xi} = \sum_{n=0}^{N} a_n \boldsymbol{\xi}_n. \tag{8.33}$$

We can then evaluate δW as a quadratic form

$$\delta W = \sum_{n=0}^{N} \sum_{m=0}^{N} A_{nm} a_m a_n, \tag{8.34}$$

where

$$A_{nm} = \delta W(\boldsymbol{\xi}_m, \boldsymbol{\xi}_n).$$

We can then minimize Eq. (8.34), subject to any convenient normalization, with respect to the amplitudes a_n. This leads to a matrix equation for the set of a_n which define a minimizing trial function through Eq. (8.33).

8.3.3 Proof of the Energy Principle

Here we consider a proof of the assertion that a system is unstable if, and only if, there exists some displacement $\boldsymbol{\xi}$ which makes δW negative. The proof which we quote is from Bernstein et al. [141]. It has more recently been shown to be flawed because it doesn't treat the continuum eigenfunctions correctly, but it is simple and intuitive. A mathematically correct proof, based on conservation of energy, is given by Laval [158].

The Bernstein et al. proof assumes that the eigenfunctions of $\mathbf{F}, \boldsymbol{\xi}_n$, form a complete set for *any functions* which satisfy the boundary conditions. Let $\boldsymbol{\xi}$ be a displacement which satisfies the boundary conditions, and which makes δW *negative*. By completeness, we can expand $\boldsymbol{\xi}$ in the $\boldsymbol{\xi}_n$,

$$\boldsymbol{\xi} = \sum_{n=0}^{\infty} a_n \boldsymbol{\xi}_n.$$

Now, we can compute

$$
\begin{aligned}
\delta W &= -\frac{1}{2} \sum_{n=0}^{\infty} \sum_{m=0}^{\infty} a_n^* a_m \int d\tau \boldsymbol{\xi}_n \cdot \mathbf{F}(\boldsymbol{\xi}_m), \\
&= \frac{1}{2} \rho \sum_{n=0}^{\infty} |a_n|^2 \omega_n^2,
\end{aligned}
\tag{8.35}
$$

where we used the orthonormality of the eigenfunctions. It therefore follows from Eq. (8.35) that δW can be made negative if, and only if, there exists at least one negative eigenvalue ω_n^2.

8.3.4 Extended Energy Principle

By starting from Eq. (8.9) and using simple vector identities, one can obtain

$$
\begin{aligned}
-\boldsymbol{\xi}^* \cdot \mathbf{F}(\boldsymbol{\xi}) = &- \boldsymbol{\xi}^* \cdot \mathbf{J} \times \mathbf{Q} + \frac{1}{\mu_0} |\mathbf{Q}|^2 + (\nabla \cdot \boldsymbol{\xi}^*)(\boldsymbol{\xi} \cdot \nabla p + \gamma p \nabla \cdot \boldsymbol{\xi}) \\
&+ \nabla \cdot \left[\frac{1}{\mu_0} \mathbf{Q} \times (\boldsymbol{\xi}^* \times \mathbf{B}) - \boldsymbol{\xi}^* (\boldsymbol{\xi} \cdot \nabla p + \gamma p \nabla \cdot \boldsymbol{\xi}) \right].
\end{aligned}
$$

We can therefore write

$$\delta W = \delta W_f + BT,$$

where

$$\delta W_f = \frac{1}{2} \int d\tau \left[\frac{1}{\mu_0} |\mathbf{Q}|^2 - \boldsymbol{\xi}^* \cdot \mathbf{J} \times \mathbf{Q} + \gamma p |\nabla \cdot \boldsymbol{\xi}|^2 + (\nabla \cdot \boldsymbol{\xi}^*) \boldsymbol{\xi} \cdot \nabla p \right].
\tag{8.36}$$

By using Eq. (8.11), the boundary term can be shown to be

$$BT = \delta W_s + \frac{1}{2\mu_0} \int_s dS \, (\hat{n} \cdot \boldsymbol{\xi}_\perp^*) \hat{\mathbf{B}} \cdot \hat{\mathbf{Q}},
\tag{8.37}$$

where

$$\delta W_s = \frac{1}{2} \int dS |\hat{n} \cdot \boldsymbol{\xi}|^2 \hat{n} \cdot \left[\left[\nabla \left(p + \frac{1}{2\mu_0} B^2 \right) \right] \right]. \tag{8.38}$$

The double bracket indicates the jump (or discontinuity) in the equilibrium quantities across the plasma/vacuum interface. It vanishes unless the equilibrium has surface currents which, as per the discussion preceding Eq. (8.12), will not be considered further here. The final integral in Eq. (8.37) is the perturbed magnetic energy in the vacuum. It can be written as

$$\delta W_v = \frac{1}{2\mu_0} \int dS \, (\hat{n} \cdot \boldsymbol{\xi}_\perp^*) \, \hat{\mathbf{B}} \cdot \hat{\mathbf{Q}} \tag{8.39}$$

$$= \frac{1}{2\mu_0} \int_v d\tau |\hat{\mathbf{Q}}|^2. \tag{8.40}$$

The last step follows from the fact that in the vacuum, we can write $\hat{\mathbf{Q}} = \nabla \times \mathbf{A}$, where $\nabla \times \nabla \times \mathbf{A} = 0$. It follows from Eq. (8.40) that

$$\delta W_v = \frac{1}{2\mu_0} \int_v d\tau \, (\nabla \times \mathbf{A}^*) \cdot \nabla \times \mathbf{A}$$

$$= -\frac{1}{2\mu_0} \int_s dS \hat{n} \cdot \mathbf{A}^* \times \hat{\mathbf{Q}}.$$

Note the minus sign because we are using the normal that points outward from the plasma region. Now, using the boundary condition, Eq. (8.13), in the form $\mathbf{A}^* = \boldsymbol{\xi}^* \times \hat{\mathbf{B}}$, we obtain Eq. (8.39).

We can now summarize the *extended energy principle*. We define the quantity

$$\delta W = \delta W_f + \delta W_s + \delta W_v \tag{8.41}$$

where δW_f, δW_s, and δW_v are defined by Eq. (8.36), (8.38), and (8.40). In the vacuum, the vector field $\hat{\mathbf{Q}}$ satisfies $\nabla \times \hat{\mathbf{Q}} = 0$ and $\nabla \cdot \hat{\mathbf{Q}} = 0$, with boundary conditions

$$\hat{n} \cdot \hat{\mathbf{Q}}|_{\text{wall}} = 0$$

$$\hat{n} \cdot \hat{\mathbf{Q}}|_{p/v} = \hat{n} \cdot \nabla \times \left(\boldsymbol{\xi} \times \hat{\mathbf{B}} \right)$$

For stability, one must show that $\delta W \geq 0$ for all allowable perturbations $\boldsymbol{\xi}$. Note that in the application of the extended energy principle, one need not impose the pressure balance condition across the plasma vacuum interface, Eq. (8.12), as it has already been incorporated into the formulation.

8.3.5   Useful Identities

There are a number of mathematical identities that follow from Eq. (8.1) and (8.2) that can be used to rewrite δW_f in more useful and intuitive forms.

(i) Curvature vector of the magnetic field

Consider the curvature vector

$$\boldsymbol{\kappa} \equiv \mathbf{b} \cdot \nabla \mathbf{b}, \tag{8.42}$$

where $\mathbf{b} = \mathbf{B}/|\mathbf{B}|$ is a unit vector in the direction of the magnetic field. The vector $\boldsymbol{\kappa}$ is of length $|\boldsymbol{\kappa}| = 1/R_c$ and points toward the center of a locally tangent circle of radius R_c. Using the vector identity $\mathbf{A} \times (\nabla \times \mathbf{A}) = \frac{1}{2}\nabla \mathbf{A}^2 - \mathbf{A} \cdot \nabla \mathbf{A}$ and noting that $\nabla \mathbf{b}^2 = 0$, we have

$$\boldsymbol{\kappa} = (\nabla \times \mathbf{b}) \times \mathbf{b}.$$

Writing the equilibrium magnetic field as $\mathbf{B} = B\mathbf{b}$, and the current density as $\mathbf{J} = \mu_0^{-1}\nabla \times B\mathbf{b}$, the equilibrium equation, Eq. (8.1), can then be written

$$\boldsymbol{\kappa} = \frac{1}{B^2}\nabla_\perp \left[\mu_0 p + \frac{1}{2}B^2 \right], \tag{8.43}$$

where we have dropped the subscript 0 from the equilibrium quantities. Here, and elsewhere, the subscript $\perp$ indicates "perpendicular to the magnetic field."

Since $\boldsymbol{\kappa} \cdot \mathbf{B} = 0$, it is often useful to project the curvature onto its two non-zero directions. If we utilize the straight field line coordinate system of Section 5.4.1, we can write the magnetic field in the form $\mathbf{B} = \mathbf{s} \times \nabla \Psi$, where $\mathbf{s} = \nabla \phi - q \nabla \theta$. A convenient form for the curvature vector is then

$$\boldsymbol{\kappa} = -\left(\kappa_\psi \nabla \Psi + \kappa_s \mathbf{s} \right). \tag{8.44}$$

Here, the normal and geodesic curvatures are defined, respectively, as

$$\kappa_\psi \equiv \boldsymbol{\kappa} \cdot \frac{\mathbf{s} \times \mathbf{B}}{B^2}, \qquad \kappa_s \equiv \boldsymbol{\kappa} \cdot \frac{\mathbf{B} \times \nabla \Psi}{B^2}.$$

It is also sometimes useful to project the curvature vector onto the reciprocal basis,

$$\boldsymbol{\kappa} = \kappa^\psi \frac{\mathbf{B} \times \mathbf{s}}{B^2} + \kappa^s \frac{\nabla \Psi \times \mathbf{B}}{B^2}, \tag{8.45}$$

where $\kappa^\psi \equiv \boldsymbol{\kappa} \cdot \nabla \Psi$ and $\kappa^s \equiv \boldsymbol{\kappa} \cdot \mathbf{s}$.

(ii) Divergence of $\boldsymbol{\xi}$

We decompose the plasma displacement vector $\boldsymbol{\xi}$ into its components parallel to and perpendicular to the magnetic field,

$$\begin{aligned} \boldsymbol{\xi} &= \boldsymbol{\xi}_\parallel + \boldsymbol{\xi}_\perp \\ &= \frac{(\boldsymbol{\xi} \cdot \mathbf{B})\mathbf{B}}{B^2} + \frac{\mathbf{B} \times (\boldsymbol{\xi} \times \mathbf{B})}{B^2} \end{aligned}$$

Consider now the divergence of the parallel part,

$$\nabla \cdot \boldsymbol{\xi}_\| = \mathbf{B} \cdot \nabla \left(\frac{\boldsymbol{\xi} \cdot \mathbf{B}}{B^2} \right), \tag{8.46}$$

and of the perpendicular part,

$$
\begin{aligned}
\nabla \cdot \boldsymbol{\xi}_\perp &= \frac{1}{B^2} \left[\boldsymbol{\xi} \times \mathbf{B} \cdot \nabla \times \mathbf{B} - \mathbf{B} \cdot \nabla \times (\boldsymbol{\xi} \times \mathbf{B}) \right] + \mathbf{B} \times (\boldsymbol{\xi} \times \mathbf{B}) \cdot \nabla \frac{1}{B^2}, \\
&= \frac{-1}{B^2} \left[\mathbf{Q} \cdot \mathbf{B} + \boldsymbol{\xi} \cdot \nabla_\perp \left(\mu_0 p + B^2 \right) \right], \\
&= \frac{-1}{B^2} \left[\mathbf{Q} \cdot \mathbf{B} - \mu_0 \boldsymbol{\xi} \cdot \nabla_\perp p \right] - 2\boldsymbol{\xi} \cdot \boldsymbol{\kappa}. \tag{8.47}
\end{aligned}
$$

(iii) Normal Component of Perturbed Magnetic Field

Consider the inner product of the equilibrium pressure and the perturbed magnetic field

$$
\begin{aligned}
\mathbf{Q} \cdot \nabla p &= \left[\nabla \times (\boldsymbol{\xi} \times \mathbf{B}) \right] \cdot \nabla p \\
&= \nabla \cdot \left[(\boldsymbol{\xi} \times \mathbf{B}) \times \nabla p \right] \\
&= \mathbf{B} \cdot \nabla (\boldsymbol{\xi} \cdot \nabla p). \tag{8.48}
\end{aligned}
$$

Since p could have been any flux function for the derivation of this identity, we have shown that the normal component of the perturbed magnetic field is proportional to the variation along $\mathbf{B}$ of the normal component of the displacement.

8.3.6 Physical Significance of Terms in δW_f

By using the identities in the last section, we can write δW_f, Eq. (8.36), as follows:

$$
\begin{aligned}
\delta W_f &= \frac{1}{2} \int d\tau \left[\frac{1}{\mu_0} Q_\perp^2 + \frac{1}{\mu_0} B^2 \left[\nabla \cdot \boldsymbol{\xi}_\perp + 2\boldsymbol{\xi}_\perp \cdot \boldsymbol{\kappa} \right]^2 \right. \\
&\quad + \left. \gamma p |\nabla \cdot \boldsymbol{\xi}|^2 - 2 \left(\boldsymbol{\xi}_\perp \cdot \nabla p \right) \left(\boldsymbol{\kappa} \cdot \boldsymbol{\xi}_\perp \right) - \sigma \boldsymbol{\xi}_\perp \times \mathbf{B} \cdot \mathbf{Q}_\perp \right]. \tag{8.49}
\end{aligned}
$$

Here, we have introduced the normalized parallel current,

$$\sigma \equiv \frac{\mathbf{J} \cdot \mathbf{B}}{B^2}. \tag{8.50}$$

In deriving Eq. (8.49) we have used the fact that terms of the form $\mathbf{B} \cdot \nabla a$, for any single-valued scalar a, integrate to zero when integrated over the plasma volume. Each of the terms in this equation has a physical interpretation which we now discuss.

(i) Shear Alfvén wave term: $\frac{1}{\mu_0}Q_\perp^2$

This is the energy necessary to bend the magnetic field lines. It is always stabilizing, and is the dominant potential energy contribution to the shear Alfvén waves.

(ii) Fast magnetoacoustic wave term: $\frac{1}{\mu_0}B^2\left[\nabla\cdot\boldsymbol{\xi}_\perp + 2\boldsymbol{\xi}_\perp\cdot\boldsymbol{\kappa}\right]^2$

This is the energy necessary to compress the magnetic field lines. It is always stabilizing, and is the dominant potential energy for compressional fast magnetosonic waves. Note that if a pure toroidal field were present, so that $\boldsymbol{\kappa} = -\hat{R}/R$ in cylindrical coordinates, this would be equivalent to $B^2\left[R^2\nabla\cdot(\boldsymbol{\xi}_\perp/R^2)\right]^2$.

(iii) Sound (slow magnetoacoustic) wave term: $\gamma p|\nabla\cdot\boldsymbol{\xi}|^2$

This is the energy required to compress the plasma (not the magnetic field). It is always stabilizing and is the potential energy for the slow magnetosonic (sound) waves. Note that it is the only term that contains the parallel displacement $\boldsymbol{\xi}_\parallel$. This is consistent with the discussion of the three wave types following Eq. (1.96).

(iv) Interchange instability term: $-2\left(\boldsymbol{\xi}_\perp\cdot\nabla p\right)\left(\boldsymbol{\kappa}\cdot\boldsymbol{\xi}_\perp\right)$

This term is not positive definite and therefore contributes to MHD instabilities. When this term is dominant, we have an interchange instability. Note that when ∇p and $\boldsymbol{\kappa}$ are parallel, this term is destabilizing. We call this *bad curvature*. Anytime that ∇p and $\boldsymbol{\kappa}$ are not parallel, it is possible to find a displacement $\boldsymbol{\xi}$ that makes this term destabilizing.

(v) Kink instability term: $-\sigma\boldsymbol{\xi}_\perp\times\mathbf{B}\cdot\mathbf{Q}_\perp$

This term is not positive definite and is proportional to the parallel current. When this term is dominant, we have a *kink* instability. We see from the definition of $\mathbf{Q}$, Eq. (8.10), that this term is proportional to the scalar product of a vector field $\boldsymbol{\xi}_\perp\times\mathbf{B}$ and its curl. It is a measure of the *twist* of the vector field.

8.3.7 Comparison Theorem

Let us compare the perturbed energy in two configurations: I. A plasma region surrounded by a vacuum region, and II. The same plasma region, but surrounded by a currentless pressureless plasma which occupies the same region as the vacuum in configuration I. The terms in the energy principle are the same, except for the one in the surrounding region. In configuration I it is:

$$\delta W_V = \frac{1}{2}\int d\tau\, |\nabla\times\mathbf{A}|^2\,, \tag{8.51}$$

whereas in configuration II, we see from Eq. (8.36) that it is:

$$\delta W_{PP} = \frac{1}{2} \int d\tau \, |\nabla \times (\boldsymbol{\xi} \times \mathbf{B})|^2 . \tag{8.52}$$

Here, $\mathbf{A}$ is the magnetic vector potential in the vacuum region, and $\boldsymbol{\xi}$ is the plasma displacement in the region with zero equilibrium pressure or current. Since $\boldsymbol{\xi} \times \mathbf{B}$ and $\mathbf{A}$ satisfy the same boundary conditions, and the same minimization conditions, if you find that configuration II is unstable, then configuration I will also be unstable, since you can just set $\mathbf{A} = \boldsymbol{\xi} \times \mathbf{B}$ and the energy terms will be the same.

However, the converse is not true. Say that you find the field $\hat{\mathbf{Q}} = \nabla \times \mathbf{A}$ that minimizes the integral in configuration I, and you want to find the corresponding displacement field $\boldsymbol{\xi}$ for configuration II. Consider the $\nabla\psi$ component of the following equation:

$$
\begin{aligned}
\nabla \times (\boldsymbol{\xi} \times \mathbf{B}) &= \hat{\mathbf{Q}} \\
\nabla\psi \cdot \nabla \times (\boldsymbol{\xi} \times \mathbf{B}) &= \nabla\psi \cdot \hat{\mathbf{Q}} \\
\mathbf{B} \cdot \nabla (\boldsymbol{\xi} \cdot \nabla\psi) &= \nabla\psi \cdot \hat{\mathbf{Q}}.
\end{aligned} \tag{8.53}
$$

It is seen that in order to solve for the $\nabla\psi$ component of the displacement, you must invert the operator $\mathbf{B} \cdot \nabla$.

We saw in Section 5.4.1 that if we define the poloidal angle θ appropriately we can form a straight field line coordinate system (ψ, θ, ϕ). If we Fourier analyze in these coordinates, the $(m,n)^{th}$ harmonic of the displacement varies like

$$\boldsymbol{\xi} \cdot \nabla\psi = \xi_\psi^{(m,n)}(\psi) e^{i(m\theta - n\phi)} .$$

Applying the result from Eq. (5.45) to Eq. (8.53) for this harmonic gives the relation

$$\frac{\Psi'}{J} i \, [m - nq(\psi)] \, \xi_\psi^{(m,n)} = \nabla\psi \cdot \hat{\mathbf{Q}}^{(m,n)} . \tag{8.54}$$

Since this operator vanishes at the rational magnetic surface where $q = m/n$, this equation cannot be solved for the displacement harmonic $\xi_\psi^{(m,n)}$ there and hence one cannot find the same minimizing field that was found for the vacuum case. We can thus say that the configurations surrounded by the vacuum region and the pressureless, currentless plasma region will have the same stability properties unless there is a rational surface in the region, and then the configuration surrounded by the vacuum region will be more unstable.

8.4 Cylindrical Geometry

Many calculations are performed in periodic "straight" circular cylindrical geometry for simplicity. If we impose periodic boundary conditions in z at $[0, 2\pi R]$, this geometry has the same periodicity properties as a torus of major radius R. We employ normal cylindrical coordinates (r, θ, z); see Figure 8.5. A circular cross section equilibrium is described by the equilibrium fields B_θ and B_z, the total field $B^2 = B_\theta^2 + B_z^2$, and by the pressure p, which are related by the equilibrium condition

$$\frac{dp}{dr} + \frac{B_z}{\mu_0}\frac{dB_z}{dr} + \frac{B_\theta}{\mu_0 r}\frac{d}{dr}(rB_\theta) = 0.$$

We decompose the displacement vector into components in cylindrical coordinates, and Fourier analyze in θ and z such that

$$\boldsymbol{\xi} = \left[\xi_r(r)\hat{r} + \xi_\theta(r)\hat{\theta} + \xi_z(r)\hat{z}\right]e^{i(m\theta - kz)}.$$

The primary simplification of the cylinder over the torus comes because the equilibrium depends only on the single coordinate r. This implies that all the linear modes with a given m and k value are uncoupled and can be examined separately.

Note that in the analogy with a torus, we would define the effective toroidal mode number n and safety factor q in this geometry as:

$$n = kR, \tag{8.55}$$

$$q(r) = \frac{rB_z}{RB_\theta}. \tag{8.56}$$

(Recall that R is just a constant that appears in the periodicity length of the cylinder as can be seen in Figure 8.5.) The safety factor can also be expressed in terms of the total longitudinal current interior to radius r, $I_p(r)$. Since Ampere's law can be integrated in this geometry to give the relation $2\pi r B_\theta = \mu_0 I_p$, we have the alternate expression

$$q(r) = \frac{2\pi r^2 B_z}{\mu_0 R I_p}. \tag{8.57}$$

The value of the safety factor on axis, $r = 0$, can be seen to be related to the longitudinal current density on axis, J_0, by

$$q(0) = \frac{2B_z}{\mu_0 R J_0}. \tag{8.58}$$

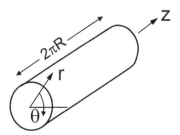

FIGURE 8.5: Straight circular cylindrical geometry with periodicity length $2\pi R$.

8.4.1 Eigenmode Equations and Continuous Spectra

To better understand the structure of the linear ideal MHD equations in the cylinder, in this section we take the time dependence to be $e^{-i\omega t}$. Five of the equations satisfied by the Fourier transforms of the quantities $p_1, \mathbf{Q}, \boldsymbol{\xi}$, Eqs. (8.5), (8.6), and (8.8), become algebraic equations. These can be used to eliminate the variables ξ_θ, ξ_z, and $\mathbf{Q}$ in favor of ξ_r and p_1. By introducing a new quantity, the perturbed fluid plus magnetic pressure $P = p_1 + \mathbf{Q} \cdot \mathbf{B}/\mu_0$, one obtains the following system of two first-order equations:

$$D \frac{d}{dr}(r\xi_r) = C_1 r\xi_r - rC_2 P, \tag{8.59}$$

$$D \frac{d}{dr}P = \frac{1}{r}C_3 r\xi_r - C_1 P. \tag{8.60}$$

Here,

$$D = \left(\rho\omega^2 - \frac{F^2}{\mu_0}\right)\left[\rho\omega^2\left(\gamma p + \frac{B^2}{\mu_0}\right) - \gamma p\frac{F^2}{\mu_0}\right],$$

$$C_1 = \frac{2B_\theta}{\mu_0 r}\left\{\rho^2\omega^4 B_\theta - \frac{m}{r}F\left[\rho\omega^2\left(\gamma p + \frac{B^2}{\mu_0}\right) - \gamma p\frac{F^2}{\mu_0}\right]\right\},$$

$$C_2 = \rho^2\omega^4 - \left(k^2 + \frac{m^2}{r^2}\right)\left[\rho\omega^2\left(\gamma p + \frac{B^2}{\mu_0}\right) - \gamma p\frac{F^2}{\mu_0}\right],$$

$$C_3 = D\left[\rho\omega^2 - \frac{F^2}{\mu_0} + \frac{2B_\theta}{\mu_0}\frac{d}{dr}\left(\frac{B_\theta}{r}\right)\right]$$

$$+ \rho\omega^2\left(\rho\omega^2 - \frac{F^2}{\mu_0}\right)\left(\frac{2B_\theta^2}{\mu_0 r}\right)^2$$

$$- \left[\gamma p\left(\rho\omega^2 - \frac{F^2}{\mu_0}\right) + \rho\omega^2\frac{B_z^2}{\mu_0}\right]\left(\frac{2B_\theta F}{\mu_0 r}\right)^2,$$

$$F = \frac{m}{r}B_\theta - kB_z = \frac{B_\theta}{r}(m - nq). \tag{8.61}$$

It is clear from this form that the system of equations (8.59) and (8.60) only has a singularity when $D = 0$, and from the form of D, it is seen that this will give rise to two continuous spectra [147]. When the first bracket in D vanishes, it is referred to as the "Alfvén continuum," and when the second bracket vanishes, it is referred to as the "sound continuum." While this form of the equations is useful in identifying the spectral properties, it is not normally the most useful for obtaining solutions since the eigenvalue ω^2 appears in a complicated non-linear way in the coefficients.

It is, however, possible to solve these equations using a "shooting" technique. The solution that is regular at the origin can be shown to have the expansion near $r = 0$

$$r\xi_r = Cr^m,\tag{8.62}$$

$$P = \frac{1}{m}\left[\frac{B_\theta^2}{\mu_0 r^2}\left(2(m-nq)-(m-nq)^2\right)+\rho\omega^2\right]r\xi_r,\tag{8.63}$$

where C is an arbitrary constant. If the plasma extends to a conducting wall at radius a, the boundary condition there is

$$\xi_r(a) = 0.\tag{8.64}$$

It can be shown that the unstable eigenvalues of Eqs. (8.59) and (8.60) behave like the eigenvalues of the Sturm–Liouville problem [162, 163]. This suggests employing an iteration scheme where one guesses an initial eigenvalue ω^2 and integrates the equations for $r\xi_r$ and P from the origin to the edge. If the edge boundary condition is not satisfied, one then changes the guess for the eigenvalue ω^2 and integrates again. This is repeated until the boundary condition is satisfied to some tolerance. The next section explains how to extend the boundary condition to that appropriate for a cylindrical plasma surrounded by a vacuum region, which is in turn surrounded by an ideal wall.

8.4.2 Vacuum Solution

If there is a vacuum region surrounding the plasma, the vacuum field can be expressed in terms of a scalar potential $\hat{Q} = \nabla\Phi_m$ that satisfies Laplaces equation,

$$\nabla^2\Phi_m = 0.\tag{8.65}$$

The solution in cylindrical coordinates for a given (m, k) mode is the sum of modified (hyperbolic) Bessel functions I_m and K_m. The vacuum fields are thus given by:

$$B_r^v = k\left[C_1 I_m'(kr)+C_2 K_m'(kr)\right],$$
$$B_\theta^v = \frac{im}{r}\left[C_1 I_m(kr)+C_2 K_m(kr)\right],$$
$$B_z^v = -ik\left[C_1 I_m(kr)+C_2 K_m(kr)\right].$$

By requiring that $B_r^v = 0$ at the radius of a perfectly conducting wall (at $r = b$), and by making B_r^v continuous at the plasma-vacuum interface, the constants C_1 and C_2 are determined in terms of the radial displacement at the edge of the plasma, ξ_a. Imposing the condition that P be continuous at the plasma vacuum interface then yields the boundary condition

$$\xi_a = \left(\frac{\mu_0 k}{F^2}\right) \frac{I'_m(ka)K'_m(kb) - K'_m(ka)I'_m(kb)}{K_m(ka)I'_m(kb) - I_m(ka)K'_m(kb)} P(a). \tag{8.66}$$

When using the extended energy principle rather than the equation of motion, it is δW_v that is needed from the vacuum region, not a relation between ξ_a and $P(a)$. Using this same analysis, we can use Eq. (8.39) to evaluate:

$$\delta W_v = \pi a \frac{F^2}{\mu_0 k} \frac{K_m(ka)I'_m(kb) - I_m(ka)K'_m(kb)}{I'_m(ka)K'_m(kb) - K'_m(ka)I'_m(kb)} \xi_a^2. \tag{8.67}$$

The expression in Eq. (8.67) is often evaluated in the large aspect ratio limit where the expansions $I_m(x) \sim x^m$ and $K_m(x) \sim x^{-m}$ are valid. This yields

$$\delta W_v = \pi B_\theta^2 \frac{(m - nq)^2}{\mu_0 m} \frac{1 + (a/b)^{2m}}{1 - (a/b)^{2m}} \xi_a^2, \tag{8.68}$$

which can be seen to be always non-negative since m here is always positive.

8.4.3   Reduction of δW_f

The quantity δW_f depends in a complicated way on all three components of the displacement field $\boldsymbol{\xi}$. Minimizing δW_f in this form is therefore equivalent to minimizing with respect to three arbitrary scalar fields, one corresponding to each of the three components of the vector field $\boldsymbol{\xi}$. Considerable simplification can be achieved if δW_f can be minimized with respect to one or more of these components analytically.

This has been shown to be possible to do algebraically [164] if we restrict consideration to cylindrical geometry. It is convenient to introduce new variables ξ, η, ζ for the displacement, related to $\xi_r(r), \xi_\theta(r), \xi_z(r)$ by

$$\begin{aligned} \xi &= \xi_r, \\ \eta &= \frac{im}{r}\xi_\theta - ik\xi_z = \nabla \cdot \boldsymbol{\xi} - \frac{1}{r}\frac{d}{dr}(r\xi_r), \\ \zeta &= i\xi_\theta B_z - i\xi_z B_\theta = i(\boldsymbol{\xi} \times \mathbf{B})_r. \end{aligned}$$

Now ξ_θ and ξ_z may be eliminated in favor of ζ and η by

$$\begin{aligned} \xi_\theta &= i\frac{kr\zeta - rB_\theta\eta}{krB_z - mB_\theta}, \\ \xi_z &= -i\frac{m\zeta - rB_z\eta}{krB_z - mB_\theta}. \end{aligned}$$

It can be shown, by straightforward though lengthy algebra, that δW_f can be written for this system as

$$\delta W_f = \int_0^a r\,dr \left\{ \frac{1}{\mu_0} \Lambda(\xi, \xi') \;+\; \gamma p \left[\eta + \frac{1}{r}(r\xi)' \right]^2 \right.$$
$$\left. + \; \frac{k^2 r^2 + m^2}{\mu_0 r^2} \left[\zeta - \zeta_0(\xi, \xi') \right]^2 \right\}. \quad (8.69)$$

Here we have introduced the quantities

$$\Lambda(\xi, \xi') \;=\; \frac{B_\theta^2}{k^2 r^2 + m^2} \left[(m - nq)\xi' - (m + nq)\frac{\xi}{r} \right]^2$$
$$+ \; \left[B_\theta^2 (m - nq)^2 - 2B_\theta (r B_\theta)' \right] \frac{\xi^2}{r^2}, \quad (8.70)$$

$$\zeta_0(\xi, \xi') = \frac{r}{k^2 r^2 + m^2} \left[(-kr B_\theta - m B_z)\xi' - (-kr B_\theta + m B_z)\frac{\xi}{r} \right]. \quad (8.71)$$

We see from Eq. (8.69) that ζ and η each occur in only one term in δW_f and in each case that term is positive definite. We can therefore minimize δW_f algebraically with respect to η and ζ by choosing

$$\eta = -\frac{1}{r}(r\xi)', \quad (8.72)$$

$$\zeta = \zeta_0(\xi, \xi'). \quad (8.73)$$

In doing this, δW_f is seen to depend only on the single scalar field $\xi(r)$, so the minimum δW_f can be obtained by minimizing the functional

$$\delta W = \delta W_f + \delta W_v, \quad (8.74)$$

where

$$\delta W_f = \int_0^a r\,dr\,\Lambda(\xi, \xi'), \quad (8.75)$$

with respect to arbitrary scalar displacements fields ξ. Note that Eq. (8.72) is equivalent to $\nabla \cdot \boldsymbol{\xi} = 0$, so that δW_f is minimized by incompressible displacements.

Another form of the energy principle is obtained by integrating Eq. (8.75) by parts to eliminate the term in $\xi\xi'$. This yields:

$$\delta W = \delta W_f + \delta W_a + \delta W_v.$$

Here, δW_a is the surface term from the integration by parts,

$$\delta W_a = \frac{B_\theta^2 \xi_a^2}{\mu_0 (k^2 r^2 + m^2)} \left. (n^2 q^2 - m^2) \right|_{r=a}, \quad (8.76)$$

and the plasma term becomes

$$\delta W_f = \frac{1}{\mu_0} \int_0^a r\, dr \left[f \left(\frac{d\xi}{dr} \right)^2 + g\xi^2 \right]. \tag{8.77}$$

The coefficients f and g are given by

$$f = \frac{r B_\theta^2 (m - nq)^2}{k^2 r^2 + m^2}, \tag{8.78}$$

$$g = \frac{1}{r} \frac{B_\theta^2 (nq + m)^2}{k^2 r^2 + m^2} \; + \; \frac{B_\theta^2}{r} (m - nq)^2 - \frac{2 B_\theta}{r} \frac{d}{dr} (r B_\theta)$$

$$- \frac{d}{dr} \left[B_\theta^2 \left(\frac{n^2 q^2 - m^2}{k^2 r^2 + m^2} \right) \right]. \tag{8.79}$$

The quantity f is seen to be never negative, but g can be either sign. Another useful expression for g is derived by using the equilibrium equation to obtain:

$$g = \frac{2 k^2 r^2 \mu_0}{k^2 r^2 + m^2} \frac{dp}{dr} \; + \; \frac{B_\theta^2}{r} (m - nq)^2 \frac{k^2 r^2 + m^2 - 1}{k^2 r^2 + m^2}$$

$$+ \frac{2 k^2 r B_\theta^2}{(k^2 r^2 + m^2)^2} \left(n^2 q^2 - m^2 \right). \tag{8.80}$$

The solution $\xi(r)$ that minimizes δW_f in Eq. (8.77) satisfies the associated Euler-Lagrange equation

$$\frac{d}{dr} \left(f \frac{d\xi}{dr} \right) - g\xi = 0. \tag{8.81}$$

This equation has a singular point, r_s wherever $f(r)$ vanishes, i.e., wherever $m - nq(r_s) = 0$. In general, it is not possible to continue an Euler-Lagrange solution past a singular point, so each subinterval bounded by the origin, a singular point, or the plasma radius, should be minimized separately. The configuration is then stable for specified values of m and k if and only if it is stable in each subinterval.

It is shown in [164] that in the vicinity of a singular point, if we define the local coordinate $x = |r - r_s|$, we can expand the functions f and g locally such that $f = \alpha x^2$ and $g = \beta$, where

$$\alpha = \frac{r_s B_\theta^2 B_z^2}{B^2} \left(\frac{q'}{q} \right)^2, \tag{8.82}$$

$$\beta = \frac{2 B_\theta^2 \mu_0}{B^2} \frac{dp}{dr}. \tag{8.83}$$

The solutions are of the form x^{-n_1} and x^{-n_2}, where n_1 and n_2 are roots of the indicial equation $n^2 - n - (\beta/\alpha) = 0$. The condition for non-oscillatory solutions, $\alpha + 4\beta > 0$ yields Suydam's condition:

$$-\frac{dp}{dr} < \frac{r}{8}\frac{B_z^2}{\mu_0}\left(\frac{q'}{q}\right)^2. \tag{8.84}$$

This can be shown to be necessary, but not sufficient, for stability. Since the pressure decreases going outward from the origin, it is a limitation on the local pressure gradient.

If Suydam's condition is satisfied, the two roots n_1 and n_2 are real and satisfy $n_1 + n_2 = 1$. If we take $n_2 > n_1$, we call x^{-n_1} the small solution at r_s. (The small solution at the origin is the one that is regular there.) It is shown in [164] that a sufficient condition for stability is the following: For specified values of m and k (or equivalently $n \equiv kR$), a cylindrical configuration is stable in a given subinterval if, and only if, Eq. (8.84) is satisfied at the endpoints, and the solution of Eq. (8.81) that is small at the left does not vanish anywhere in the interval. This provides a straightforward prescription for evaluating stability of a cylindrical configuration.

8.5   Toroidal Geometry

In toroidal geometry as illustrated in Figure 8.1 the equilibrium depends on the poloidal angle θ and so one cannot Fourier analyze in the poloidal angle and obtain a separate, uncoupled equation for each poloidal harmonic as is possible in the cylinder. However, if the configuration has an axisymmetric equilibrium, it is still possible to Fourier analyze in the toroidal angle ϕ so that

$$\boldsymbol{\xi}(\psi,\theta,\phi) = \sum_{n=-N}^{N} \boldsymbol{\xi}_n(\psi,\theta)e^{-in\phi}, \tag{8.85}$$

and each n harmonic $\boldsymbol{\xi}_n(\psi,\theta)$ is linearly independent of the others.

The finite element method, presented in Chapter 11, has been used [160] to evaluate δW_f as given by Eq. (8.49) along with δW_v, Eq. (8.39), and the kinetic energy K, Eq. (8.27). However, for many applications, some analytic simplifications can be used to obtain an asymptotic form of the stability equations that are appropriate for most applications. We discuss two such simplifications that target global and localized eigenfunctions, respectively, in Sections 8.5.3 and 8.5.4. However, as with Section 8.4 in cylindrical geometry, we start by deriving a form for solving the toroidal eigenmode equations directly that displays the role of the continuous spectra and follow this by a discussion of the vacuum calculation.

8.5.1   Eigenmode Equations and Continuous Spectra

The toroidal geometry analogue of the eigenmode equations presented for the cylinder in Section 8.4.1 employs the variables $\nabla \cdot \boldsymbol{\xi}$, ξ_s, ξ_ψ, and $P = p_1 + \mathbf{Q} \cdot \mathbf{B}/\mu_0$ [165]. We again take the time dependence $\boldsymbol{\xi}(\mathbf{x}, t) = \boldsymbol{\xi}(\mathbf{x})e^{-i\omega t}$. The displacement vector and perturbed magnetic field are decomposed as

$$\boldsymbol{\xi} = \frac{\xi_\psi}{|\nabla\Psi|^2}\nabla\Psi + \frac{\xi_s}{B^2}(\mathbf{B} \times \nabla\Psi) + \frac{\xi_b}{B^2}\mathbf{B}, \tag{8.86}$$

and

$$\mathbf{Q} = \frac{Q_\psi}{|\nabla\Psi|^2}\nabla\Psi + \frac{Q_s}{|\nabla\Psi|^2}(\mathbf{B} \times \nabla\Psi) + \frac{Q_b}{B^2}\mathbf{B}. \tag{8.87}$$

The components are seen to be defined by the projections

$$\xi_\psi = \boldsymbol{\xi} \cdot \nabla\Psi, \qquad \xi_s = \boldsymbol{\xi} \cdot \frac{\mathbf{B} \times \nabla\Psi}{|\nabla\Psi|^2}, \qquad \xi_b = \boldsymbol{\xi} \cdot \mathbf{B},$$

$$Q_\psi = \mathbf{Q} \cdot \nabla\Psi, \qquad Q_s = \mathbf{Q} \cdot \frac{\mathbf{B} \times \nabla\Psi}{B^2}, \qquad Q_b = \mathbf{Q} \cdot \mathbf{B}.$$

After eliminating the time dependence, the three components of the equation of motion, Eq. (8.14), can be written

$$\nabla\Psi \cdot \nabla P = \omega^2 \rho \xi_\psi + |\nabla\Psi|^2 \mathbf{B} \cdot \nabla\left(\frac{\mathbf{B} \cdot \nabla\xi_\psi}{|\nabla\Psi|^2}\right) \tag{8.88}$$

$$- \left(S\frac{|\nabla\Psi|^2}{B^2} + \sigma\right)|\nabla\Psi|^2(\mathbf{B} \cdot \nabla\xi_s + S\xi_\psi) + 2\boldsymbol{\kappa} \cdot \nabla\Psi Q_b,$$

$$(\mathbf{B} \times \nabla\Psi) \cdot \nabla P = \omega^2 \rho |\nabla\Psi|^2 \xi_s + B^2 \sigma \mathbf{B} \cdot \nabla\xi_\psi \tag{8.89}$$

$$+ B^2 \mathbf{B} \cdot \nabla\left[\frac{|\nabla\Psi|^2}{B^2}(\mathbf{B} \cdot \nabla\xi_s + S\xi_\psi)\right] + 2\boldsymbol{\kappa} \cdot (\mathbf{B} \times \nabla\Psi) Q_b,$$

$$\omega^2 \rho \xi_b = \mathbf{B} \cdot \nabla(p_1 + p'\xi_\psi). \tag{8.90}$$

Here, $p' = dp/d\Psi$, $\boldsymbol{\kappa}$ and σ are defined in Eqs. (8.42) and (8.50), and

$$S = -\left(\frac{\mathbf{B} \times \nabla\Psi}{|\nabla\Psi|^2}\right) \cdot \nabla \times \left(\frac{\mathbf{B} \times \nabla\Psi}{|\nabla\Psi|^2}\right) \tag{8.91}$$

is the local magnetic shear. The three components of the induction equation, Eq. (8.10), can similarly be written as

$$Q_\psi = \mathbf{B} \cdot \nabla \xi_\psi, \tag{8.92}$$

$$Q_s = \left(\frac{|\nabla \Psi|^2}{B^2} \right) (\mathbf{B} \cdot \nabla \xi_s + S\xi_\psi), \tag{8.93}$$

$$Q_b = B^2 \mathbf{B} \cdot \nabla \left(\frac{\xi_b}{B^2} \right) - B^2 \nabla \cdot \boldsymbol{\xi} - 2\boldsymbol{\kappa} \cdot (\mathbf{B} \times \nabla \Psi) \xi_s$$
$$-2 (\boldsymbol{\kappa} \cdot \nabla \Psi) \frac{B^2}{|\nabla \Psi|^2} \xi_\psi + p' \xi_\psi, \tag{8.94}$$

and $\nabla \cdot \boldsymbol{\xi}$ can be explicitly expressed as

$$\nabla \cdot \boldsymbol{\xi} = \frac{\nabla \Psi \cdot \nabla \xi_\psi}{|\nabla \Psi|^2} + \left[\nabla \cdot \left(\frac{\nabla \Psi}{|\nabla \Psi|^2} \right) \right] \xi_\psi + \frac{\mathbf{B} \times \nabla \Psi \cdot \nabla \xi_s}{B^2}$$
$$-2\boldsymbol{\kappa} \cdot (\mathbf{B} \times \nabla \Psi) \xi_s + \mathbf{B} \cdot \nabla \left(\frac{\xi_b}{B^2} \right). \tag{8.95}$$

The next step is to eliminate ξ_b, Q_ψ, Q_s, and Q_b by using Eqs. (8.90) and (8.92)–(8.94). Then, by using Eqs. (8.5), (8.88), (8.89), and (8.95) the toroidal form of the linearized ideal MHD eigenmode equations can be written [165]

$$\nabla \Psi \cdot \nabla \begin{bmatrix} P \\ \xi_\psi \end{bmatrix} = C \begin{bmatrix} P \\ \xi_\psi \end{bmatrix} + D \begin{bmatrix} \xi_s \\ \nabla \cdot \boldsymbol{\xi} \end{bmatrix}, \tag{8.96}$$

and

$$E \begin{bmatrix} \xi_s \\ \nabla \cdot \boldsymbol{\xi} \end{bmatrix} = F \begin{bmatrix} P \\ \xi_\psi \end{bmatrix}, \tag{8.97}$$

where C, D, E, F are 2×2 matrix operators involving only surface derivatives $\mathbf{B} \cdot \nabla$ and $(\mathbf{B} \times \nabla \Psi) \cdot \nabla$. The matrix operators are given by

$$C = \begin{bmatrix} 2\kappa^\psi & G \\ 0 & -|\nabla \Psi|^2 \nabla \cdot \left(\frac{\nabla \Psi}{|\nabla \Psi|^2} \right) \end{bmatrix},$$

$$D = \begin{bmatrix} -\left(S\frac{|\nabla \Psi|^2}{B^2} + \sigma \right) |\nabla \Psi|^2 \mathbf{B} \cdot \nabla & 2\gamma p \kappa^\psi \\ |\nabla \Psi|^2 \left(2\kappa_s - \frac{\mathbf{B} \times \nabla \Psi}{B^2} \cdot \nabla \right) & |\nabla \Psi|^2 \left[1 + \frac{\gamma p}{\omega^2 \rho} \mathbf{B} \cdot \nabla \left(\frac{\mathbf{B} \cdot \nabla}{B^2} \right) \right] \end{bmatrix},$$

$$E = \begin{bmatrix} \frac{\omega^2 \rho |\nabla \Psi|^2}{B^2} + \mathbf{B} \cdot \nabla \left(\frac{|\nabla \Psi|^2 \mathbf{B} \cdot \nabla}{B^2} \right) & 2\gamma p \kappa_s \\ 2\kappa_s & \frac{\gamma p + B^2}{B^2} + \frac{\gamma p}{\omega^2 \rho} \mathbf{B} \cdot \nabla \left(\frac{\mathbf{B} \cdot \nabla}{B^2} \right) \end{bmatrix},$$

$$
F = \begin{bmatrix} -2\kappa_s + \frac{\mathbf{B}\times\nabla\Psi}{B^2}\cdot\nabla & -\mathbf{B}\cdot\nabla\frac{|\nabla\Psi|^2}{B^2}S - \sigma\mathbf{B}\cdot\nabla - 2p'\kappa_s \\ -\frac{1}{B^2} & -2\frac{\kappa^\psi}{|\nabla\Psi|^2} \end{bmatrix}.
$$

In these matrix operator expressions, $\mathbf{B}\cdot\nabla$ operates on all the quantities to its right. The curvature projections κ^ψ and κ_s are defined in Eqs. (8.45) and (8.44) and we have introduced the quantities

$$
\nabla\Psi\cdot\nabla = |\nabla\Psi|^2\frac{\partial}{\partial\Psi} + (\nabla\Psi\cdot\nabla\theta)\frac{\partial}{\partial\theta},
$$

$$
G = \omega^2\rho + 2p'\kappa^\psi + |\nabla\Psi|^2\mathbf{B}\cdot\nabla\left(\frac{\mathbf{B}\cdot\nabla}{|\nabla\Psi|^2}\right) - \left(\sigma + S\frac{|\nabla\Psi|^2}{B^2}\right)S|\nabla\Psi|^2.
$$

The boundary condition at the axis is that ξ_ψ be regular there. For fixed boundary modes, the boundary condition is $\xi_\psi = 0$ at the plasma-wall interface. For free boundary modes, the boundary condition at the plasma-vacuum interface, Eq. (8.13), becomes $\nabla\Psi\cdot\hat{\mathbf{Q}} = \mathbf{B}\cdot\nabla\xi_\psi$, where $\hat{\mathbf{Q}}$ is the vacuum magnetic field which must be solved from the vacuum equations $\nabla\cdot\hat{\mathbf{Q}} = 0$ and $\nabla\times\hat{\mathbf{Q}} = 0$. Once this is solved, the pressure balance boundary condition, Eq. (8.12), is applied. This effectively relates ξ_ψ and P at the interface.

For a given equilibrium we first solve for ξ_s and $\nabla\cdot\boldsymbol{\xi}$ in terms of P and ξ_ψ from Eq. (8.97) by inverting the surface matrix operator E. Equation (8.96) then reduces to an equation for P and ξ_ψ. Admissible solutions must be regular and satisfy the appropriate boundary conditions. This procedure fails if the inverse of the surface matrix operator E does not exist for a given ω^2 at some Ψ surface. Then, only non-square-integrable solutions with spatial singularities at the singular surface are possible. If at each surface nontrivial single-valued periodic solutions in θ and ϕ can be found for the equation

$$
E\begin{bmatrix} \xi_s \\ \nabla\cdot\boldsymbol{\xi} \end{bmatrix} = 0, \tag{8.98}
$$

the corresponding set of eigenvalues ω^2 forms the continuous spectrum for the equilibrium. Equation (8.98) represents the coupling of the sound waves and the shear Alfvén waves through the surface component of the magnetic curvature and the plasma pressure. It is the toroidal generalization of the condition $D = 0$ in the cylindrical equations, Eq. (8.61).

Because the formulation presented here is not variational, it does not make use of the self-adjointness property of the MHD operator. It therefore can and has been extended to included non-ideal effects such as plasma resistivity [166], wall resistivity [167], and certain kinetic effects associated with a small energetic particle population [168].

8.5.2 Vacuum Solution

In the vacuum region, which may be surrounded by a conducting wall, there are no currents and so both $\nabla \times \hat{\mathbf{Q}} = 0$ and $\nabla \cdot \hat{\mathbf{Q}} = 0$. We can therefore express the perturbed magnetic field in terms of a scalar potential that satisfies Laplace's equation,

$$\hat{\mathbf{Q}} = \nabla \chi, \tag{8.99}$$

where

$$\nabla^2 \chi = 0. \tag{8.100}$$

We shall assume in this section that the scalar potential χ is single valued. However, Lüst and Martensen [169] have shown that the most general solution in a toroidal annulus requires that χ be multivalued. They showed that one can decompose the gradient of a multivalued scalar potential into the gradient of a single-valued piece χ^* and a linear combination of vectors $\mathcal{D}_i; i = 1, 2$ such that

$$\hat{\mathbf{Q}} = \nabla \chi = \nabla \chi^* + \sum_{i=1}^{2} \gamma_i \mathcal{D}_i. \tag{8.101}$$

The coefficients of the non-periodic parts of the potential can be uniquely determined by the path-independent integrals [170]

$$\gamma_i = \int_{\mathcal{C}_i} \nabla \chi \cdot d\mathbf{S}, \tag{8.102}$$

where $\mathcal{C}_i; i = 1, 2$ are closed curves which thread the vacuum region in the poloidal and toroidal sense. The vectors $\mathcal{D}_i; i = 1, 2$ lie point in the poloidal and toroidal directions, respectively, and are used to determine a set of inductance matrices from which the coefficients $\gamma_i; i = 1, 2$ can be found.

Green's second identity for the Laplacian with the observer points inside the vacuum region and the source points on the plasma and conductor surfaces surrounding it gives

$$4\pi \chi(\mathbf{x}) = \int d\mathbf{S}' \cdot [G(\mathbf{x}, \mathbf{x}') \nabla' \chi(\mathbf{x}') - \chi(\mathbf{x}) \nabla' G(\mathbf{x}, \mathbf{x}')], \tag{8.103}$$

where the free-space Green's function satisfies

$$\nabla^2 G(\mathbf{x}, \mathbf{x}') = -4\pi \delta(\mathbf{x} - \mathbf{x}'). \tag{8.104}$$

The solution that vanishes at infinity is given by

$$G(\mathbf{x}, \mathbf{x}') = \frac{1}{|\mathbf{x} - \mathbf{x}'|}. \tag{8.105}$$

For linear modes with ϕ dependence $e^{in\phi}$, we can perform the ϕ integration analytically to give [170]

$$2\pi\bar{G}_n(R,Z;R',Z') \equiv \int_0^{2\pi} G(\mathbf{x},\mathbf{x}')e^{in(\phi-\phi')} = \frac{2\pi^{1/2}\Gamma(1/2-n)}{H}P_{-1/2}^n(s),$$

(8.106)

where $P_{-1/2}^n$ is the associated Legendre function of the first kind, $\rho^2 = (R - R')^2 + (Z - Z')^2$, $H^4 = \rho^2\left(\rho^2 + 4RR'\right)$, and

$$s = \frac{R^2 + R'^2 + (Z - Z')^2}{H^2}.$$

A given Fourier harmonic $\chi_n(R,Z)$, defined by $\chi(\mathbf{x}) = \chi_n e^{in\phi}$, then satisfies

$$2\chi_n + \int_{p+c}\mathbf{dS}\cdot\nabla\bar{G}_n\chi_n = \int_p \mathbf{dS}\cdot\nabla\chi_n\bar{G}_n,$$

(8.107)

where the first integral is over both the plasma/vacuum boundary (p) and the conductor (c), but the second does not include the conductor since $\hat{n}\cdot\nabla\chi_n = 0$ there.

Equation (8.107) is best solved by the method of collocation. By taking a finite number of points along the plasma/vacuum boundary and the wall, the integral becomes a sum and the integral equation a matrix equation relating χ_n and $\hat{n}\cdot\nabla\chi_n$ at points on the boundary curves.

If we define a coordinate system (ψ,θ,ϕ) on each contour such that ψ=const. on a given integration contour, then the normal to that boundary contour can be written $\hat{n} = \nabla\psi/|\nabla\psi|$ and the Jacobian is $J = [\nabla\psi\times\nabla\theta\cdot\nabla\phi]^{-1}$. On the plasma/vacuum boundary we can take $\psi = \Psi$, the equilibrium flux function.

The integrals in Eq. (8.107) are then only over the poloidal angle θ, which is defined separately on each of the two contours. Since the integrals are singular, care must be taken to take the appropriate analytic continuation. We can write this as

$$\chi_n(\theta) \quad + \quad P\int_{p+c} d\theta'\, J\nabla\psi\cdot\nabla\bar{G}_n(\theta,\theta')\chi_n(\theta')$$

$$= \int_p d\theta'\, J\bar{G}_n(\theta,\theta')\nabla\psi\cdot\nabla\chi_n(\theta').$$

(8.108)

Here P denotes the principal value of the integral where the singular term appears. Note that the residue from the analytic continuation has canceled one factor of χ from the first term on the left side of the equation. This integral equation is evaluated at N discrete points on the the plasma/vacuum boundary and N discrete points on the conductor. If the integrals are each replaced by a sum over these same N points on each contour, Eq. (8.108) becomes a $2N\times 2N$ matrix equation that can be solved for χ_n on the boundary as a function of $\hat{n}\cdot\nabla\chi_n$.

8.5.3 Global Mode Reduction in Toroidal Geometry

It has been shown [171] that a reduction similar to that described in Section 8.4.3 is possible in toroidal geometry, although it is no longer strictly algebraic. It involves solving for the minimizing in-surface displacements as functions of the displacements normal to the surfaces by solving a set of in-surface differential operator equations. This reduction forms the basis for the PEST-II [160] stability program. The derivation uses straight field line coordinates as introduced in Section 5.4.1, so that the magnetic field is represented as

$$
\begin{aligned}
\mathbf{B} &= \nabla\phi \times \nabla\Psi + g(\Psi)\nabla\phi \\
&= \nabla\phi \times \nabla\Psi + q(\Psi)\nabla\Psi \times \nabla\theta.
\end{aligned}
$$

Here $q(\Psi)$ is the safety factor profile. If ϕ is the cylindrical coordinate angle, then this implies that the Jacobian is given by

$$
J = R^2 \frac{q(\Psi)}{g(\Psi)}.
$$

It is shown in [171] that the kinetic and potential energies can be written as

$$
K = \frac{1}{2}\int_p \rho|\xi|^2 d\tau,
\tag{8.109}
$$

$$
\delta W = \frac{1}{2}\int_P d\tau \left\{ \left[\mathbf{Q} + (\xi\cdot\nabla\Psi)\frac{\mathbf{J}\times\nabla\Psi}{|\nabla\Psi|^2}\right]^2 + \gamma p|\nabla\cdot\xi|^2 - 2U\,|\xi\cdot\nabla\Psi|^2 \right\}
$$
$$
+ \int_V d\tau |\nabla\times\mathbf{A}|^2,
\tag{8.110}
$$

where

$$
2U = -2p'\kappa_\psi + \frac{\sigma^2 B^2}{|\nabla\Psi|^2} - \mathbf{B}\cdot\nabla\left[\frac{\sigma\mathbf{s}\cdot\nabla\Psi}{|\nabla\Psi|^2}\right] + \sigma\frac{q'}{J}.
$$

Here, we have defined

$$
\mathbf{s} = \nabla\phi - q\nabla\theta,
$$

and the normalized parallel current σ and the field line curvature κ and its projections are defined in Eqs. (8.50) and (8.44). Noting that $\mathbf{B} = \mathbf{s}\times\nabla\Psi$, the displacement vector is decomposed as

$$
\xi = \xi^\psi\frac{\mathbf{B}\times\mathbf{s}}{B^2} + \xi_s\frac{\mathbf{B}\times\nabla\Psi}{B^2} + \xi_b\frac{\mathbf{B}}{B^2}.
\tag{8.111}
$$

In analogy with the Newcomb method in cylindrical geometry, the procedure is to analytically eliminate the $\mathbf{B}$ and $\mathbf{B}\times\nabla\Psi$ components of ξ from the

δW functional. Operationally, this is accomplished by employing the model normalization

$$2\tilde{K}(\boldsymbol{\xi}^*, \boldsymbol{\xi}) = \int_p d\tau \boldsymbol{\xi}^* \cdot \overleftrightarrow{\rho} \cdot \boldsymbol{\xi}, \tag{8.112}$$

with the anisotropic density tensor $\overleftrightarrow{\rho} = \rho \nabla \Psi \nabla \Psi$, and extremizing the functional

$$L = \omega^2 \tilde{K}(\boldsymbol{\xi}^*, \boldsymbol{\xi}) - \delta W(\boldsymbol{\xi}^*, \boldsymbol{\xi}). \tag{8.113}$$

Since only ξ^ψ enters into the modified kinetic energy $\tilde{K}$, equations for the components ξ_s and ξ_b are obtained from the Euler equation that results from extremizing δW. The field-aligned displacement ξ_b appears only in the positive definite term $\gamma p |\nabla \cdot \boldsymbol{\xi}|^2$. This term can be made to vanish by extremizing with respect to ξ_b. To extremize with respect to the other surface component ξ_s, we note that

$$\begin{aligned}
\boldsymbol{\xi} \times \mathbf{B} &= \xi^\psi \mathbf{s} + \xi_s \nabla \Psi, \\
\mathbf{Q} &= \nabla \times (\boldsymbol{\xi} \times \mathbf{B}) = \nabla \times (\xi^\psi \mathbf{s}) + \nabla \times (\xi_s \nabla \Psi).
\end{aligned}$$

Insertion of this expression for $\mathbf{Q}$ into Eq. (8.110) and extremizing δW with respect to ξ_s gives the Euler equation

$$\nabla \cdot |\nabla \Psi|^2 \nabla_s \xi_s = \nabla \cdot \left[\nabla \Psi \times \nabla \times (\xi^\psi \mathbf{s}) + \xi_\psi \mathbf{J} \right], \tag{8.114}$$

where ∇_s is the surface operator

$$\nabla_s \equiv \left[I - \frac{\nabla \Psi \nabla \Psi}{|\nabla \Psi|^2} \right] \cdot \nabla.$$

Equation (8.114) is then solved surface by surface to relate ξ_s to ξ^ψ. This is used to eliminate ξ_s from Eq. (8.113) so that the final extremization needed to be carried out using numerical basis functions (such as finite elements, see Chapter 11) is only with respect to the single component of the displacement vector ξ^ψ.

8.5.4 Ballooning Modes

There is an important class of instabilities in toroidal geometry that have displacements that are localized about a small band of magnetic surfaces for which they are locally resonant, but that have significant variation in the poloidal angle on those surfaces. There is an extensive literature on the analytic properties of these modes [172, 173]. Here we present a short derivation of the ballooning equation that can be solved numerically to determine the stability of this class of modes.

The derivation starts with the straight field line representation of the magnetic field given by Eq. (5.43) which we write as

$$\begin{aligned} \mathbf{B} &= \nabla(\phi - q\theta) \times \nabla\Psi, \\ &= \Psi'\nabla(\phi - q\theta) \times \nabla\psi, \end{aligned} \qquad (8.115)$$

where $q(\psi)$ is the safety factor. The ballooning formalism uses a new coordinate that follows the field lines, $\beta \equiv \phi - q\theta$, instead of ϕ, so that we construct a (ψ, β, θ) coordinate system. In this system, Eq. (8.115) becomes

$$\mathbf{B} = \Psi'\nabla\beta \times \nabla\psi.$$

For any scalar function f, the operator $\mathbf{B} \cdot \nabla$ in these coordinates becomes simply

$$\mathbf{B} \cdot \nabla f = \frac{\Psi'}{J}\frac{\partial f}{\partial\theta}. \qquad (8.116)$$

Note that the gradient of β is given by

$$\nabla\beta = \nabla\phi - q\nabla\theta - \theta q'\nabla\psi. \qquad (8.117)$$

The last term is seen to be non-periodic because θ appears explicitly (as opposed to just $\nabla\theta$).

It is convenient to express the displacement vector in terms of three scalar quantities (ξ, ξ_s, ξ_b) as follows:

$$\boldsymbol{\xi} = \xi\Psi'\frac{\mathbf{B} \times \nabla\beta}{B^2} + \xi_s\frac{\mathbf{B} \times \nabla\Psi}{B^2} + \xi_b\frac{\mathbf{B}}{B^2}. \qquad (8.118)$$

These quantities represent the plasma displacement across surfaces, $\xi = \boldsymbol{\xi}\cdot\nabla\psi$, within surfaces, $\xi_s = \boldsymbol{\xi}\cdot\nabla\beta$, and parallel to the field, $\xi_b = \boldsymbol{\xi}\cdot\mathbf{B}$.

Ballooning mode displacements have the property that gradients of the displacement are large compared to gradients of equilibrium quantities. We therefore introduce a small ordering parameter, ϵ, and order the displacement gradients large,

$$\left[\frac{1}{\xi}|\nabla\xi|\right]^{-1} \sim \left[\frac{1}{\beta}|\nabla\beta|\right]^{-1} \sim \left[\frac{1}{\xi_b}|\nabla\xi_b|\right]^{-1} \sim \epsilon, \qquad (8.119)$$

whereas the gradients of all equilibrium quantities are order unity. We see from examining Eq. (8.49) that the first three terms in δW_f will be of higher order (ϵ^{-2}) than the others in this ordering. To order ϵ^{-2} we therefore have

$$\delta W^{(-2)} = \frac{1}{2}\int dV \left[\frac{1}{\mu_0}|\mathbf{Q}_\perp|^2 + \frac{1}{\mu_0}B^2|\nabla\cdot\boldsymbol{\xi}_\perp|^2 + \gamma p|\nabla\cdot\boldsymbol{\xi}|^2\right], \qquad (8.120)$$

where we have dropped the lower-order curvature piece of the second term. The third term is the only one that involves the parallel displacement ξ_b, and so we can choose ξ_b to make this term vanish and proceed to minimize with respect to ξ and ξ_s.

It is readily verified that for the first two terms to vanish to leading order, we require a first-order stream function $\Phi^{(1)} \sim \epsilon$ that varies along the field on the same scale as the equilibrium, i.e.,

$$\xi_\perp^{(0)} = \frac{\mathbf{B} \times \nabla \Phi^{(1)}}{B^2}, \qquad \mathbf{B} \cdot \nabla \Phi^{(1)} \sim \Phi^{(1)} \sim \epsilon, \qquad (8.121)$$

or, in component form

$$\xi^{(0)} = \frac{1}{\Psi'} \frac{\partial \Phi^{(1)}}{\partial \beta}, \qquad \xi_s^{(0)} = \frac{1}{\Psi'} \frac{\partial \Phi^{(1)}}{\partial \psi}, \qquad \frac{\partial \Phi^{(1)}}{\partial \theta} \sim \epsilon. \qquad (8.122)$$

We now turn our attention to the terms that appear in $\delta W_f^{(0)}$. The compressive terms can again be minimized to zero by the first-order terms in the displacement, $\xi^{(1)}$ and $\xi_s^{(1)}$, so we do not need to consider them further. We seek an equation determining $\Phi^{(1)}$. Returning to Eq. (8.49), we see that there will be two terms giving order unity contributions when the substitution from Eq. (8.121) is made. These are the field line bending and pressure-curvature terms,

$$|\mathbf{Q}_\perp|^2 = \frac{1}{B^2} \left| \nabla \left(\mathbf{B} \cdot \nabla \Phi^{(1)} \right) \right|^2,$$

and

$$-2 \left(\xi \cdot \nabla p \right) \left(\xi \cdot \kappa \right) = -2 \left[\frac{\nabla p \times \mathbf{B}}{B^2} \cdot \nabla \Phi^{(1)} \right] \left[\frac{\kappa \times \mathbf{B}}{B^2} \cdot \nabla \Phi^{(1)} \right].$$

The final term in Eq. (8.49), associated with kink modes, vanishes to this order when integrated by parts since the parallel current σ is an equilibrium function with no fast spatial variation. The zero-order energy can therefore be written

$$\delta W^{(0)} = \frac{1}{2} \int dV \left\{ \frac{1}{\mu_0 B^2} \left| \tilde{\nabla} \left(\mathbf{B} \cdot \nabla \Phi^{(1)} \right) \right|^2 \right.$$
$$\left. - 2 \left[\frac{\nabla p \times \mathbf{B}}{B^2} \cdot \tilde{\nabla} \Phi^{(1)} \right] \left[\frac{\kappa \times \mathbf{B}}{B^2} \cdot \tilde{\nabla} \Phi^{(1)} \right] \right\}. \qquad (8.123)$$

Note that the plasma displacement is described in terms of a single scalar function $\Phi^{(1)}$. In Eq. (8.123) we denote the fast derivatives, of order ϵ^{-1}, by $\tilde{\nabla}$, whereas the equilibrium scale derivatives are denoted by ∇. The Euler equation associated with Eq. (8.123) is

$$\mathbf{B} \cdot \nabla \left[\frac{1}{\mu_0 B^2} \tilde{\nabla}^2 (\mathbf{B} \cdot \nabla \Phi) \right] + 2 \frac{\kappa \times \mathbf{B}}{B^2} \cdot \tilde{\nabla} \left[\frac{\nabla p \times \mathbf{B}}{B^2} \cdot \tilde{\nabla} \Phi \right] = 0, \qquad (8.124)$$

where here and in what follows we have dropped the superscript (1) on Φ.

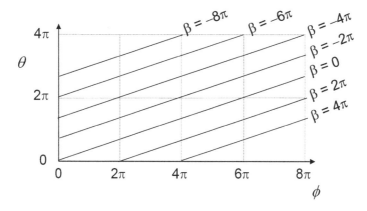

FIGURE 8.6: All physical quantities must satisfy periodicity requirements $\Phi(\psi, \theta, \beta) = \Phi(\psi, \theta, \beta + 2\pi) = \Phi(\psi, \theta + 2\pi, \beta - 2\pi q)$.

Ballooning Representation

To solve Eq. (8.124) one must deal with the periodicity properties of Φ, the θ-secularity in $\nabla\beta$, and the multiple spatial scales. Consider first the periodicity. Any physical quantity must be single valued and thus have periodicity properties with the conventional angles

$$\Phi(\psi, \theta, \phi) = \Phi(\psi, \theta, \phi + 2\pi) = \Phi(\psi, \theta + 2\pi, \phi). \qquad (8.125)$$

Now, since $\beta = \phi - q(\psi)\theta$, the periodicity requirements become

$$\Phi(\psi, \theta, \beta) = \Phi(\psi, \theta, \beta + 2\pi) = \Phi(\psi, \theta + 2\pi, \beta - 2\pi q). \qquad (8.126)$$

This is illustrated in Figure 8.6. Since this periodicity condition would be difficult to implement, one is led to a different approach in which we first suspend all periodicity conditions and look for an aperiodic solution $\hat{\Phi}$ with boundary conditions that it vanish at θ equals plus infinity or minus infinity, i.e.,

$$\hat{\Phi}(\psi, \theta, \beta) = 0 \qquad \theta \to \pm\infty. \qquad (8.127)$$

Having obtained an aperiodic solution $\hat{\Phi}$, we can construct a periodic solution by taking linear superpositions of solutions as illustrated in Figure 8.7.

$$\Phi(\psi, \theta, \beta) = \sum_{l=-\infty}^{\infty} \hat{\Phi}(\psi, \theta - 2\pi l, \beta + 2\pi l q). \qquad (8.128)$$

This manifestly satisfies the periodicity conditions

$$\Phi(\psi, \theta, \beta) = \Phi(\psi, \theta, \beta + 2\pi) = \Phi(\psi, \theta + 2\pi, \beta - 2\pi q). \qquad (8.129)$$

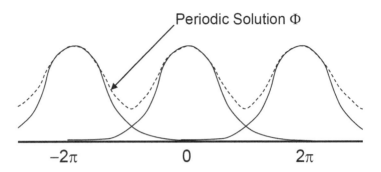

FIGURE 8.7: A solution with the correct periodicity properties is constructed by taking a linear superposition of an infinite number of offset aperiodic solutions.

To deal with the multiple spatial scales, we Fourier analyze in ϕ, the ignorable coordinate for the equilibrium. For toroidal mode number n, Eq. (8.128) thus becomes

$$\Phi(\psi, \theta, \beta) = \sum_{l=-\infty}^{\infty} \hat{\Phi}(\psi, \theta - 2\pi l)\, e^{-in(\beta + 2\pi lq)}. \qquad (8.130)$$

For $n \gg 1$, all the rapid phase variation is in the exponent. We therefore have

$$\tilde{\nabla} \hat{\Phi}(\psi, \theta - 2\pi l, \beta + 2\pi lq)$$
$$= -in\hat{\Phi}(\psi, \theta - 2\pi l)(\nabla\beta + 2\pi lq'\nabla\psi)\, e^{-in(\beta + 2\pi lq)},$$
$$= -in\hat{\Phi}(\psi, \theta - 2\pi l)\nabla(\Phi - q(\theta - 2\pi l))\, e^{-in(\beta + 2\pi lq)},$$

where we made use of Eq. (8.117). We can now make the substitution $\theta - 2\pi l \to \theta$ to obtain

$$\tilde{\nabla} \hat{\Phi}(\psi, \theta, \beta + 2\pi lq) = -in\hat{\Phi}(\psi, \theta)\nabla\beta e^{-in(\beta + 2\pi lq)}. \qquad (8.131)$$

Since all the fast variation is in the exponent, we can return to Eq. (8.124) and make the substitutions $\tilde{\nabla} \to \nabla\beta$ and $\mathbf{B} \cdot \nabla \to (\Psi'/J)\partial/\partial\theta$. Finally, we evaluate

$$\frac{\nabla p \times \mathbf{B}}{B^2} \cdot \nabla\beta = -\frac{p'}{\Psi'}, \qquad \frac{\kappa \times \mathbf{B} \cdot \nabla\beta}{B^2} = -(\kappa_\psi - q'\theta\kappa_s), \qquad (8.132)$$

where the geodesic curvature, κ_s, and the normal curvature, κ_ψ, are defined in Eq. (8.44). These substitutions lead to the canonical form for the balloon equation,

$$\frac{\Psi'}{J}\frac{\partial}{\partial\theta}\left[\frac{|\nabla\beta|^2}{\mu_0 B^2}\frac{\Psi'}{J}\frac{\partial\hat{\Phi}}{\partial\theta}\right] + 2\frac{p'}{J}(\kappa_\psi - q'\theta\kappa_s)\hat{\Phi} = 0, \qquad (8.133)$$

where, from Eq. (8.117), we have

$$|\nabla\beta|^2 = q'^2|\nabla\psi|^2\theta^2 + 2qq'\,(\nabla\psi\cdot\nabla\theta)\,\theta + |\nabla\phi - q\nabla\theta|^2.$$

It can be shown that the configuration is unstable with an eigenfunction localized near the surface ψ if the function $\hat\Phi(\psi,\theta)$ that satisfies Eq. (8.133) vanishes twice in the interval $-\infty < \theta < \infty$. The determination of stability with respect to this class of modes is thus reduced to solving a single second-order ordinary differential equation on each magnetic surface and looking for zero crossings.

Mercier Criterion
By analyzing the asymptotic behavior of the ballooning equation, Eq. (8.133), as θ approaches infinity and requiring that the solution not be oscillatory there, we obtain a necessary condition for stability known as the *Mercier criterion* [97, 174, 175]. If we define an arbitrary magnetic surface label ψ and denote the derivatives with respect to ψ of the volume within a surface as V', the poloidal magnetic flux as Ψ', the poloidal current flux as K', the toroidal magnetic flux as Φ', and the toroidal current flux as I', and then define the function (here we adopt the standard convention for this equation that we use *rationalized* MKS units in which $\mu_0 \to 1$)

$$
\begin{aligned}
F &= \frac{1}{4V'^2}\left(\Phi'\Psi'' - \Psi'\Phi''\right)^2 - \frac{1}{V'}\left(\Phi'\Psi'' - \Psi'\Phi''\right)\left\langle\frac{\sigma B^2}{|\nabla\psi|^2}\right\rangle \\
&+ \left[\left\langle\frac{\sigma B^2}{|\nabla\psi|^2}\right\rangle^2 - \left\langle\frac{\sigma^2 B^2}{|\nabla\psi|^2}\right\rangle\left\langle\frac{B^2}{|\nabla\psi|^2}\right\rangle\right] - (p')^2\left\langle\frac{1}{B^2}\right\rangle\left\langle\frac{B^2}{|\nabla\psi|^2}\right\rangle \\
&+ \frac{1}{V'}\left(p'V'' + I'\Psi'' - K'\Phi''\right)\left\langle\frac{B^2}{|\nabla\psi|^2}\right\rangle,
\end{aligned}
\tag{8.134}
$$

the Mercier criterion is then that you must have $F > 0$ on all surfaces for stability. Evaluation of this at the magnetic axis for a circular cross section toroidal plasma (now using MKS units) gives the criterion for stability [176]

$$-\frac{dp}{dr}(1 - q^2) < \frac{r\,B_z^2}{8\,\mu_0}\left(\frac{q'}{q}\right)^2.\tag{8.135}$$

Note now that $q' = dq/dr$. Comparison of this criterion with the Suydam criterion in a cylinder, Eq. (8.84), we see that the left side has picked up a factor $(1 - q^2)$ which will be stabilizing if $q > 1$ on the axis since the pressure is normally decreasing away from the magnetic axis.

Edge Localized Instabilities The ballooning mode formalism presented above has been extended by a number of authors [177, 178] to include higher-order terms in the expansion. The ballooning equation, Eq. (8.133), can be regarded as the lowest-order result in a formal expansion in the inverse toroidal mode number n^{-1}. The next order in this expansion still has the property that two components of the displacement can be eliminated. However, the terms that provide the kink drive, proportional to the radial derivative of the parallel current density, are retained in this order, as are cross-surface derivative terms that determine the global mode structure, and the vacuum region. This is the basis for the ELITE [177] stability code which is able to obtain very high accuracy for modes with high toroidal mode number n that are localized near the plasma edge.

8.6 Summary

The ideal MHD equations play a very important role in the stability analysis of magnetically confined plasmas since the terms retained (from the full magnetohydrodynamic equations) in the ideal MHD description are normally the dominant terms in the equations. The analysis of ideal MHD linear stability is a very mature field. The self-adjointness property of the linear ideal MHD operator (in the absence of equilibrium flow) leads one to a variational formulation. If concerned only with the determination of stability or instability, the energy principle provides a powerful framework in which significant simplifications are possible by minimizing the quadratic form for the energy analytically or quasi-analytically for the surface components of the displacement in terms of the cross-surface component. Further expansions in which multiple spatial scale orderings are applied offer additional important simplifications for important classes of localized modes. We will discuss in the remaining chapters methods for solving some of the sets of equations derived here.

Problems

8.1: Show that Eq. (8.16) is satisfied if we replace the perturbed density, ρ_1, by Eq. (8.21) and the perturbed velocity, $\mathbf{v}_1$, by Eq. (8.20). Let $\partial/\partial t \to -i\omega$ and make use of the equilibrium condition $\nabla \cdot \rho \mathbf{V} = 0$.

8.2: Show that Eq. (8.24) follows from Eq. (8.19) once ρ_1 and $\mathbf{v}_1$ are eliminated using Eqs. (8.21) and (8.20).

8.3: Derive the form for δW_f given in Eq. (8.49), starting from Eq. (8.36).

8.4: Derive the form of the displacement field $\boldsymbol{\xi}$ that minimizes the fast magnetosonic wave term in Eq. (8.49) for a torus with a purely toroidal magnetic field.

8.5: Derive the cylindrical normal mode equations, Eqs. (8.59) and (8.60).

8.6: Derive the forms for the plasma displacement ξ and perturbed pressure P at the origin as given in Eqs. (8.62) and (8.63).

Chapter 9

Numerical Methods for Hyperbolic Equations

9.1 Introduction

In this chapter we begin the study of finite difference methods for solving hyperbolic systems of equations. Some excellent general references for this material are books by Hirsch [179], Leveque [180], and Roache [24]. If we denote by $\mathbf{U}$ the vector of unknowns, then a conservative system of equations can be written (in one dimension) either as

$$\frac{\partial \mathbf{U}}{\partial t} + \frac{\partial \mathbf{F}}{\partial x} = 0, \tag{9.1}$$

or, in the non-conservative form

$$\frac{\partial \mathbf{U}}{\partial t} + \mathbf{A} \cdot \frac{\partial \mathbf{U}}{\partial x} = 0. \tag{9.2}$$

The matrix $\mathbf{A} = \partial \mathbf{F}/\partial \mathbf{U}$ has real eigenvalues for a pure hyperbolic system. Since the stability analysis we perform is linear, the stability properties of the same method as applied to either Eq. (9.1) or Eq. (9.2) will be identical. Systems of linear equations will be of the form Eq. (9.2) from the outset. We begin by examining the numerical stability of several finite difference methods applied to the system Eq. (9.2).

9.2 Explicit Centered-Space Methods

Perhaps the simplest method one could apply to Eq. (9.2) is to finite difference forward in time and centered in space (FTCS). However, this is unstable as can be shown by simple stability analysis. Consider the FTCS method

$$\frac{\mathbf{U}_j^{n+1} - \mathbf{U}_j^n}{\delta t} + \mathbf{A} \cdot \left[\frac{\mathbf{U}_{j+1}^n - \mathbf{U}_{j-1}^n}{2\delta x} \right] = 0. \tag{9.3}$$

We saw in Section 2.5.2 that von Neumann stability analysis corresponds to making the substitution

$$\mathbf{U}_j^n \rightarrow \mathbf{V}_k r^n e^{ij\theta_k} \tag{9.4}$$

and using a transformation where the matrix $\mathbf{A}$ becomes diagonal with the eigenvalues λ_A along the diagonal. Here $\theta_k = k\delta x$, with k being the wavenumber.

Substitution of Eq. (9.4) into Eq. (9.3), making the matrix transformation, and canceling common terms yields

$$r = 1 - i\frac{\delta t}{\delta x}\lambda_A \sin(\theta_k). \tag{9.5}$$

For stability, the magnitude of the amplification factor r must be less than or equal to unity for all values of θ_k. Since the eigenvalues of $\mathbf{A}$ are real for a hyperbolic system, we see that $|r| > 1$ for all combinations of δt and δx, making the FTCS method unconditionally unstable for hyperbolic equations. Note that this is in contrast to what was found in Chapter 7, that FTCS is conditionally stable for parabolic equations.

9.2.1 Lax–Friedrichs Method

A simple modification to the difference method Eq. (9.3) will lead to a stable difference scheme. In the Lax–Friedrichs method [181, 182], the old time value is replaced by the average of its two nearest neighbors,

$$\frac{\mathbf{U}_j^{n+1} - \frac{1}{2}\left(\mathbf{U}_{j+1}^n + \mathbf{U}_{j-1}^n\right)}{\delta t} + \mathbf{A} \cdot \left[\frac{\mathbf{U}_{j+1}^n - \mathbf{U}_{j-1}^n}{2\delta x}\right] = 0. \tag{9.6}$$

Substitution of Eq. (9.4) into Eq. (9.6) and making the matrix transformation yields for the amplification factor

$$r = \cos(\theta_k) - i\frac{\delta t}{\delta x}\lambda_A \sin(\theta_k). \tag{9.7}$$

For stability, we require that $|r| \leq 1$, which implies that the ellipse in the complex plane defined by Eq. (9.7) as θ_k increases from 0 to 2π lie inside the unit circle. Since the eigenvalues λ_A are real, this stability condition implies

$$\delta t \leq \frac{\delta x}{|\lambda_A|}, \tag{9.8}$$

for all eigenvalues of the matrix $\mathbf{A}$. This is just the Courant–Friedrichs–Lewy (CFL) condition [183, 184] that we first saw in Section 2.5.2. It is typical for an explicit difference method applied to a hyperbolic system.

Let us examine the truncation error of the Lax–Friedrichs method,

Eq. (9.6). Expanding in a Taylor's series about $\mathbf{U}_j^n$ we find the leading order truncation error to be

$$T_\Delta = -\frac{1}{2}\delta t \frac{\partial^2 \mathbf{U}}{\partial t^2} + \frac{1}{2}\frac{\delta x^2}{\delta t}\frac{\partial^2 \mathbf{U}}{\partial x^2} + \mathcal{O}\left(\delta x^2, \delta t^2\right). \tag{9.9}$$

It is important to notice that, as $\delta t \to 0$ for fixed δx, the truncation error becomes unbounded and so, in contrast to most methods, one will not obtain a more accurate solution simply by reducing the time step.

We can apply the stability analysis of Hirt [28] to this method by using the differential equation, Eq. (9.2), to eliminate time derivatives in Eq. (9.9) in favor of spatial derivatives. This converts the leading order truncation error, Eq. (9.9), into a diffusion-like term,

$$T_\Delta = -\frac{1}{2}\left[\delta t \mathbf{A}^2 - \frac{\delta x^2}{\delta t}\mathbf{I}\right]\cdot\frac{\partial^2 \mathbf{U}}{\partial x^2} + \mathcal{O}\left(\delta x^2, \delta t^2\right).$$

The requirement that the eigenvalues of the effective diffusion matrix be positive gives back the stability condition Eq. (9.8).

The Lax–Friedrichs method is mostly of historical importance because of the low-order truncation error, and of the accuracy problem in the limit as $\delta t \to 0$. However, because it is easily programmed and dependable, it is often used to advantage in the early stages of the development of a program after which it is replaced by a higher-order more complex method.

9.2.2 Lax–Wendroff Methods

One of the most popular methods for solving a hyperbolic system of equations, especially the compressible fluid equations, is the Lax–Wendroff method [185]. The two-step version, as applied to Eq. (9.1), is as follows:

$$\mathbf{U}_{j+\frac{1}{2}}^{n+\frac{1}{2}} = \frac{1}{2}\left[\mathbf{U}_{j+1}^n + \mathbf{U}_j^n\right] - \frac{\delta t}{2\delta x}\left[\mathbf{F}_{j+1}^n - \mathbf{F}_j^n\right], \tag{9.10}$$

$$\mathbf{U}_j^{n+1} = \mathbf{U}_j^n - \frac{\delta t}{\delta x}\left[\mathbf{F}_{j+\frac{1}{2}}^{n+\frac{1}{2}} - \mathbf{F}_{j-\frac{1}{2}}^{n+\frac{1}{2}}\right]. \tag{9.11}$$

The first step, Eq. (9.10), which is seen to be equivalent to the Lax–Friedrichs method, Eq. (9.6), applied to a single zone spacing, defines provisional values which are used in the second step. If we linearize the equation so that $\mathbf{F} = \mathbf{A}\cdot\mathbf{U}$, with $\mathbf{A}$ a constant, then Eqs. (9.10) and (9.11) can be combined to give the single step difference equation

$$\begin{aligned}
\mathbf{U}_j^{n+1} = {} & \mathbf{U}_j^n - \frac{\delta t}{2\delta x}\mathbf{A}\cdot\left[\mathbf{U}_{j+1}^n - \mathbf{U}_{j-1}^n\right] \\
& + \frac{1}{2}\left(\frac{\delta t}{\delta x}\right)^2 \mathbf{A}^2\cdot\left[\mathbf{U}_{j+1}^n - 2\mathbf{U}_j^n + \mathbf{U}_{j-1}^n\right],
\end{aligned} \tag{9.12}$$

or, in a conservative form:

$$\mathbf{U}_j^{n+1} = \mathbf{U}_j^n - \frac{\delta t}{2\delta x}\left[\mathbf{F}_{j+1}^n - \mathbf{F}_{j-1}^n\right] \tag{9.13}$$

$$+ \quad \frac{1}{2}\left(\frac{\delta t}{\delta x}\right)^2\left[\mathbf{A}_{j+\frac{1}{2}}^n \cdot \left(\mathbf{F}_{j+1}^n - \mathbf{F}_j^n\right) - \mathbf{A}_{j-\frac{1}{2}}^n \cdot \left(\mathbf{F}_j^n - \mathbf{F}_{j-1}^n\right)\right].$$

Von Neumann stability analysis of Eq. (9.12) leads to the amplification factor

$$r = 1 - i\lambda_A \frac{\delta t}{\delta x}\sin\theta_k + \left(\lambda_A\frac{\delta t}{\delta x}\right)^2\left[\cos\theta_k - 1\right]. \tag{9.14}$$

This is seen to be stable with $|r| \leq 1$ for $\lambda_A \delta t/\delta x \leq 1$. The amplification factor r in Eq. (9.14) can be written

$$|r|^2 = 1 - 4s^2\left(1 - s^2\right)\sin^4\frac{1}{2}\theta_k,$$

where $s = \delta t\lambda_A/\delta x$ so that the difference scheme is dissipative of fourth order in the long-wavelength limit of small θ_k [181]. The truncation error is seen to be second-order in both space and time, $\mathcal{O}(\delta x^2, \delta t^2)$.

To gain some insight into the origin of Eqs. (9.12) and (9.13), let us start with a Taylor series expansion in time up to third order so as to achieve second-order accuracy:

$$\mathbf{U}_j^{n+1} = \mathbf{U}_j^n + \delta t\frac{\partial\mathbf{U}}{\partial t}\bigg|_j + \frac{1}{2}\delta t^2\frac{\partial^2\mathbf{U}}{\partial t^2}\bigg|_j + \mathcal{O}(\delta t^3).$$

Now, make the substitution from the PDE, Eq. (9.2), $\frac{\partial}{\partial t} \rightarrow -\mathbf{A}\cdot\frac{\partial}{\partial x}$ to obtain up to order δt^2:

$$\mathbf{U}_j^{n+1} = \mathbf{U}_j^n - \delta t\mathbf{A}\cdot\frac{\partial\mathbf{U}}{\partial x}\bigg|_j + \frac{1}{2}\delta t^2\mathbf{A}^2\cdot\frac{\partial^2\mathbf{U}}{\partial x^2}\bigg|_j. \tag{9.15}$$

When centered spatial finite difference operators replace the spatial derivatives in Eq. (9.15), we obtain Eq. (9.12). We see that the second-order spatial difference term in brackets is just that required to provide second-order accuracy in time.

9.2.3   MacCormack Differencing

An interesting variation on the two step Lax–Wendroff method which has been used for aerodynamic calculations is known as *MacCormack differencing*. MacCormack's method [186] applied to Eq. (9.1) is as follows:

$$\mathbf{U}_j^{\overline{n+1}} = \mathbf{U}_j^n - \frac{\delta t}{\delta x}\left[\mathbf{F}_{j+1}^n - \mathbf{F}_j^n\right],$$

$$\mathbf{U}_j^{n+1} = \frac{1}{2}\left\{\mathbf{U}_j^n + \mathbf{U}_j^{\overline{n+1}} - \frac{\delta t}{\delta x}\left[\mathbf{F}_j^{\overline{n+1}} - \mathbf{F}_{j-1}^{\overline{n+1}}\right]\right\}. \tag{9.16}$$

A provisional value of the solution is computed at the new time using a one-sided spatial difference, and that provisional value is used to compute a one-sided spatial difference in the opposite direction and is also averaged into the final solution. This can also be shown to be second-order accurate in space and time and, in fact, to be equivalent to the Lax–Wendroff method if $\mathbf{A}$ is a constant matrix.

9.2.4 Leapfrog Method

The leapfrog method applied to Eq. (9.1) can be written as follows:

$$\mathbf{U}_j^{n+1} = \mathbf{U}_j^{n-1} - \frac{\delta t}{\delta x}\left[\mathbf{F}_{j+1}^n - \mathbf{F}_{j-1}^n\right]. \tag{9.17}$$

Note that this involves three time levels, and is centered in both time and space, making its truncation error $\mathcal{O}\left(\delta t^2, \delta x^2\right)$. Letting $\mathbf{F} = \mathbf{A}\cdot\mathbf{U}$, we find for the amplification factor

$$r = is\sin\theta_k \pm \left[1 - s^2\sin^2\theta_k\right]^{1/2}, \tag{9.18}$$

where $s = \lambda_A\delta t/\delta x$. The method has $|r| = 1$ identically for $s \leq 1$ and is therefore non-dissipative. It is unstable for $s > 1$.

We have seen in Problem 1 of Chapter 7 that the leapfrog method is unstable when applied to diffusion equations. However, the DuFort–Frankel method combines naturally with the leapfrog method when dissipative terms are present. For example, the mixed type equation

$$\frac{\partial\phi}{\partial t} + u\frac{\partial\phi}{\partial x} = D\frac{\partial^2\phi}{\partial x^2} \tag{9.19}$$

would be differenced using the combined leapfrog/DuFort–Frankel method as

$$\frac{\phi_j^{n+1} - \phi_j^{n-1}}{2\delta t} + u\left[\frac{\phi_{j+1}^n - \phi_{j-1}^n}{2\delta x}\right] = D\left[\frac{\phi_{j+1}^n + \phi_{j-1}^n - \phi_j^{n+1} - \phi_j^{n-1}}{\delta x^2}\right]. \tag{9.20}$$

Since ϕ_j^{n+1} is the only new time value in Eq. (9.20), it can be solved for algebraically without matrix inversion.

The combined method, Eq. (9.20), has many advantages and is used often, but it has some shortcomings. One is that the grid points at even and odd values of time and space indices $n + j$ are completely decoupled, as shown in Figure 9.1, and can drift apart after many time cycles. Using an FTCS method instead for the diffusive term will couple the grids together. This implies evaluating the diffusion term at the $n - 1$ time level:

$$\frac{\phi_j^{n+1} - \phi_j^{n-1}}{2\delta t} + u\left[\frac{\phi_{j+1}^n - \phi_{j-1}^n}{2\delta x}\right] = D\left[\frac{\phi_{j+1}^{n-1} - 2\phi_j^{n-1} + \phi_{j-1}^{n-1}}{\delta x^2}\right]. \tag{9.21}$$

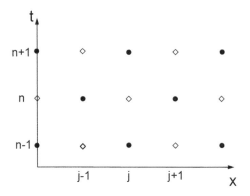

FIGURE 9.1: In the combined DuFort–Frankel/leapfrog method, the space-time points at odd and even values of $n + j$ are completely decoupled.

If the time step restriction from the explicit diffusion term in Eq. (9.21), $\delta t \leq \delta x^2 / 4D$, is too severe, one can take a weighted combination of the right sides of Eqs. (9.20) and (9.21) where the diffusion term D in the first is multiplied by some parameter $(1 - \alpha)$ and the D in the second by α. The value of α can be chosen small enough so that the associated time step restriction $\delta t < \delta x^2 / 4\alpha D$ does not limit the time step.

Often, a slight variation of the leapfrog method involving a staggering in space and/or in time offers significant advantages in removing the decoupled grid problem, or in incorporating boundary conditions more easily. For example, consider the wave equation

$$\frac{\partial^2 \phi}{\partial t^2} = c^2 \frac{\partial^2 \phi}{\partial x^2}, \tag{9.22}$$

where the wave propagation velocity c is a constant. This can be written in terms of the variables $v = \partial \phi / \partial t$ and $\omega = c \partial \phi / \partial x$ as

$$\frac{\partial v}{\partial t} = c \frac{\partial \omega}{\partial x},$$
$$\frac{\partial \omega}{\partial t} = c \frac{\partial v}{\partial x}. \tag{9.23}$$

If we adopt the convention, as shown in Figure 9.2, that v is defined at half-integer time points and integer space points, and ω is defined at integer time and half-integer space points as shown, then the leapfrog method applied to Eq. (9.22) becomes

$$v_j^{n+\frac{1}{2}} = v_j^{n-\frac{1}{2}} + \frac{c\delta t}{\delta x}\left(\omega_{j+\frac{1}{2}}^n - \omega_{j-\frac{1}{2}}^n\right), \tag{9.24}$$

$$\omega_{j+\frac{1}{2}}^{n+1} = \omega_{j+\frac{1}{2}}^n + \frac{c\delta t}{\delta x}\left(v_{j+1}^{n+\frac{1}{2}} - v_j^{n+\frac{1}{2}}\right). \tag{9.25}$$

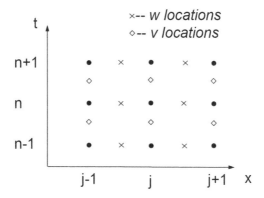

FIGURE 9.2: Staggering variables in space and time can remove the decoupled grid problem from leapfrog.

Stability analysis of the system given by Eqs. (9.24)–(9.25) again shows that the amplification factor is given by

$$r^2 - (2 - 4a^2)r + 1 = 0, \qquad (9.26)$$

where $a = c\delta t/\delta x \sin(\theta_k/2)$. Both roots of Eq. (9.26) have absolute value unity for $a^2 \leq 1$, which implies $c\delta t/\delta x \leq 1$ for stability.

9.2.5   Trapezoidal Leapfrog

A popular method used to overcome the grid decoupling problem with leapfrog is the leapfrog-trapezoidal method [187]. Instead of Eq. (9.17), the two-step finite difference approximation to Eq. (9.1) is

$$
\begin{aligned}
\mathbf{U}_j^{n+\frac{1}{2}} &= \frac{1}{2}\left(\mathbf{U}_j^{n-1} + \mathbf{U}_j^n\right) - \frac{\delta t}{\delta x}\left[\mathbf{F}_{j+\frac{1}{2}}^n - \mathbf{F}_{j-\frac{1}{2}}^n\right], \\
\mathbf{U}_j^{n+1} &= \mathbf{U}_j^n - \frac{\delta t}{\delta x}\left[\mathbf{F}_{j+\frac{1}{2}}^{n+\frac{1}{2}} - \mathbf{F}_{j-\frac{1}{2}}^{n+\frac{1}{2}}\right].
\end{aligned}
\qquad (9.27)
$$

Here, $\mathbf{F}_{j+\frac{1}{2}} = (1/2)(\mathbf{F}_j + \mathbf{F}_{j+1})$, etc. The first step, called the trapezoidal step, effectively couples the two leapfrog grids by strongly damping any associated computational mode. Time and space centering in both steps guarantees second-order accuracy in both time and space. Variants of this that require spatial differencing over only a single zone spacing can also be applied by staggering the spatial locations of the different variables.

Von Neumann stability analysis of this difference equation is complicated by the need to solve a quadratic equation with complex coefficients, but it can readily be verified numerically that the algorithm given by Eq. (9.27) is stable for $s = \lambda_A \delta t / \delta x < \sqrt{2}$. If explicit diffusive terms are to be added to Eq. (9.27), they should be evaluated at the $(n-1)$ time level on the first step and the n time level on the second step so they would effectively be FTCS for each step.

9.3   Explicit Upwind Differencing

Suppose we are concerned with solving a simple one-dimensional conservation equation such as the continuity equation,

$$\frac{\partial \rho}{\partial t} + \frac{\partial}{\partial x}(\rho u) = 0. \tag{9.28}$$

This is of the form of Eq. (9.1) if we identify $\mathbf{U} = \rho$, $\mathbf{F} = \rho u$, $\mathbf{A} = u$. An explicit, two-time-level method which is conditionally stable involves the use of one-sided, rather than space-centered, differencing. Backward spatial differences are used if the velocity u is positive, and forward spatial differences are used if u is negative.

The basic method is then

$$\frac{\rho_j^{n+1} - \rho_j^n}{\delta t} = -\left[\frac{(\rho u)_j^n - (\rho u)_{j-1}^n}{\delta x}\right] \text{ for } u > 0$$

$$= -\left[\frac{(\rho u)_{j+1}^n - (\rho u)_j^n}{\delta x}\right] \text{ for } u < 0. \tag{9.29}$$

This method is known as "upwind," "upstream," or "donor cell" differencing. It has a truncation error $T_\Delta = \mathcal{O}(\delta t, \delta x)$, and is stable for $\delta t \leq \delta x / u$. It is instructive to examine the leading order truncation error for the constant u case. Expanding in Taylor series and using Eq. (9.28) to eliminate time derivatives in favor of space derivatives gives for the truncation error

$$T_\Delta = \left(\frac{1}{2}|u|\delta x - \frac{1}{2}u^2\delta t\right)\frac{\partial^2 \rho}{\partial x^2} + \mathcal{O}\left(\delta x^2, \delta t^2\right). \tag{9.30}$$

We see that the upwind difference method introduces an artificial (or numerical) diffusion term which makes the method stable if $\delta t \leq \delta x / u$ so that the effective diffusion coefficient in brackets is positive.

To gain further insight as to why the upwind method works and is robustly stable, let us apply it to the simplified case where the velocity u is constant and positive, $u > 0$. Eq. (9.29) then reduces to

$$\rho_j^{n+1} = s\rho_{j-1}^n + (1-s)\rho_j^n, \tag{9.31}$$

where $s \equiv u\delta t / \delta x$. This is illustrated schematically in Figure 9.3. We see that

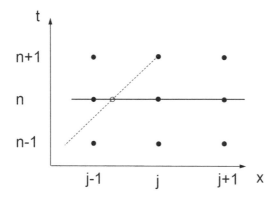

FIGURE 9.3: The upwind method corresponds to linear interpolation for the point where the characteristic curve intersects time level n as long as $s \equiv u\delta t/\delta x \leq 1$

for $s \leq 1$, the new value ρ_j^{n+1} is obtained by following the characteristic curve back to the previous time level n and using linear interpolation to evaluate the intersection point. This also implies that there will be no new minima or maxima since the interpolated value will aways be bounded by the values at the two endpoints. This guarantees stability of the upwind method for $s \leq 1$.

As a generalization of Eq. (9.31), we see that it can be written in the form

$$\rho_j^{n+1} = \rho_j^n + \sum_{j \neq k} C_{j,k} \left(\rho_k^n - \rho_j^n \right), \tag{9.32}$$

where the coefficients are all non-negative and are zero except for nearest neighbors, i.e.,

$$C_{j,k} = \begin{cases} \geq 0 & k = j \pm 1 \\ = 0 & \text{else} \end{cases} \tag{9.33}$$

If ρ_j^n is a local maximum, then $\left(\rho_k^n - \rho_j^n \right) \leq 0$, and so $\rho_j^{n+1} \leq \rho_j^n$. Conversely, if ρ_j^n is a local minimum, then $\left(\rho_k^n - \rho_j^n \right) \geq 0$, and so $\rho_j^{n+1} \geq \rho_j^n$. The scheme is therefore *local extremum diminishing*. It is clear that a method built on these principles can readily be extended to multiple dimensions.

The difference method described by Eq. (9.29) is conservative if the velocity u doesn't change sign in the computational domain. If it does change sign, there will be a non-conservative error which introduces an artificial source or sink of mass, although it will be proportional to the local u which would be small since it is passing through zero near that point.

There is a variant of the upwind differencing method which is fully conservative. We first define average interface velocities on each of the mesh points.

The sign of these velocities determines, by upwind differencing, which cell values of ρ to use. The method can be expressed as

$$\frac{\rho_j^{n+1} - \rho_j^n}{\delta t} = -\left[\frac{u_{j+\frac{1}{2}}^n \rho_{j+\frac{1}{2}}^n - u_{j-\frac{1}{2}}^n \rho_{j-\frac{1}{2}}^n}{\delta x}\right], \qquad (9.34)$$

where

$$u_{j+\frac{1}{2}} = \frac{1}{2}\left[u_{j+1} + u_j\right],$$
$$u_{j-\frac{1}{2}} = \frac{1}{2}\left[u_j + u_{j-1}\right]. \qquad (9.35)$$

Alternatively, because of the staggered locations, the $u_{j+\frac{1}{2}}$ may be the fundamental variables. In either case, we would define the ρ values as follows:

$$
\begin{aligned}
\rho_{j+\frac{1}{2}} &= \rho_j & \text{for } u_{j+\frac{1}{2}} > 0, \\
&= \rho_{j+1} & \text{for } u_{j+\frac{1}{2}} < 0, \\
\rho_{j-\frac{1}{2}} &= \rho_{j-1} & \text{for } u_{j-\frac{1}{2}} > 0, \\
&= \rho_j & \text{for } u_{j-\frac{1}{2}} < 0.
\end{aligned}
$$

Because the method of Eq. (9.34) is conservative and also takes centered derivatives of u, it is generally preferred to that of Eq. (9.29).

9.3.1 Beam–Warming Upwind Method

We can use the analysis of Eq. (9.15), introduced in conjunction with the Lax–Wendroff method, to construct a second-order accurate upwind method. Suppose all the eigenvalues of $\mathbf{A}$ are positive. In place of the centered difference formulas used in Eq. (9.12), we could use the one-sided difference formulas:

$$
\begin{aligned}
\frac{\partial \mathbf{U}}{\partial x} &= \frac{1}{2\delta x}\left(3\mathbf{U}_j - 4\mathbf{U}_{j-1} + \mathbf{U}_{j-2}\right) + \mathcal{O}(\delta x^2), \\
\frac{\partial^2 \mathbf{U}}{\partial x^2} &= \frac{1}{\delta x^2}\left(\mathbf{U}_j - 2\mathbf{U}_{j-1} + \mathbf{U}_{j-2}\right) + \mathcal{O}(\delta x).
\end{aligned}
$$

Substitution of these one-sided difference operators into Eq. (9.15) gives the second-order accurate upwind method known as the *Beam–Warming method* [188],

$$
\begin{aligned}
\mathbf{U}_j^{n+1} &= \mathbf{U}_j^n - \frac{\delta t}{2\delta x}\mathbf{A} \cdot \left(3\mathbf{U}_j^n - 4\mathbf{U}_{j-1}^n + \mathbf{U}_{j-2}^n\right) \\
&+ \frac{1}{2}\left(\frac{\delta t}{\delta x}\right)^2 \mathbf{A}^2 \cdot \left(\mathbf{U}_j^n - 2\mathbf{U}_{j-1}^n + \mathbf{U}_{j-2}^n\right). \qquad (9.36)
\end{aligned}
$$

This can also be expressed in conservative form as

$$\mathbf{U}_j^{n+1} = \mathbf{U}_j^n - \frac{\delta t}{\delta x}\left[\mathbf{F}_{j+\frac{1}{2}}^n - \mathbf{F}_{j-\frac{1}{2}}^n\right], \qquad (9.37)$$

where the fluxes are defined as

$$\mathbf{F}^n_{j+\frac{1}{2}} = \mathbf{F}^n_j + \frac{1}{2}\left(\mathbf{I} - \frac{\delta t}{\delta x}\mathbf{A}_{j-\frac{1}{2}}\right)\left[\mathbf{F}^n_j - \mathbf{F}^n_{j-1}\right]. \tag{9.38}$$

9.3.2 Upwind Methods for Systems of Equations

The simple upwind method described in the previous sections will normally not work for systems of equations with multiple characteristic velocities which could be of either sign. However, there are extensions of this method that form the basis for many modern methods of exceptional stability and accuracy. The basic idea of these methods is to locally decompose the solution into right-going and left-going subproblems, and then to use an appropriate upwind method for each of the subproblems. This idea goes back to a paper by Courant, Isaacson, and Rees [189] and is sometimes called the CIR method.

As an example, consider the linear system of m equations given by Eq. (9.2):

$$\frac{\partial \mathbf{U}}{\partial t} + \mathbf{A} \cdot \frac{\partial \mathbf{U}}{\partial x} = 0, \tag{9.39}$$

where $\mathbf{U}$ is an m vector and $\mathbf{A}$ is an $m \times m$ matrix.

We first construct two diagonal matrices from $\mathbf{A}$, each of which has either the positive, $\mathbf{D}^+$, or negative, $\mathbf{D}^-$, eigenvalues of $\mathbf{A}$ along the diagonal. Let the p-th eigenvalue of the matrix $\mathbf{A}$ be denoted λ_p. We introduce the compact notation

$$\lambda_p^+ = \max(\lambda_p, 0), \quad \mathbf{D}^+ = \mathrm{diag}(\lambda_1^+, \cdots, \lambda_m^+), \tag{9.40}$$

$$\lambda_p^- = \min(\lambda_p, 0), \quad \mathbf{D}^- = \mathrm{diag}(\lambda_1^-, \cdots, \lambda_m^-). \tag{9.41}$$

Note that $\mathbf{D}^+ + \mathbf{D}^- = \mathbf{D}$, where $\mathbf{D} = \mathbf{T} \cdot \mathbf{A} \cdot \mathbf{T}^{-1}$ is the diagonal matrix with the same eigenvalues as $\mathbf{A}$. In terms of the vector $\mathbf{V} = \mathbf{T} \cdot \mathbf{U}$, the upwind method can be written

$$\mathbf{V}^{n+1}_j = \mathbf{V}^n_j - \frac{\delta t}{\delta x}\mathbf{D}^+ \cdot (\mathbf{V}^n_j - \mathbf{V}^n_{j-1}) - \frac{\delta t}{\delta x}\mathbf{D}^- \cdot (\mathbf{V}^n_{j+1} - \mathbf{V}^n_j). \tag{9.42}$$

Note that for the ideal MHD equations, the unknown vector $\mathbf{U}$ can be thought of as the sum of eight scalar amplitudes times the eight orthonormal eigenvectors found in Eq. (1.95),

$$\mathbf{U} = \sum_{i=1}^{8} \alpha_i \mathbf{S}_i. \tag{9.43}$$

The matrix $\mathbf{T}$ is constructed by making each row one of the eigenvectors. Then, using the orthonormal property, we see that the vector $\mathbf{V}$ just contains the eight amplitudes and that the inverse matrix $\mathbf{T}^{-1}$ is just the transpose of $\mathbf{T}$.

We can transform Eq. (9.42) back to the physical $\mathbf{U}$ variables by multiplying on the left by $\mathbf{T}^{-1}$,

$$\mathbf{U}_j^{n+1} = \mathbf{U}_j^n - \frac{\delta t}{\delta x}\mathbf{A}^+ \cdot (\mathbf{U}_j^n - \mathbf{U}_{j-1}^n) - \frac{\delta t}{\delta x}\mathbf{A}^- \cdot (\mathbf{U}_{j+1}^n - \mathbf{U}_j^n), \qquad (9.44)$$

where

$$\mathbf{A}^+ = \mathbf{T}^{-1} \cdot \mathbf{D}^+ \cdot \mathbf{T}, \qquad \mathbf{A}^- = \mathbf{T}^{-1} \cdot \mathbf{D}^- \cdot \mathbf{T}. \qquad (9.45)$$

Note that $\mathbf{A} = \mathbf{A}^+ + \mathbf{A}^-$, and if all of the eigenvalues have the same sign, then either $\mathbf{A}^+$ or $\mathbf{A}^-$ is zero, and the method given by Eq. (9.44) reduces to a standard upwind method.

There have been several efforts to extend this method to the conservative form, Eq. (9.1). Godunov [190] is credited with being the first to do so for a nonlinear system of equations. His approach was to treat each computational zone as if it had uniform properties for the conserved quantities, and thus "discontinuities" would exist at the zone boundaries. An initial value problem with discontinuities such as this is known as a *Riemann problem*. . The method consists of solving the Riemann problem for a small time, δt, evolving each eigenmode according to its characteristic velocity as in the CIR method, and then reconstructing the conserved variables in each zone by averaging the time-evolved variables over each zone volume. When applied to a linear problem, it yields just the CIR method discussed above.

In general, this Riemann problem cannot be solved exactly over the time interval δt and some approximations need to be made. Consider the space-centered conservative differencing form in one dimension:

$$\mathbf{U}_j^{n+1} = \mathbf{U}_j^n - \frac{1}{\delta x}\left[\bar{\mathbf{F}}_{j+\frac{1}{2}}^n - \bar{\mathbf{F}}_{j-\frac{1}{2}}^n\right]. \qquad (9.46)$$

Let us define the upwind differenced fluxes as

$$\bar{\mathbf{F}}_{j+\frac{1}{2}}^n \equiv \frac{1}{2}(\mathbf{F}_j + \mathbf{F}_{j+1}) - \mathbf{H}_{j+\frac{1}{2}},$$

where

$$\mathbf{H}_{j+\frac{1}{2}} = \frac{1}{2}\left[(\mathbf{F}_{j+1}^+ - \mathbf{F}_j^+) - (\mathbf{F}_{j+1}^- - \mathbf{F}_j^-)\right].$$

Here, the signed fluxes $\mathbf{F}_j^+$ and $\mathbf{F}_j^-$ have the property that

$$\mathbf{F}_j = \mathbf{F}_j^+ + \mathbf{F}_j^-, \qquad (9.47)$$

and the corresponding eigenvalues of $\partial\mathbf{F}^+/\partial\mathbf{U}$ are all positive or zero, and those of $\partial\mathbf{F}^-/\partial\mathbf{U}$ are all negative or zero. This is seen to lead to a conservative analogue of Eq. (9.44).

There still remains a question of how $\mathbf{F}_j^+$ and $\mathbf{F}_j^-$ are to be evaluated.

We saw in Eq. (9.45) how $\mathbf{A} = \partial \mathbf{F}/\partial \mathbf{U}$ can be split in the linearized form. However, for the conservative form, we need to find matrices $\bar{\mathbf{A}}^+$ and $\bar{\mathbf{A}}^-$ such that, in addition to Eq. (9.47), the fluxes $\mathbf{F}_j^+$ and $\mathbf{F}_j^-$ satisfy

$$
\begin{aligned}
\mathbf{F}_j^+ - \mathbf{F}_{j-1}^+ &= \bar{\mathbf{A}}^+ \cdot (\mathbf{U}_j - \mathbf{U}_{j-1}), \\
\mathbf{F}_{j+1}^- - \mathbf{F}_j^- &= \bar{\mathbf{A}}^- \cdot (\mathbf{U}_{j+1} - \mathbf{U}_j),
\end{aligned}
$$

For the Euler equations, these matrices were derived by Roe by showing that there is a unique interface state (the Roe-averaged state) that has certain desired properties [191]. For the MHD equations, several authors have shown that by applying suitable averaging techniques, acceptable solutions can be found [192, 193, 194]. Various limiter schemes, as described in the next section, have also been combined with the method presented here to improve the accuracy away from discontinuities [195, 196].

9.4 Limiter Methods

Second- or higher-order accurate methods such as Lax–Wendroff or Beam–Warming give much better accuracy on smooth solutions than does the simple upwind method, but they can fail near discontinuities where oscillations are generated. This is because the Taylor series expansion upon which these methods are based breaks down when the solution is not smooth.

The first-order upwind method has the advantage of keeping the solution monotonic in regions where it should be and not producing spurious maxima or minima, but the accuracy is not very good. The idea of high-accuracy limiter methods is to combine the best features of both the high-order and the upwind methods. They try to obtain second-order accuracy away from discontinuities, but also to avoid oscillations in regions where the solution is not smoothly varying.

To better understand some of the concepts involved, let us apply the Lax–Wendroff method, Eq. (9.12), to the simple linear scalar equation

$$
\frac{\partial u}{\partial t} + a \frac{\partial u}{\partial x} = 0, \tag{9.48}
$$

where a is a constant, assumed here to be positive. It is readily verified that the Lax–Wendroff method applied to this equation can be written as

$$
u_j^{n+1} = u_j^n - s \left(u_j^n - u_{j-1}^n \right) - \frac{1}{2} s (1-s) \left(u_{j+1}^n - 2u_j^n + u_{j-1}^n \right), \tag{9.49}
$$

where $s = a \delta t / \delta x$.

The first term in brackets in Eq. (9.49) is seen to be just upwind differencing, while the second term can be considered to be a diffusive correction.

However, note that since the CFL condition requires $s < 1$, the second term will have a negative coefficient which corresponds to *anti-diffusion*. The way to interpret this is that the upwind method has so much intrinsic *numerical diffusion* associated with it that the Lax–Wendroff method needs to subtract off some of the diffusion term in order to get a higher-accuracy method.

The idea of the limiter methods is to modify the last term in Eq. (9.49) in a way that modifies this anti-diffusion term so that it does not produce undesirable results in regions with sharp gradients. This idea goes back to the 1970s with the *flux corrected transport (FCT)* method [187, 197], the *hybrid method* [198], and others [199]. An excellent discussion of limiter methods can be found in Leveque [180].

The most common methods can be described by writing Eq. (9.49) in the form:

If $a > 0$:

$$u_j^{n+1} = u_j^n - s\left(u_j^n - u_{j-1}^n\right)$$
$$- \frac{1}{2}s(1-s)\left[\phi\left(\theta_{j+\frac{1}{2}}^n\right)\left(u_{j+1}^n - u_j^n\right) - \phi\left(\theta_{j-\frac{1}{2}}^n\right)\left(u_j^n - u_{j-1}^n\right)\right].$$

Else, if $a < 0$:

$$u_j^{n+1} = u_j^n - s\left(u_{j+1}^n - u_j^n\right)$$
$$+ \frac{1}{2}s(1+s)\left[\phi\left(\theta_{j+\frac{1}{2}}^n\right)\left(u_{j+1}^n - u_j^n\right) - \phi\left(\theta_{j-\frac{1}{2}}^n\right)\left(u_j^n - u_{j-1}^n\right)\right].$$

The argument of the limiter function is defined as follows:

$$\theta_{j-\frac{1}{2}}^n = \begin{cases} \dfrac{u_{j-1}^n - u_{j-2}^n}{u_j - u_{j-1}} & \text{if } a > 0, \\[2ex] \dfrac{u_{j+1}^n - u_j^n}{u_j - u_{j-1}} & \text{if } a < 0. \end{cases} \tag{9.50}$$

The different methods are defined by their choice for the limiter function $\phi(\theta)$. (Note that the upwind method is recovered if we let $\phi(\theta) = 0$, the Lax–Wendroff method is recovered if we let $\phi(\theta) = 1$ and the Beam–Warming method is recovered for $\phi(\theta) = \theta$.)

Four popular high-resolution limiter methods can be expressed as:

$$\begin{aligned} \text{minmod [180]} \quad &: \quad \phi(\theta) = \text{minmod}(1,\theta), \\ \text{superbee [200]} \quad &: \quad \phi(\theta) = \max\left(0, \min(1, 2\theta), \min(2, \theta)\right), \\ \text{MC [180]} \quad &: \quad \phi(\theta) = \max\left(0, \min\left((1+\theta)/2, 2, 2\theta\right)\right), \\ \text{van Leer [201]} \quad &: \quad \phi(\theta) = \frac{\theta + |\theta|}{1 + |\theta|} \end{aligned}$$

Here the minmod function of two arguments is defined as

$$\text{minmod}(a, b) = \begin{cases} a & \text{if } |a| < |b| \text{ and } ab > 0, \\ b & \text{if } |b| < |a| \text{ and } ab > 0, \\ 0 & \text{if } ab \leq 0. \end{cases} \tag{9.51}$$

If a and b have the same sign, this selects the one that is smaller in modulus, otherwise it returns zero.

These prescriptions can all be shown to have the property that they are *total variation diminishing (TVD)* [202]. If we define the *total variation* of a grid function u_j by

$$\text{TV}(u) \equiv \sum_{j=-\infty}^{\infty} |u_j - u_{j-1}|, \tag{9.52}$$

then the TVD property is that

$$\text{TV}(u^{n+1}) \leq \text{TV}(u^n). \tag{9.53}$$

It can be shown that any TVD method is also *monotonicity preserving*, that is, if $u_j^n \geq u_{j+1}^n$ for all j, then this implies that $u_j^{n+1} \geq u_{j+1}^{n+1}$ for all j.

The limiting process described above is, of course, complicated by the fact that a hyperbolic system of equations such as described by Eq. (9.2) actually consists of a superposition of more than one type of wave. In this case the correction term must be decomposed into eigencomponents as discussed in Section 9.3.2 and each scalar eigencoefficient must be limited separately, based on the algorithm for scalar advection. A more complete description of the implementation of flux-limiter methods to systems of equations, as well as some subtleties involving non-linear systems, can be found in Leveque [180].

9.5 Implicit Methods

If all the eigenvalues of the matrix $\mathbf{A}$, λ_A, are close together in magnitude, it is generally not worthwhile to use an implicit method for a purely hyperbolic problem. The reasons for this are as follows: It requires more computational work per time step using an implicit method compared to that required for an explicit method. This difference becomes greater as the rank of the matrix $\mathbf{A}$ increases or especially when working in two or higher dimensions. Also, to take advantage of the unconditional stability of the implicit method, we would need to significantly violate the CFL condition so that $\delta t \gg \delta x / |\lambda_A|$. Since the truncation error is normally of the same order in both δt and δx, this would imply that the $\mathcal{O}(\delta t^2)$ term would dominate in the truncation error and thus solutions using larger time steps would be less accurate than those with smaller time steps. This is in contrast to a Crank–Nicolson implicit solution of a diffusion equation where we saw in Section 7.2.1 that the time step can be increased from $\delta t \sim \delta x^2$ to $\delta t \sim \delta x$ without substantially increasing the overall truncation error.

Another possible disadvantage of implicit methods applied to hyperbolic

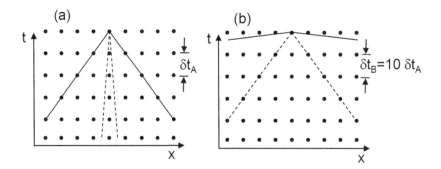

FIGURE 9.4: Characteristics for the Alfvén wave (dotted) and the fast magnetoacoustic wave (solid) are shown in this schematic diagram. (a) Space-time plot for fully explicit method. (b) Same for partially implicit method where the time step is at CFL limit for the Alfvén wave.

equations is that they result in an infinite signal propagation speed. Physically, all information should be propagated along the characteristics, which corresponds to a finite velocity of propagation. An explicit finite difference method normally has a domain of dependence that very closely resembles the domain of dependence of the physical system, while the domain of dependence of the implicit method is totally different.

Explicit finite difference methods are also ideally suited to pure hyperbolic problems with with a wide range of characteristic velocities but where interest lies in phenomena associated with the fastest wave speed (or equivalently, the largest eigenvalue λ_A). However, in a strongly magnetized plasma we normally have the situation where there are multiple characteristic velocities that differ widely, and where primary interest lies not with the fastest wave speed (which is the fast magnetoacoustic wave which compresses the strong background field), but rather with the slower Alfvén wave, which is nearly incompressible [203]. The situation is then as depicted in Figure 9.4, where we draw the characteristics associated with the fast wave as a solid line, and those associated with the slower shear Alfvén wave with a dotted line.

Figure 9.4a illustrates the space-time points in an explicit calculation where the time step is determined by the CFL condition for the fast wave. It is seen that it takes many time steps for the shear Alfvén wave to traverse one zone spacing. In contrast, Figure 9.4b shows the same characteristics but with a time step chosen so that the CFL condition is just satisfied for the shear Alfvén wave. It is seen that the CFL condition for the fast wave is strongly violated and so an implicit treatment is needed, at least for the eigencomponents associated with the fast wave.

There are also many situations where there are additional terms in the equations, such as resistivity, that lead to even longer time scales that need to be followed, for example the slow growth of a resistive instability in a fusion device [11, 12, 13]. In these cases, the CFL time step restriction based on the fast wave would be much too restrictive and some form of implicit solution is required. The next few sections describe some different approaches to solving the implicit system in multiple dimensions. We discuss both implicit and partially implicit methods that are all aimed at producing efficient algorithms for studying plasma motion that is associated with other than the fastest characteristic.

9.5.1 θ-Implicit Method

We first consider an implicit method for the hyperbolic system of Eq. (9.2) that is similar to the θ-implicit and Crank–Nicolson methods for parabolic equations discussed in Section 7.2.1,

$$\frac{\mathbf{U}_j^{n+1} - \mathbf{U}_j^n}{\delta t} + \mathbf{A} \cdot \left[\theta \left(\frac{\mathbf{U}_{j+1}^{n+1} - \mathbf{U}_{j-1}^{n+1}}{2\delta x} \right) + (1 - \theta) \left(\frac{\mathbf{U}_{j+1}^n - \mathbf{U}_{j-1}^n}{2\delta x} \right) \right], \quad (9.54)$$

where $0 \leq \theta \leq 1$ is the implicit parameter. Von Neumann analysis yields for the amplification factor

$$r = \frac{1 + is(\theta - 1)}{1 + is\theta}, \quad (9.55)$$

where $s = (\delta t \lambda_A / \delta x) \sin \theta_k$, with λ_A being each of the eigenvalues of the matrix $\mathbf{A}$. We see that $|r| \leq 1$, implying unconditional stability, for $0.5 \leq \theta \leq 1.0$. For $\theta = 0.5$ we have $|r| = 1$, implying no dissipation, and the method described by Eq. (9.54) is centered in both space and time and therefore has a truncation error $T_\Delta = \mathcal{O}(\delta x^2, \delta t^2)$. In this sense, it is analogous to the Crank–Nicolson method for parabolic equations.

The θ-implicit method as described by Eq. (9.54) uses the quasi-linear form of conservation equations, and does not deal with any non-linearity present in $\mathbf{A}$. If $\mathbf{A}$ is a strongly non-linear function of $\mathbf{U}$, or if we wish to solve the problem in conservation form, the method can be modified in several ways. One obvious way is to update the solution to a provisional time, $(n+1)^*$, to evaluate the matrix $\mathbf{A}$ at this time, and then repeat the update of the solution from time n to $(n+1)$ but using some linear combination of $\mathbf{A}$ at time n and $(n+1)^*$.

A straightforward linearization technique can also be used to apply the θ-implicit method to the conservative formulation of a hyperbolic system as given by Eq. (9.1) in one dimension [204]. We first Taylor expand $\mathbf{F}^{n+1}$ in time about $\mathbf{F}^n$

$$\mathbf{F}^{n+1} = \mathbf{F}^n + \mathbf{A}^n \cdot \left(\mathbf{U}^{n+1} - \mathbf{U}^n \right) + \mathcal{O}(\delta t^2). \quad (9.56)$$

Applying the θ-implicit method to the system of conservation equations,

Eq. (9.1), and rearranging terms gives

$$\left[\mathbf{I} + \theta \delta t \frac{\partial}{\partial x} \mathbf{A}^n \right] \cdot \mathbf{U}^{n+1} = \left[\mathbf{I} + \theta \delta t \frac{\partial}{\partial x} \mathbf{A}^n \right] \cdot \mathbf{U}^n - \delta t \left. \frac{\partial \mathbf{F}}{\partial x} \right|^n . \tag{9.57}$$

Note that the derivative is acting on $\mathbf{A} \cdot \mathbf{U}$ for both bracketed terms. If $\mathbf{A}$ is a constant matrix, Eq. (9.57) is equivalent to Eq. (9.54) if centered spatial difference operators are used. For $\theta = 1/2$, the method is seen to be time and space centered and thus have a truncation error $\mathcal{O}(\delta x^2, \delta t^2)$.

9.5.2   Alternating Direction Implicit (ADI)

In one dimension, the operator in brackets on the left in Eq. (9.57) leads to a block tridiagonal matrix which is easily and efficiently inverted using the methods of Section 3.3.3. Consider now the two-dimensional equation

$$\frac{\partial \mathbf{U}}{\partial t} + \frac{\partial \mathbf{F}}{\partial x} + \frac{\partial \mathbf{G}}{\partial y} = 0. \tag{9.58}$$

We introduce the second Jacobian matrix $\mathbf{C} = \partial \mathbf{G}/\partial \mathbf{U}$. A straightforward generalization of the one-dimensional method leads to the time advance

$$\left[\mathbf{I} + \theta \delta t \left(\frac{\partial}{\partial x} \mathbf{A}^n + \frac{\partial}{\partial y} \mathbf{C}^n \right) \right] \cdot \mathbf{U}^{n+1} = \left[\mathbf{I} + \theta \delta t \left(\frac{\partial}{\partial x} \mathbf{A}^n + \frac{\partial}{\partial y} \mathbf{C}^n \right) \right] \cdot \mathbf{U}^n$$
$$- \delta t \left(\frac{\partial \mathbf{F}}{\partial x} + \frac{\partial \mathbf{G}}{\partial y} \right)^n . \tag{9.59}$$

The two-dimensional operator in brackets on the left in Eq. (9.59) needs to be inverted each time step. This is more problematic than the operator in Eq. (9.57) because it no longer leads to block tridiagonal matrices with small block size, and the difference terms involving $\mathbf{A}^n$ and $\mathbf{C}^n$ are not diagonally dominant. Some type of good preconditioner is needed if one of the iterative methods of Chapter 3 is to be used.

However, in two dimensions, we can consider use of the ADI scheme introduced in Section 7.3.5 [204, 205]. Specializing to $\theta = 1/2$, we establish a factorable operator by adding to each side of Eq. (9.59) a higher-order term:

$$\frac{1}{4} \delta t^3 \frac{\partial}{\partial x} \mathbf{A}^n \cdot \frac{\partial}{\partial y} \mathbf{C} \cdot \left(\frac{\mathbf{U}^{n+1} - \mathbf{U}^n}{\delta t} \right) . \tag{9.60}$$

Then, a factored scheme with the same order of accuracy as Eq. (9.59) can be written as

$$\left[\mathbf{I} + \frac{1}{2} \delta t \frac{\partial}{\partial x} \mathbf{A}^n \right] \cdot \left[\mathbf{I} + \frac{1}{2} \delta t \frac{\partial}{\partial y} \mathbf{C}^n \right] \cdot \mathbf{U}^{n+1}$$
$$= \left[\mathbf{I} + \frac{1}{2} \delta t \frac{\partial}{\partial x} \mathbf{A}^n \right] \cdot \left[\mathbf{I} + \frac{1}{2} \delta t \frac{\partial}{\partial y} \mathbf{C}^n \right] \cdot \mathbf{U}^n$$
$$- \delta t \left(\frac{\partial \mathbf{F}}{\partial x} + \frac{\partial \mathbf{G}}{\partial y} \right)^n . \tag{9.61}$$

This is seen to be similar to the ADI method which breaks the full time step into two half-steps, in each of which only a single direction is treated implicitly which leads to just tridiagonal matrices. This can be solved in two steps, each involving the solution of block tridiagonal systems of equations as follows:

$$
\left[\mathbf{I} + \frac{1}{2}\delta t \frac{\partial}{\partial x} \mathbf{A}^n \right] \mathbf{U}^* = \left[\mathbf{I} + \frac{1}{2}\delta t \frac{\partial}{\partial x} \mathbf{A}^n \right] \cdot \left[\mathbf{I} + \frac{1}{2}\delta t \frac{\partial}{\partial y} \mathbf{C}^n \right] \cdot \mathbf{U}^n
$$
$$
- \delta t \left(\frac{\partial \mathbf{F}}{\partial x} + \frac{\partial \mathbf{G}}{\partial y} \right)^n,
$$
$$
\left[\mathbf{I} + \frac{1}{2}\delta t \frac{\partial}{\partial y} \mathbf{C}^n \right] \cdot \mathbf{U}^{n+1} = \mathbf{U}^*. \tag{9.62}
$$

9.5.3 Partially Implicit 2D MHD

As discussed at the start of Section 9.5, a fully explicit solution of a system of hyperbolic equations can be very inefficient if the range of propagation velocities, or eigenvalues of $\mathbf{A}$, $|\lambda_A^{max}|/|\lambda_A^{min}|$ is large. If this is the case, we say that the system of equations is "stiff," or contains multiple time scales. This is only a difficulty when the fastest time scale in the problem is not the one that is dominantly contributing to the physical phenomena of interest.

This is exactly the situation for the case of the MHD equations being applied to tokamak geometry. We have seen in Chapter 1 that the fast compressible wave, the shear Alfvén wave, and the slow compressible wave can have characteristic wave transit times that can each be separated by an order of magnitude in a strongly magnetized plasma for propagation directions that are nearly perpendicular to the background magnetic field.

It is the fast compressible wave propagating perpendicular to the magnetic field that will normally set the limit on the time step for a fully explicit method, while the physical phenomena of interest normally develop on the shear Alfvén time scale, or slower. Since the fast compressible wave will manifest itself in small amplitude stable compressible oscillations, one strategy is to isolate it and treat only this phenomenon with implicit differences, and to use explicit differencing on the remaining terms. The goal would be to end up with a set of difference equations that form a relatively small and well-conditioned sparse matrix equation which can be solved efficiently and which allows the time step to be determined only by the CFL condition based on the shear Alfvén wave.

To illustrate the application of these methods to a tractable but non-trivial system of equations, let us consider the two-dimensional resistive MHD equations of Section 1.2.2. We will allow propagation only in the $\hat{\mathbf{x}}, \hat{\mathbf{y}}$ plane by taking $\partial/\partial z = 0$ and adopt a notation that incorporates $\nabla \cdot \mathbf{B} = 0$ and thus allows us to represent the magnetic field with just two scalar variables. We also anticipate different propagation velocities for compressible and incompressible motions. These considerations lead us to represent the magnetic field in terms of a poloidal flux ψ and a longitudinal field b,

$$\mathbf{B} = \hat{\mathbf{z}} \times \nabla\psi + b\hat{\mathbf{z}}, \tag{9.63}$$

and to represent the velocity vector in terms of a stream function, a potential, and a longitudinal part as follows:

$$\mathbf{u} = \hat{\mathbf{z}} \times \nabla A + \nabla\Omega + v\hat{\mathbf{z}}. \tag{9.64}$$

The current density is given by the curl of the magnetic field:

$$\mu_0 \mathbf{J} = \nabla^2\psi\hat{\mathbf{z}} + \nabla b \times \hat{\mathbf{z}}.$$

Note that in terms of the velocity stream function A and potential Ω, the Cartesian components of the velocity are given by

$$u_x = -\partial A/\partial y + \partial\Omega/\partial x,$$
$$u_y = \partial A/\partial x + \partial\Omega/\partial y,$$
$$u_z = v.$$

In what follows we neglect the density variation (set $\rho = 1$) and also assume that the convective derivative $(\mathbf{u}\cdot\nabla\mathbf{u})$ term, the pressure gradient term, and the viscous term in the equation of the motion are negligible for simplicity. We take the following projections of the momentum equation, Eq. (1.54):

$$\nabla\cdot\left[\frac{\partial\mathbf{u}}{\partial t} = \mathbf{J}\times\mathbf{B}\right],$$

$$\hat{\mathbf{z}}\cdot\nabla\times\left[\frac{\partial\mathbf{u}}{\partial t} = \mathbf{J}\times\mathbf{B}\right],$$

$$\mathbf{z}\cdot\left[\frac{\partial\mathbf{u}}{\partial t} = \mathbf{J}\times\mathbf{B}\right],$$

and of the field evolution equation, Eq. (1.56):

$$\hat{\mathbf{z}}\cdot\left[\frac{\partial\mathbf{B}}{\partial t} = \nabla\times(\mathbf{u}\times\mathbf{B} - \eta\mathbf{J})\right],$$

$$\hat{\mathbf{z}}\cdot[\nabla\times]^{-1}\cdot\left[\frac{\partial\mathbf{B}}{\partial t} = \nabla\times(\mathbf{u}\times\mathbf{B} - \eta\mathbf{J})\right].$$

This gives the five scalar evolution equations

$$\frac{\partial}{\partial t}\nabla^2\Omega = \frac{-1}{\mu_0}\frac{\partial}{\partial x}\left[b\frac{\partial b}{\partial x} + \frac{\partial \psi}{\partial x}\nabla^2\psi\right] - \frac{1}{\mu_0}\frac{\partial}{\partial y}\left[b\frac{\partial b}{\partial y} + \frac{\partial \psi}{\partial y}\nabla^2\psi\right], \quad (9.65)$$

$$\frac{\partial}{\partial t}\nabla^2 A = -\frac{1}{\mu_0}\frac{\partial \psi}{\partial y}\frac{\partial}{\partial x}\left[\nabla^2\psi\right] + \frac{1}{\mu_0}\frac{\partial \psi}{\partial x}\frac{\partial}{\partial y}\left[\nabla^2\psi\right], \quad (9.66)$$

$$\frac{\partial v}{\partial t} = \frac{\partial \psi}{\partial x}\frac{\partial b}{\partial y} - \frac{\partial \psi}{\partial y}\frac{\partial b}{\partial x}, \quad (9.67)$$

$$\frac{\partial b}{\partial t} + u_x\frac{\partial b}{\partial x} + u_y\frac{\partial b}{\partial y} = -b\nabla^2\Omega + \frac{\partial \psi}{\partial x}\frac{\partial v}{\partial y} - \frac{\partial \psi}{\partial y}\frac{\partial v}{\partial x} + \frac{\eta}{\mu_0}\nabla^2 b, \quad (9.68)$$

$$\frac{\partial \psi}{\partial t} + u_x\frac{\partial \psi}{\partial x} + u_y\frac{\partial \psi}{\partial y} = \frac{\eta}{\mu_0}\nabla^2\psi. \quad (9.69)$$

We are interested in solving Eqs. (9.65)–(9.69) in the strong longitudinal field regime where $|\nabla\Psi|/b \sim \epsilon \ll 1$. Here we have introduced a small dimensionless parameter ϵ. It follows from equilibrium considerations [106] that the gradients of the longitudinal field will be of second-order in this parameter, $\nabla b/b \sim \epsilon^2$. There are a multiplicity of time scales present in the equations associated with the fast wave, the shear Alfvén wave, and with field diffusion, which will be much slower than the wave propagation time of the slowest MHD wave for Lundquist number $S \gg 1$.

We could perform a purely explicit time advance, for example, using the combined leapfrog and DuFort–Frankel methods. However, the time step criterion becomes unacceptable since the fast wave will be controlling the time step while the shear Alfvén wave will be dominating the physics. A fully implicit solution would be inefficient due to the size and the lack of diagonal dominance of the resulting matrices. It is often better to select only certain terms (which in this case are the order ϵ^0 terms that contribute to the fast wave) and treat them implicitly, thereby allowing the time step to be set by the slower shear Alfvén wave.

If we define new variables for the divergence of the velocity,

$$U \equiv \nabla^2\Omega$$

and the total pressure of the longitudinal magnetic field,

$$P \equiv \frac{1}{2\mu_0}b^2,$$

we can effectively isolate the fast wave dynamics. We also define the (higher order in ϵ) auxiliary variables

$$R = \frac{-1}{\mu_0}\left\{\frac{\partial}{\partial x}\left(\frac{\partial \psi}{\partial x}\nabla^2\psi\right) + \frac{\partial}{\partial y}\left(\frac{\partial \psi}{\partial y}\nabla^2\psi\right)\right\},$$

$$S = -\left[u_x\frac{\partial P}{\partial x} + u_y\frac{\partial P}{\partial y}\right] + b\left[\frac{\partial \psi}{\partial x}\frac{\partial v}{\partial y} - \frac{\partial \psi}{\partial y}\frac{\partial v}{\partial x}\right] + \frac{\eta}{\mu_0}b\nabla^2 b.$$

Equations (9.65) and (9.68) then take the form

$$\frac{\partial U}{\partial t} + \nabla^2 P = R, \tag{9.70}$$

$$\frac{\partial P}{\partial t} + b^2 U = S. \tag{9.71}$$

We see that the order ϵ^0 terms have been isolated on the left in Eqs. (9.70) and (9.71). Consider now the BTCS partially implicit time advance:

$$\frac{U^{n+1} - U^n}{\delta t} + \nabla^2 P^{n+1} = R^n, \tag{9.72}$$

$$\frac{P^{n+1} - P^n}{\delta t} + b^2 U^{n+1} = S^n. \tag{9.73}$$

We can eliminate U^{n+1} from the second equation to give

$$\left[I - \delta t^2 b^2 \nabla^2\right] P^{n+1} = P^n + \delta t S^n - \delta t b^2 \left[U^n + \delta t R^n\right]. \tag{9.74}$$

The elliptic operator in Eq. (9.74), which is now symmetric and positive definite, can either be inverted as is, or the semi-implicit method of Section 7.3.3 can be used to replace it with a similar constant coefficient elliptic operator. We return to this in Section 9.5.6. Once P^{n+1} is obtained, we can obtain U^{n+1} from Eq. (9.72), and then Ω^{n+1} from solving the elliptic equation $\nabla^2 \Omega^{n+1} = U^{n+1}$. The first-order equations, Eqs. (9.66), (9.67), and (9.69) are then advanced using the new time values of $b^{n+1} = \sqrt{2\mu_0 P^{n+1}}$ and Ω^{n+1}, but with the other terms evaluated using an explicit method such as leapfrog plus DuFort–Frankel. This would involve introducing a new variable for the vorticity, $\omega \equiv \nabla^2 A$, advancing this in time using Eq. (9.66) and then obtaining A from solving $\nabla^2 A^{n+1} = \omega^{n+1}$.

9.5.4 Reduced MHD

We include a brief discussion of reduced MHD here since it can be thought of as taking the limit of the equations in the last section as $|\nabla \psi|/b \sim \epsilon \to 0$. The strong longitudinal field b becomes a spatial constant and is not solved for. It follows that $U = 0$ and $v = 0$, and the velocity is incompressible and in the (x, y) plane. The remaining equations (now including the convective term and adding a viscosity term with coefficient μ) can be written in the compact form:

$$\frac{\partial \omega}{\partial t} + [A, \omega] = \frac{1}{\mu_0} [\psi, \nabla^2 \psi] + \mu \nabla^2 \omega, \tag{9.75}$$

$$\frac{\partial \psi}{\partial t} + [A, \psi] = \frac{\eta}{\mu_0} \nabla^2 \psi, \tag{9.76}$$

$$\nabla^2 A = \omega. \tag{9.77}$$

Here we have introduced the *Poisson bracket*, which for any two scalar variables a and b is defined as

$$[a, b] \equiv \frac{\partial a}{\partial x} \frac{\partial b}{\partial y} - \frac{\partial a}{\partial y} \frac{\partial b}{\partial x}. \tag{9.78}$$

The two-variable reduced equation set given by Eqs. (9.75)–(9.77) has been extended to 3D Cartesian geometry by Strauss [18] and to 3D toroidal geometry by Breslau [30].

The Poisson bracket operator, or some equivalent to it, appears in most all formulations of reduced MHD. The most obvious centered difference approximation to it on an equally spaced rectangular grid would be

$$
\begin{aligned}
[a, b]_{i,j}^{++} &= \frac{1}{4h^2} [(a_{i+1,j} - a_{i-1,j})(b_{i,j+1} - b_{i,j-1}) \\
&- (a_{i,j+1} - a_{i,j-1})(b_{i+1,j} - b_{i-1,j})],
\end{aligned} \tag{9.79}
$$

where $\delta x = \delta y = h$ is the grid spacing.

Arakawa [206] has shown that if a time-centered difference scheme is used together with this form of spatial differencing of this operator, a non-linear computational instability can develop. He attributed this instability to the fact that the finite difference operator in Eq. (9.79) does not maintain certain integral constraints that the differential operator does. He considered two other second-order approximations to the Poisson bracket:

$$
\begin{aligned}
[a, b]_{i,j}^{+\times} &= \frac{1}{4h^2} [a_{i+1,j}(b_{i+1,j+1} - b_{i+1,j-1}) - a_{i-1,j}(b_{i-1,j+1} - b_{i-1,j-1}) \\
&- a_{i,j+1}(b_{i+1,j+1} - b_{i-1,j+1}) + a_{i,j-1}(b_{i+1,j-1} - b_{i-1,j-1})],
\end{aligned}
$$

$$
\begin{aligned}
[a, b]_{i,j}^{\times+} &= \frac{1}{4h^2} [a_{i+1,j+1}(b_{i,j+1} - b_{i+1,j}) - a_{i-1,j-1}(b_{i-1,j} - b_{i,j-1}) \\
&- a_{i-1,j+1}(b_{i,j+1} - b_{i-1,j}) + a_{i+1,j-1}(b_{i+1,j} - b_{i,j-1})].
\end{aligned}
$$

By taking the average of these three approximations,

$$[a, b]_{i,j} = \frac{1}{3} \left\{ [a, b]_{i,j}^{++} + [a, b]_{i,j}^{+\times} + [a, b]_{i,j}^{\times+} \right\}, \tag{9.80}$$

it can be shown [206] that the integral constraints that correspond to conservation of the mean vorticity, the mean kinetic energy, and the mean square vorticity in an incompressible fluid flow calculation would be conserved and the non-linear instability will not occur. This form of differencing of the Poisson bracket, or a fourth-order analogue of it, is now widely used in plasma physics as well as in meteorology and other disciplines.

9.5.5 Method of Differential Approximation

Here we describe an approach for solving stiff hyperbolic systems that is based on the method of differential approximations [207] and is now used in several contemporary multidimensional MHD codes [7, 208]. We start with again considering the second-order hyperbolic system in one dimension given by Eq. (9.23),

$$\frac{\partial v}{\partial t} = c\frac{\partial \omega}{\partial x}, \tag{9.81}$$

$$\frac{\partial \omega}{\partial t} = c\frac{\partial v}{\partial x}. \tag{9.82}$$

These can be combined to give the single second-order equation,

$$\frac{\partial^2 v}{\partial t^2} = \mathcal{L}(v), \tag{9.83}$$

where we have introduced the linear second-order spatial operator,

$$\mathcal{L}(v) \equiv c^2 \frac{\partial^2}{\partial x^2} v. \tag{9.84}$$

Consider now the following algorithm for advancing the system in time:

$$\left(1 - \theta^2 \delta t^2 \mathcal{L}\right) v^{n+1} = \left(1 - \alpha \delta t^2 \mathcal{L}\right) v^n + \delta t c \left(\frac{\partial \omega}{\partial x}\right)^m \tag{9.85}$$

$$\omega^{m+1} = \omega^m + \delta t c \phi \left(\frac{\partial v}{\partial x}\right)^{n+1} + \delta t c (1 - \phi) \left(\frac{\partial v}{\partial x}\right)^n. \tag{9.86}$$

We have introduced the implicit parameters θ, α, and ϕ and the time step δt. The derivatives are to be replaced by centered finite difference (or spectral or finite element) operators. For $\theta = \alpha = 0$ and $\phi = 1$, the algorithm described by Eqs. (9.85) and (9.86) is just the explicit leapfrog method, equivalent to that described by Eqs. (9.24) and (9.26).

To examine the numerical stability of the general case, we make the substitutions

$$\frac{\partial}{\partial x} \rightarrow -ik_e,$$

$$\mathcal{L} \rightarrow -c^2 k_e^2,$$

where $k_e = -(1/\delta x)\sin\theta_k$ is the effective wavenumber when the centered finite difference operator is substituted for the spatial derivative. Letting $v^{n+1} = rv^n$ and $\omega^{m+1} = r\omega^m$, the amplification factor r can then be determined from the quadratic equation

$$(1 + \theta^2 D)(r - 1)^2 + D(\theta^2 + \phi - \alpha)(r - 1) + D = 0, \tag{9.87}$$

where we have defined $D \equiv \delta t^2 c^2 k_e^2$. Equation (9.87) has roots

$$r = \frac{1 + \frac{1}{2}D\left(\theta^2 - \phi + \alpha\right) \pm i\sqrt{D + \left[\theta^2 - \frac{1}{4}\left(\theta^2 + \phi - \alpha\right)^2\right]D^2}}{1 + \theta^2 D}, \tag{9.88}$$

for which

$$|r|^2 = \frac{1 + (1 + \alpha - \phi)D}{1 + \theta^2 D} \tag{9.89}$$

when the quantity within the square root in Eq. (9.90) is non-negative. There are two cases that are of interest.

For the *Caramana method* [207], we set $\phi = 1$ and $\alpha = \theta^2$. From Eq. (9.89), this yields $|r|^2 = 1$ for any D, as long as $\theta \geq \frac{1}{2}$. Thus the method is linearly stable and non-dissipative for $\theta \geq \frac{1}{2}$. Truncation error analysis shows the time discretization error to be second-order in δt for any stable value of θ. This method has the additional feature that the multiplier of the operator $\mathcal{L}$ is the same on both sides of Eq. (9.86), so that in steady state, when $v^{n+1} = v^n$, the operator will have no affect on the solution.

For the *split θ-implicit method*, we set $\phi = \theta$ and $\alpha = \theta(\theta - 1)$. This is what one would obtain if both Eqs. (9.81) and (9.82) were time differenced using the θ-implicit method, and then the differenced form of Eq. (9.82) was used to algebraically eliminate ω^{n+1} from the differenced form of Eq. (9.81). The amplification factor is $|r|^2 = 1 + (1 - 2\theta)D/(1 + \theta^2 D)$, which is less than or equal to 1 (and hence stable) when $\theta \geq \frac{1}{2}$. The quantity within the square root in Eq. (9.88) is exactly D for this method, which is assumed to be positive. Note that $|r|^2 = 1$ only when $\theta = \frac{1}{2}$ for this method. The truncation error analysis for this method is exactly the same as it would be for the θ-implicit method, yielding second-order accuracy only for $\theta = \frac{1}{2}$.

In each of these algorithms, the operator $(1 - \theta^2 \delta t^2 \mathcal{L})$ needs to be inverted each time step. However, this is a well-conditioned symmetric diagonally dominant operator, and so the iterative methods discussed in Chapter 3 should perform well on this. Also, we see that the two equations, Eqs. (9.85) and (9.86), can be solved sequentially, and so the associated sparse matrix equation is only for a single scalar variable. Both of these features offer clear advantages over the unsplit θ-implicit method discussed in Section 9.5.1.

To obtain the operator $\mathcal{L}$ for the ideal MHD equations, we start with the momentum equation, again ignoring for simplicity the density and convective derivative terms, assumed small. Denoting time derivatives with a dot, i.e., $\dot{\mathbf{u}} \equiv \partial \mathbf{u}/\partial t$, we have

$$\rho_0 \dot{\mathbf{u}} + \nabla p = \frac{1}{\mu_0}[(\nabla \times \mathbf{B}) \times \mathbf{B}]. \tag{9.90}$$

Time differentiating gives

$$\rho_0 \ddot{\mathbf{u}} + \nabla \dot{p} = \frac{1}{\mu_0}\left[\left(\nabla \times \dot{\mathbf{B}}\right) \times \mathbf{B} + (\nabla \times \mathbf{B}) \times \dot{\mathbf{B}}\right]. \tag{9.91}$$

Next, we take the ideal MHD components of the magnetic field and pressure equations:

$$\dot{\mathbf{B}} = \nabla \times [\mathbf{u} \times \mathbf{B}], \tag{9.92}$$
$$\dot{p} = -\mathbf{u} \cdot \nabla p - \gamma p \nabla \cdot \mathbf{u}. \tag{9.93}$$

We substitute for $\dot{\mathbf{B}}$ and $\dot{p}$ from Eqs. (9.92) and (9.93) into Eq. (9.90), whereby in analogy with Eq. (9.83), we can identify

$$
\begin{aligned}
\mathcal{L}(\mathbf{u}) \;\equiv\; & \frac{1}{\mu_0} \left\{ \nabla \times \left[\nabla \times (\mathbf{u} \times \mathbf{B}) \right] \right\} \times \mathbf{B} \\
+ \;& \frac{1}{\mu_0} \left\{ (\nabla \times \mathbf{B}) \times \nabla \times (\mathbf{u} \times \mathbf{B}) \right\} \\
+ \;& \nabla \left(\mathbf{u} \cdot \nabla p + \gamma p \nabla \cdot \mathbf{u} \right).
\end{aligned}
\tag{9.94}
$$

We note that this is equivalent to the ideal MHD operator $\mathbf{F}$ used in Eq. (8.26).

Applying the Caramana method to this equation gives the following equation to advance the velocity to the new time level

$$
\begin{aligned}
\left\{ \rho_0 - \theta^2 \delta t^2 \mathcal{L} \right\} \mathbf{u}^{n+1} \;=\; & \left\{ \rho_0 - \theta^2 \delta t^2 \, \mathcal{L} \right\} \mathbf{u}^n \\
+ \;& \delta t \left\{ -\nabla p + \frac{1}{\mu_0} \left[(\nabla \times \mathbf{B}) \times \mathbf{B} \right] \right\}^{n+\frac{1}{2}}.
\end{aligned}
\tag{9.95}
$$

When finite difference or finite element methods are used to discretize the spatial operators in Eq. (9.95), it becomes a sparse matrix equation for the advanced time velocity $\mathbf{u}^{n+1}$. When this matrix equation is solved using the methods of Chapter 3, the new velocity can be used to advance the magnetic field and pressure according to Eqs. (9.92) and (9.93), which now take the form:

$$
\begin{aligned}
\dot{\mathbf{B}} \;=\; & \nabla \times \left[\left(\theta \mathbf{u}^{n+1} + (1-\theta)\mathbf{u}^n \right) \times \mathbf{B} \right], \tag{9.96} \\
\dot{p} \;=\; & -\left(\theta \mathbf{u}^{n+1} + (1-\theta)\mathbf{u}^n \right) \cdot \nabla p \\
& -\gamma p \nabla \cdot \left(\theta \mathbf{u}^{n+1} + (1-\theta)\mathbf{u}^n \right). \tag{9.97}
\end{aligned}
$$

A von Neumann stability analysis shows this method to be linearly stable for all δt as long as $\theta \geq 1/2$. We also note that it gives accurate solutions in steady state when the terms in Eq. (9.95) on the two sides of the equation which involve the operator $\mathcal{L}$ cancel.

Non-ideal terms such as resistivity and viscosity can be added to Eqs. (9.95)–(9.97) by treating them implicitly and this will not affect the form of the operator in Eq. (9.94) or the numerical stability. The challenge in this method is to find an efficient method for inverting the matrix corresponding to the operator on the left in Eq. (9.95). This is presently an area of active research.

9.5.6   Semi-Implicit Method

The semi-implicit method as applied to hyperbolic problems is very similar to the method of differential approximation presented in the last section, except that it approximates the operator $\mathcal{L}$ such as is defined in Eq. (9.84) or Eq. (9.94) with a simpler form that is easier to invert but that still provides numerical stability [209].

As an almost trivial example, let us consider the system given by Eqs. (9.85) and (9.86) with $\alpha = \theta^2$ and $\phi = 1$, but instead of defining the operator $\mathcal{L}$ by Eq. (9.84), we define it by

$$\mathcal{L}(v) \equiv a_0^2 \frac{\partial^2}{\partial x^2} v, \qquad (9.98)$$

for some constant a_0. In this case, the analogue of Eq. (9.89) gives stability for $4\theta^2 a_0^2 > c^2$. There is no advantage in using a_0^2 instead of c^2 in this case, since it can be compensated by the choice of θ used. However, if c was not constant but had spatial variation, then replacing the operator $c\frac{\partial}{\partial x}c\frac{\partial}{\partial x}$ by $a_0^2 \frac{\partial^2}{\partial x^2}$ could offer advantages as it would allow use of a transform-based solver as discussed in Section 3.8.2.

Now, let us consider semi-implicit approximations to the ideal MHD operator, Eq. (9.94). Harned and Schnack [210] proposed replacing the ideal MHD operator in Eq. (9.94) with the simplified operator

$$\mathcal{L}(\mathbf{u}) \equiv \frac{1}{\mu_0}\{\nabla \times [\nabla \times (\mathbf{u} \times \mathbf{C}_0)]\} \times \mathbf{C}_0, \qquad (9.99)$$

where $\mathbf{C}_0$ is a vector with constant coefficients, but with a crucial modification in the evaluation of Eq. (9.99). In Cartesian coordinates, they require that each component of $\mathbf{C}_0$ be larger than each component of $\mathbf{B}$ in magnitude, *and that all terms of the form C_iC_j be replaced by $C_iC_j\delta_{ij}$* where δ_{ij} is the Kronecker delta. They present a two-dimensional stability analysis that shows stability for this modified operator, but instability if the C_iC_j cross terms are retained in the semi-implicit operator.

Harned and Mikic [211] have also developed a semi-implicit operator for the Hall term in the magnetic field advance equation for two-fluid MHD. If we insert Eq. (1.32) into Eq. (1.24), the term they were concerned with is

$$\frac{\partial \mathbf{B}}{\partial t} = \nabla \times \left[-\frac{1}{ne}\mathbf{J} \times \mathbf{B} + \cdots \right]. \qquad (9.100)$$

By time differentiating this equation, and substituting in for the time derivative of $\mathbf{J}$ from this same equation, using Eqs. (1.25) and (1.26) and some vector identities, they proposed the semi-implicit operator

$$\mathcal{L}(\mathbf{B}) = -\frac{1}{(ne\mu_0)^2}(\mathbf{B}_0 \cdot \nabla)^2 \nabla^2 \mathbf{B}. \qquad (9.101)$$

Using this, Eq. (9.100) would be time differenced as

$$\{1 - \delta t^2 \mathcal{L}\}\mathbf{B}^{n+1} = \{1 - \delta t^2 \mathcal{L}\}\mathbf{B}^n$$
$$+ \delta t \nabla \times \left[-\frac{1}{ne}\mathbf{J} \times \mathbf{B} + \cdots \right]^n \qquad (9.102)$$

They were able to show numerical stability and second-order accuracy with analysis in simplified geometry and in numerical tests for $\mathbf{B}_0$ comparable to the largest value of $\mathbf{B}$ in the domain.

The semi-implicit method can also be combined with the partially implicit method of Section 9.5.3 to change the implicit equation, Eq. (9.74), to

$$
\begin{aligned}
\left[I - \delta t^2 b_0^2 \nabla^2\right] P^{n+1} &= \left[I - \delta t^2 (b_0^2 - b^2)\nabla^2\right] P^n \\
&+ \delta t S^n - \delta t b^2 \left[U^n + \delta t R^n\right],
\end{aligned} \tag{9.103}
$$

where b_0 is some constant satisfying $b_0^2 \geq b^2$ everywhere. In a simple geometry, this would allow the operator on the left to be inverted by fast transform techniques such as discussed in Section 3.8.2.

9.5.7 Jacobian-Free Newton–Krylov Method

A technique that is gaining in popularity is the Jacobian-free Newton–Krylov method [212]. This is a *non-linear implicit* algorithm that consists of at least two levels of iterations. Suppose that we have used spatial finite differences or finite elements to express the MHD (or any other) system of equations as the vector system,

$$
\mathbf{F}(\mathbf{U}) = 0, \tag{9.104}
$$

where $\mathbf{F}$ is a large vector function representing all of the discretized equations at all of the grid points, and $\mathbf{U}$ represents all of the unknowns. For the MHD equations, these would normally be the velocities, densities, pressures, and magnetic field components at all of the grid points. If there were M equations and M unknowns at each grid point, and N grid points, then the system given by Eq. (9.104) would be $M \times N$ non-linear equations with the same number of unknowns. In a time-dependent problem, the unknown $\mathbf{U}$ would be the complete solution vector at the new time level $(n+1)$.

The multivariate Newton iteration for $\mathbf{F}(\mathbf{U}) = 0$ is based on a first-order Taylor expansion about a current point $\mathbf{U}^k$:

$$
\mathbf{F}(\mathbf{U}^{k+1}) \simeq \mathbf{F}(\mathbf{U}^k) + \mathbf{F}'(\mathbf{U}) \cdot (\mathbf{U}^{k+1} - \mathbf{U}^k). \tag{9.105}
$$

Setting the right side to zero yields the basic Newton method

$$
\mathbf{J}(\mathbf{U}^k) \cdot (\delta \mathbf{U}^k) = -\mathbf{F}(\mathbf{U}^k). \tag{9.106}
$$

Here, we have introduced the notation

$$
\mathbf{U}^{k+1} = \mathbf{U}^k + \delta \mathbf{U}^k, \qquad k = 0, 1, \dots \tag{9.107}
$$

given $\mathbf{U}^0$, $\mathbf{J} \equiv \mathbf{F}' \equiv \partial \mathbf{F}/\partial \mathbf{U}$ is the Jacobian matrix, and k is the non-linear iteration index. The Newton iteration is terminated when the solution stops changing to some tolerance.

Forming the Jacobian matrix $\mathbf{J}$ requires taking analytic or discrete derivatives of the system of equations with respect to $\mathbf{U}$: $J_{ij} = \partial F_i(\mathbf{U})/\partial U_j$. For a complicated system like the MHD equations, this can be both error-prone and time consuming, and since $\mathbf{J}$ is a large sparse matrix, some sparse storage convention is required. However, we will see that it is not actually necessary to form $\mathbf{J}$ explicitly in order to solve Eq. (9.106).

Recall from Section 3.7 the Krylov subspace approach for solving equations of the form of Eq. (9.106), which we generically write as $\mathbf{Ax} = \mathbf{b}$. This involves forming an initial residual $\mathbf{r}_0 = \mathbf{b} - \mathbf{Ax}_0$ and then constructing a Krylov subspace, $\mathbf{K}_j$,

$$\mathbf{K}_j = span\left\{\mathbf{r}_0, \mathbf{Ar}_0, \mathbf{A}^2\mathbf{r}_0, ..., \mathbf{A}^{j-1}\mathbf{r}_0\right\}.$$

A property of Krylov methods is that they require only matrix-vector products to carry out the iteration. This is the key feature that makes them compatible with the Newton method, Eq. (9.106).

The Krylov method applied to Eq. (9.106) defines the initial linear residual, $\mathbf{r}_0$, given the initial guess, $\delta\mathbf{U}_0$, for the Newton correction,

$$\mathbf{r}_0 = -\mathbf{F}(\mathbf{U}) - \mathbf{J} \cdot \delta\mathbf{U}_0. \tag{9.108}$$

The non-linear iteration index, k, has been dropped since the Krylov iteration is performed at a fixed k. Let us introduce j as the Krylov iteration index. The jth approximate to the solution is obtained from the subspace spanned by the Krylov vectors, $\{\mathbf{r}_0, \mathbf{Jr}_0, \mathbf{J}^2\mathbf{r}_0, ..., \mathbf{J}^{j-1}\mathbf{r}_0\}$, and can be written

$$\delta\mathbf{U}_j = \delta\mathbf{U}_0 + \sum_{i=0}^{j-1} \beta_i \mathbf{J}^i \mathbf{r}_0, \tag{9.109}$$

where the scalars β_i are chosen to minimize the residual. Krylov methods such as GMRES, described in Section 3.7.2, thus require the action of the Jacobian only in the form of matrix-vector products, which can be approximated by [212, 213, 214]

$$\mathbf{Jv} \approx [\mathbf{F}(\mathbf{U} + \epsilon\mathbf{v}) - \mathbf{F}(\mathbf{U})]/\epsilon, \tag{9.110}$$

where ϵ is a small perturbation. The entire process of forming and storing the Jacobian matrix and multiplying it by a vector is replaced by just two evaluations of the function $\mathbf{F}(\mathbf{U})$, Eq. (9.104), hence the name "Jacobian-free." While the Jacobian-free Newton–Krylov method is potentially very powerful, there are a number of practical issues in their application. Convergence of the Newton iteration in Eq. (9.105) is not guaranteed. The Krylov iteration, Eq. (9.109), requires a good preconditioner. Techniques for addressing both of these are topics of current research, and [212] presents an excellent review of progress in these areas.

9.6 Summary

The most straightforward differencing, forward in time and centered in space, is numerically unstable for hyperbolic equations. Explicit time advance methods can be made stable by including additional terms, such as the Lax–Wendroff class of methods, or by time-centering such as is done in the leapfrog class of methods. The other possibility is to use one-sided differences, with the side chosen by the sign of the corresponding characteristic velocity. These one-sided methods are of lower-accuracy than centered difference methods, but limiter techniques can be used to effectively use the lower-accuracy differencing only where it is required for numerical stability, and to use the higher-order differencing elsewhere. Many problems in MHD require implicit time differencing because the time step restriction associated with the fastest of the MHD characteristics can be unacceptably small. Both the method of differential approximation and the Jacobian-free Newton–Krylov method are being used in contemporary simulation efforts.

Problems

9.1: Consider the Beam–Warming method, Eq. (9.36).
(a) Perform a Taylor series analysis to find the leading order truncation error.
(b) Perform von Neumann stability analysis to calculate the amplification factor and stability limits.

9.2: The equations of linear acoustics in a medium with equilibrium velocity u_0 in one dimension can be written in terms of the pressure p and the velocity u as

$$\frac{\partial}{\partial t} \begin{bmatrix} p \\ u \end{bmatrix} + \begin{bmatrix} u_0 & K_0 \\ 1/\rho_0 & u_0 \end{bmatrix} \cdot \frac{\partial}{\partial x} \begin{bmatrix} p \\ u \end{bmatrix} = 0. \tag{9.111}$$

Here ρ_0 is the equilibrium density and $K_0 = \rho_0 \partial p / \partial \rho_0$.
(a) Determine the matrices $\mathbf{A}^+$ and $\mathbf{A}^-$ as defined in Eq. (9.45).
(b) What are the associated eigenvectors?

9.3: Show that the conservative upwind method of Eq. (9.44) is equivalent to the non-conservative form, Eq. (9.46), if the matrices $\mathbf{A}^+$ and $\mathbf{A}^-$ have constant coefficients.

9.4: Show that Eq. (9.49) is equivalent to the Lax–Wendroff method when s is constant.

9.5: Consider the *implicit upwind* method applied to Eq. (9.48),

$$U_j^{n+1} = U_j^n - s \left[\theta \left(U_j^{n+1} - U_{j-1}^{n+1} \right) + (1 - \theta) \left(U_j^n - U_{j-1}^n \right) \right], \qquad (9.112)$$

where $0 \leq \theta \leq 1$ is a parameter and $s = a\delta t/\delta x$.
(a) Show that the finite difference equation is consistent with the differential equation. What is the leading order truncation error?
(b) Use von Neumann stability analysis to calculate the range of values of δt for which the method is stable as a function of θ and a.

9.6: Show that the MacCormack difference method, Eq. (9.16), is equivalent to the Lax–Wendroff method, Eq. (9.12), if the matrix $\mathbf{A}$ is constant.

9.7: Consider the *Wendroff method* applied to Eq. (9.48):

$$U_{j+\frac{1}{2}}^{n+1} = U_{j+\frac{1}{2}}^n - s \left(U_{j+1}^{n+\frac{1}{2}} - U_j^{n+\frac{1}{2}} \right). \qquad (9.113)$$

The half-integer subscripts refer to evaluating the quantity halfway between two grid points by simple averaging, i.e.,

$$U_{j+\frac{1}{2}}^n = \frac{1}{2} \left(U_j^n + U_{j+1}^n \right), \qquad U_j^{n+\frac{1}{2}} = \frac{1}{2} \left(U_j^n + U_j^{n+1} \right), \qquad (9.114)$$

etc. Substitute the definitions of the half-integer points into the finite difference formula and calculate the numerical stability using the von Neumann method.

9.8: Consider the one-dimensional electron-MHD equation for the time evolution of the (x, y) Cartesian components of the magnetic field which vary in the z dimension:

$$\frac{\partial \mathbf{B}}{\partial t} = -\nabla \times \left[\alpha \left(\nabla \times \mathbf{B} \right) \times \mathbf{B} \right].$$

Here, the magnetic field is given by

$$\mathbf{B} = b_x(z)\hat{\mathbf{x}} + b_y(z)\hat{\mathbf{y}} + B_0\hat{\mathbf{z}},$$

with B_0 and α being real constants.
(a) Find the 2×2 matrix $\mathbf{A}$ for which the linearized form of the electron-MHD equation can be written as:

$$\frac{\partial}{\partial t} \begin{bmatrix} b_x \\ b_y \end{bmatrix} = \alpha B_0 \mathbf{A} \cdot \frac{\partial^2}{\partial z^2} \begin{bmatrix} b_x \\ b_y \end{bmatrix}. \qquad (9.115)$$

(b) Write down a forward in time and centered in space finite difference method for solving Eq. (9.115), and analyze the stability using the von Neumann stability analysis.

(c) Devise an implicit method for solving Eq. (9.115) and show that it is unconditionally stable.

9.9: Linearize the reduced MHD equations, Eqs. (9.75)–(9.77), around an equilibrium state with $A = 0$, $\nabla^2 \psi_0 = 0$, $\frac{\partial \psi_0}{\partial x} = 0$, $\frac{\partial \psi_0}{\partial y} = B_0$, where B_0 is a spatial constant. Show that you obtain a single wave equation. What direction does the wave propagate? What is its velocity?

9.10: Derive the split θ-implicit method, corresponding to Eqs. (9.85) and (9.86) with $\phi = \theta$ and $\alpha = \theta(\theta - 1)$, by starting with both Eqs. (9.81) and (9.82) and time differencing using the θ-implicit method, and then using the differenced form of Eq. (9.82) to algebraically eliminate ω^{n+1} from the differenced form of Eq. (9.81).

Chapter 10

Spectral Methods for Initial Value Problems

10.1 Introduction

Spectral methods are fundamentally different from finite difference methods for obtaining approximate numerical solutions to partial differential equations. Instead of representing the solution by a truncated Taylor's series expansion, we expand the solution in a truncated series using some set of basis functions, or trial functions. To obtain the discrete equations to be solved numerically, we minimize the error in the differential equation, produced by using the truncated expansion, with respect to some measure. A set of test functions are used to define the minimization criterion.

This same description could be applied to finite element methods, which are discussed in the next chapter. The choice of trial functions is the main feature that distinguishes spectral methods from finite element methods. Spectral methods utilize infinitely differentiable global functions. Finite element methods, on the other hand, utilize local trial functions with derivatives only up to some order.

In general, we will see that the finite element method leads to sparse matrices which do not require much storage and are relatively easy to invert, but have convergence properties such that the error decreases only as h^p or as N^{-p}, where h is the typical element size, N is the number of elements, and p is a small integer which depends on the type of finite element being used. Spectral methods, on the other hand, will generally lead to full matrices, albeit with lower dimension, since a small number of global basis functions may be able to adequately represent the solution. Also, the asymptotic convergence rates of spectral methods can be very good, faster than any power of N, where N is the number of basis functions. This is normally referred to as exponential convergence. However, for finite values of N, spectral methods are not necessarily superior to either finite element or finite difference methods. Enough basis functions as needed to represent all the essential structure in the solution before the superior asymptotic convergence rates begin to apply.

10.1.1   Evolution Equation Example

Let us consider the one-dimensional first-order evolution equation

$$\frac{\partial u}{\partial t} = \mathcal{L}(u), \tag{10.1}$$

where $u(x,t)$ is the solution, to be determined, and $\mathcal{L}(u)$ is an operator which contains spatial derivatives of u. For example, for a simple constant coefficient convection equation, $\mathcal{L}(u) = -v\partial u/\partial x$, where v is a constant. We also need to supply initial conditions $u(x,0)$ and boundary conditions, which for now we will take to be periodic over the interval $[0, 2\pi]$.

We represent the approximate solution to Eq. (10.1) as

$$u^N(x,t) = \sum_{k=-N/2}^{N/2} u_k(t)\phi_k(x), \tag{10.2}$$

where the $\phi_k(x)$ are a set of known trial functions, or basis functions, which are compatible with the boundary conditions, and the $u_k(t)$ are expansion coefficients which remain to be determined. To obtain equations for the $u_k(t)$, we multiply Eq. (10.1) by a test function $\psi_k(x)$ and integrate over all space,

$$\int_0^{2\pi} \left[\frac{\partial u^N}{\partial t} - \mathcal{L}\left(u^N\right) \right] \psi_k(x)\, dx = 0, \tag{10.3}$$

for $k = -N/2, \cdots, N/2$. This is effectively enforcing the condition that the error be orthogonal to each of the test functions with respect to the inner produce defined by Eq. (10.3).

For a concrete example, let us take as trial and test functions

$$\phi_k(x) = e^{ikx}, \qquad\qquad \psi_k(x) = e^{-ikx}. \tag{10.4}$$

These satisfy the orthogonality relation

$$\int_0^{2\pi} \phi_k(x)\,\psi_\ell(x)dx = 2\pi\delta_{k\ell}, \tag{10.5}$$

where $\delta_{k\ell}$ is the Kronecker delta. Substituting Eqs. (10.2) and (10.4) into Eq. (10.3), we have

$$\int_0^{2\pi} e^{-ikx} \left[\left(\frac{\partial}{\partial t} + v\frac{\partial}{\partial x} \right) \sum_{\ell=-N/2}^{N/2} u_\ell(t)\, e^{i\ell x} \right] dx = 0.$$

After analytic differentiation and integration, and using Eq. (10.5), this yields

$$\frac{d}{dt}u_k + ikvu_k = 0, \qquad k = -N/2, \cdots, N/2. \tag{10.6}$$

Equation (10.6) is a system of ordinary differential equations. The initial conditions are the coefficients for the expansion of the initial condition $u(x, 0)$,

$$u_k(0) = \frac{1}{2\pi} \int_0^{2\pi} u(x, 0)e^{-ikx}\,dx.$$

10.1.2 Classification

There are several types of spectral methods which differ with respect to which type of test functions are used. In the *pure Galerkin method*, as illustrated above, the test functions are the same as (or the complex conjugate of) the trial functions, and they each individually satisfy the boundary conditions. In the *collocation method*, the test functions are translated Dirac delta functions centered at special collocation points. There are also variations, such as the so-called *tau methods* which are similar to Galerkin's method; however, none of the test functions needs to satisfy the boundary conditions. With tau methods, there is a supplementary set of equations which enforce the boundary conditions.

10.2 Orthogonal Expansion Functions

The theory underlying the spectral method is built upon the expansion of a function $u(x)$ over a finite interval (here taken to be $[0, 2\pi]$) in terms of an infinite sequence of orthogonal functions $\phi_k(x)$,

$$u(x) = \sum_{k=-\infty}^{\infty} \hat{u}_k \phi_k(x). \tag{10.7}$$

The coefficients $\hat{u}_k$ are determined by the integral

$$\hat{u}_k = \frac{1}{2\pi} \int_0^{2\pi} u(x)\bar{\phi}_k(x)dx, \tag{10.8}$$

where the expansion function and its complex conjugate satisfy the orthogonality relation which we now take to have the normalization

$$\int_0^{2\pi} \phi_k(x)\bar{\phi}_\ell(x)dx = 2\pi\delta_{k\ell}. \tag{10.9}$$

Certain classes of expansion functions yield what is called *spectral accuracy*, by which is meant exponentially rapid decay of coefficients for sufficiently large number of expansion functions N. Trigonometric functions exhibit this for periodic boundary conditions, but it can also be proven for the eigen-functions of any singular Sturm–Liouville operator with no restriction on the boundary conditions [215].

The transform indicated in Eqs. (10.7)–(10.9) is called the *finite transform* of u between physical space and transform space because the integral is over a finite interval. If the system of orthogonal functions is complete, the transform can be inverted. Functions can therefore be described either by their values in physical space or by their coefficients in transform space. Note that the expansion coefficients, Eq. (10.8), depend on all the values of u in physical space, and hence they cannot in general be computed exactly.

If a finite number of approximate expansion coefficients are instead com-puted using the values of u at a finite number of selected points, perhaps the nodes of a high-precision quadrature formula, we call the procedure a *discretetransform* between the set of values of u at the quadrature points and the set of approximate, or discrete coefficients. For the most common orthog-onal systems, Fourier and Chebyshev polynomials, the discrete transform of a function defined at N points can be computed in a "fast" way, as we saw in Section 3.8.1, requiring on the order of $N \log_2 N$ operations rather than N^2.

10.2.1   Continuous Fourier Expansion

This is the most common expansion and is natural for problems with periodic boundary conditions. The basis functions are proportional to those introduced in Eq. (10.4),

$$\phi_k \left(x \right) = e^{ikx}. \tag{10.10}$$

Here, we consider several properties associated with the Fourier expansion, namely (i) What is the relation between the series and the function u? and (ii) How rapidly does the series converge?

The basic issue we address is how well the function $u(x)$ is approximated by the sequence of trigonometric polynomials

$$P_N u(x) = \sum_{k=-N/2}^{N/2} \hat{u}_k e^{ikx}, \tag{10.11}$$

as $N \to \infty$. The function $P_N u(x)$ is the Nth order truncated Fourier series of u. We define the L^2 inner product and norm in the standard way, so that for any two functions $u(x)$ and $v(x)$,

$$(u, v) = \int_0^{2\pi} u(x)\overline{v(x)}dx, \tag{10.12}$$

and

$$\|u\|^2 = \int_0^{2\pi} |u(x)|^2 dx. \tag{10.13}$$

From inserting Eq. (10.7) into (10.13), and making use of the orthogonality relation, Eq. (10.9), we obtain the Parseval identity

$$\|u\|^2 = 2\pi \sum_{k=-\infty}^{\infty} |\hat{u}_k|^2. \tag{10.14}$$

From this, it follows, using (10.11), that

$$\|u - P_N u\|^2 = 2\pi \sum_{|k| \geq N/2} |\hat{u}_k|^2. \tag{10.15}$$

Hence, the size of the error created by replacing u with its Nth order truncated Fourier series depends upon how fast the Fourier coefficients of u decay to zero. We can show that this depends on the regularity of u in the domain $[0, 2\pi]$. Thus, integrating Eq. (10.8) by parts, we obtain

$$\begin{aligned} \hat{u}_k &= \frac{1}{2\pi} \int_0^{2\pi} u(x) e^{-ikx} dx, \\ &= \frac{1}{2\pi i k} \int_0^{2\pi} u'(x) e^{-ikx} dx, \end{aligned}$$

where $()'$ denotes differentiation with respect to x, and use was made of the periodic boundary conditions. Iterating on this result, we find that if u is m times continuously differentiable in $[0, 2\pi]$ and if the jth derivative of u, $u^{(j)}$, is periodic for all $j \leq m - 2$, then

$$\hat{u}_k = O(k^{-m}). \tag{10.16}$$

It follows that the kth Fourier coefficient of a function which is infinitely differentiable and periodic decays faster than any negative power of k.

As an example, consider the function

$$U(x) = \frac{3}{5 - 4\cos x}.$$

It is infinitely differentiable and periodic with all its derivatives in $[0, 2\pi]$. We can calculate directly that its Fourier coefficients are

$$\hat{u}_k = 2^{-|k|}; \qquad k = 0, \pm 1, \cdots,$$

which is an example of exponential decay, or spectral accuracy.

It is to be emphasized that the asymptotic rate of decay of the Fourier coefficients does not convey the entire story of the error made in a given truncation. If a series has an asymptotic rate of decay as given by Eq. (10.16), this decay is observed only for $k > k_0$, where k_0 is indicative of the smoothness of the function. If the series is truncated below k_0, then the approximation can be quite bad.

10.2.2 Discrete Fourier Expansion

In most applications, the discrete Fourier expansion is preferred to the continuous Fourier expansion described in the last section since it provides an efficient way to compute the Fourier coefficients of a function that is only known at discrete points, and to recover in physical space the information that is calculated in transform space. As we have seen in Section 2.5, the discrete Fourier transform makes use of the points

$$x_j = \frac{2\pi j}{N}, \qquad j = 0, 1, \cdots, N-1,$$

normally referred to as grid points, to define the transform pair

$$\tilde{u}_k = \frac{1}{N} \sum_{j=0}^{N-1} u(x_j) e^{-ikx_j}, \tag{10.17}$$

$$u(x_j) = \sum_{k=-N/2}^{N/2-1} \tilde{u}_k e^{ikx_j}. \tag{10.18}$$

Equations (10.17) and (10.18) define an exact transform pair between the N grid values $u(x_j)$ and the N discrete Fourier coefficients $\tilde{u}_k$. We can transform from one representation to the other without loss of information. This can be shown by inserting Eq. (10.17) into Eq. (10.18) and making use of the orthogonality property

$$\frac{1}{N} \sum_{k=-N/2}^{N/2-1} e^{-2\pi ik(j-j')/N} = \delta_{jj'}. \tag{10.19}$$

We can find the relation between the discrete Fourier coefficients $\tilde{u}_k$ and the exact Fourier coefficients as given in Eq. (10.8) by equating $u(x_j)$ from Eq. (10.7) to that given by Eq. (10.18) to obtain

$$\tilde{u}_k = \hat{u}_k + \sum_{\substack{m=-\infty \\ m \neq 0}}^{\infty} \hat{u}_{k+Nm}. \tag{10.20}$$

We see that the kth mode of the discrete transform depends not only on the kth mode of u, but also on all the modes that "alias" the kth one on the discrete grid. This is illustrated in Figure 10.1. The error associated with the difference between the discrete transform representation and the truncated Fourier series is called the *aliasing error*. It is orthogonal to the truncation error, but can be shown to be of the same asymptotic order for N large.

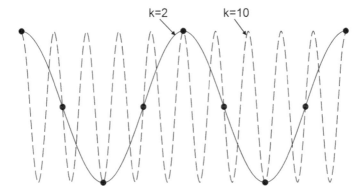

FIGURE 10.1: Example of a $k = 2$ and $k = 2 + N$ mode aliasing on a grid with $N = 8$. The two modes take on the same values at the grid points.

Note that for the transform pair defined by Eqs. (10.17) and (10.18), the derivatives of u defined at the grid points are given by

$$u'(x_j) = \sum_{k=-N/2}^{N/2-1} u_k e^{ikx_j}, \tag{10.21}$$

where

$$u_k - ik\tilde{u}_k - \frac{ik}{N} \sum_{j=0}^{N-1} u(x_j) e^{-ikx_j}. \tag{10.22}$$

Thus, if a function is known in real space, to evaluate the derivative one must evaluate the discrete Fourier coefficients, multiply by ik, and transform back to physical space.

It has been noted by several authors [217] that the wavenumber $k = -N/2$ appears asymmetrically in the representation given by Eq. (10.18) and those that follow. This can cause a problem, for example in a time-dependent problem, if $\tilde{u}_{-N/2}$ has a non-zero imaginary part, since $u(x_j)$ would then not be a real-valued function. The recommended solution is to enforce the condition that $\tilde{u}_{-N/2}$ be zero throughout the time dependence. The problem occurs because N is normally taken to be even in order to utilize FFT algorithms, but an even number does not result in a symmetric distribution about 0.

10.2.3 Chebyshev Polynomials in $(-1, 1)$

The eigenfunctions of a suitable Sturm–Liouville problem in the interval $(-1, 1)$ form a complete set of orthogonal basis functions. The ones of special interest are those such that the expansion of a smooth infinitely differentiable function in terms of their eigenfunctions guarantees spectral accuracy. This is the case if the associated Sturm–Liouville problem is *singular*, i.e., of the form

$$\begin{cases} (-pu')' + qu = \lambda wu & \text{in } (-1,1), \\ +\text{boundary conditions for u}, \end{cases}$$

(10.23)

where λ is the eigenvalue. The coefficients p, q, and w are each prescribed, real-valued functions with the following properties: p is continuously differentiable, positive in $(-1,1)$, and zero at the boundaries $x = \pm 1$; q is continuous, non-negative, and bounded in $(-1,1)$, and the weight function w is is continuous, non-negative, and integrable over the interval $(-1,1)$. Equations whose eigenfunctions are algebraic polynomials are of particular importance because of the efficiency with which they are evaluated.

There are several sets of eigenfunctions that have these desirable properties, but we will consider here the Chebyshev polynomials [216], denoted $\{T_k(x), k = 0, 1, \cdots\}$, which are the eigenfunctions of the singular Sturm–Liouville problem

$$\left[\sqrt{1 - x^2} T_k'(x) \right]' + \frac{k^2}{\sqrt{1 - x^2}} T_k(x) = 0,$$

(10.24)

or

$$\left(1 - x^2\right) T_k'' - x T_k' + k^2 T_k = 0.$$

(10.25)

Equation (10.24) is of the form of Eq. (10.23) if we identify

$$\begin{aligned} p(x) &= (1 - x^2)^{\frac{1}{2}}, \\ q(x) &= 0, \\ w(x) &= (1 - x^2)^{-\frac{1}{2}}. \end{aligned}$$

For any k, $T_k(x)$ is even if k is even, and odd if k is odd. If $T_k(x)$ is such that $T_k(1) = 1$, then

$$T_k(x) = \cos k\theta, \qquad \theta = \arccos x.$$

(10.26)

Hence, $T_k(\cos \theta) = \cos k\theta$. The Chebyshev polynomials are therefore just cosine functions after a change of independent variables. The transformation $x = \cos \theta$ allows many results of the Fourier system to be adapted to the Chebyshev system.

The Chebyshev polynomials can be expanded in power series as

$$T_k(x) = \frac{k}{2} \sum_{\ell=0}^{\lfloor k/2 \rfloor} (-1)^k \frac{(k - \ell - 1)!}{\ell!(k - 2\ell)!} (2x)^{k-2\ell},$$

where $[k/2]$ denotes the integral part of $k/2$. One can make use of the trigono-metric relation $\cos(k+1)\theta+\cos(k-1)\theta = 2\cos\theta\cos k\theta$ to obtain the recurrence relation

$$T_{k+1}(x) = 2xT_k(x) - T_{k-1}(x). \tag{10.27}$$

For the Chebyshev polynomials of the first kind, this recurrence relation is initialized with $T_0(x) = 1, T_1(x) = x$. This gives for the first few polynomials:

$$
\begin{aligned}
T_2(x) &= -1 + 2x^2, \\
T_3(x) &= -3x + 4x^3, \\
T_4(x) &= 1 - 8x^2 + 8x^4, \\
T_5(x) &= 5x - 20x^3 + 16x^5, \\
T_6(x) &= -1 + 18x^2 - 48x^4 + 32x^6.
\end{aligned}
$$

The Chebyshev polynomials satisfy the orthonormalization condition

$$\int_{-1}^{1} T_k(x)T_{k'}(x)\frac{dx}{\sqrt{1-x^2}} = \delta_{kk'}c_k\frac{\pi}{2}, \tag{10.28}$$

where

$$c_k = \begin{cases} 2 & \text{if } k = 0, \\ 1 & \text{if } k \geq 1. \end{cases} \tag{10.29}$$

Some other useful integral relations involving the Chebyshev polynomials are

$$\int_{-1}^{1} T_n(x)T_m(x)dx - \begin{cases} 0, & n+m \text{ odd}, \\ \frac{1}{1-(n+m)^2} + \frac{1}{1-(n-m)^2} & n+m \text{ even}, \end{cases} \tag{10.30}$$

$$\int_{-1}^{1} T_n'(x)T_m'(x)dx = \begin{cases} 0, & n+m \text{ odd}, \\ \frac{nm}{2}\left[J_{|(n-m)/2|} - J_{|(n+m)/2|}\right] & n+m \text{ even}, \end{cases} \tag{10.31}$$

where

$$J_k = \begin{cases} -4\sum_{p=1}^{k}\frac{1}{2p-1}, & k \geq 1, \\ 0, & k = 0. \end{cases} \tag{10.32}$$

The Chebyshev expansion of a function u defined on $(-1,1)$ is given by

$$u(x) = \sum_{k=0}^{\infty} \hat{u}_k T_k(x), \tag{10.33}$$

$$\hat{u}_k = \frac{2}{\pi c_k}\int_{-1}^{1} u(x)T_k(x)w(x)dx. \tag{10.34}$$

The result in Eq. (10.34) follows from Eq. (10.33) by using the orthogonality condition, Eq. (10.28). By the same integration by parts argument developed for the Fourier series, the Chebyshev coefficients can be shown to decay faster than algebraically.

The derivative of a function $u(x)$ expanded in Chebyshev polynomials can be written as

$$u'(x) = \sum_{m=0}^{\infty} \hat{u}_m^{(1)} T_m(x). \tag{10.35}$$

A recurrence relation for the $\hat{u}_m^{(1)}$ can be obtained by making use of the relation

$$2T_k(x) = \frac{1}{k+1}T'_{k+1}(x) - \frac{1}{k-1}T'_{k-1}(x), \qquad k \geq 1. \tag{10.36}$$

This follows from the trigonometric identity

$$2\sin\theta\cos k\theta = \sin(k+1)\theta - \sin(k-1)\theta.$$

By differentiating Eq. (10.33), equating it to Eq. (10.35), and making use of the recurrence relation in Eq. (10.36), we obtain the relation

$$2k\hat{u}_k = c_{k-1}\hat{u}_{k-1}^{(1)} - \hat{u}_{k+1}^{(1)}; \qquad k \geq 1. \tag{10.37}$$

This provides an efficient way of differentiating a polynomial of degree N in Chebyshev space. Since $\hat{u}_k^{(1)} = 0$ for $k \geq N$, the non-zero coefficients can be computed in decreasing order by the recurrence relation

$$c_k\hat{u}_k^{(1)} = \hat{u}_{k+2}^{(1)} + 2(k+1)\hat{u}_{k+1}; \qquad 0 \leq k \leq N-1. \tag{10.38}$$

The generalization of this to higher derivatives is

$$c_k\hat{u}_k^{(q)} = \hat{u}_{k+2}^{(q)} + 2(k+1)\hat{u}_{k+1}^{(q-1)}. \tag{10.39}$$

The expressions for $\hat{u}_k^{(1)}$ and $\hat{u}_k^{(2)}$ can also be expressed as

$$\hat{u}_k^{(1)} = \frac{2}{c_k} \sum_{\substack{p=k+1 \\ p+k \text{ odd}}}^{\infty} p\hat{u}_p, \tag{10.40}$$

and

$$\hat{u}_k^{(2)} = \frac{1}{c_k} \sum_{\substack{p=k+2 \\ p+k \text{ even}}}^{\infty} p(p^2 - k^2)\hat{u}_p. \tag{10.41}$$

10.2.4 Discrete Chebyshev Series

As for the Fourier series, a discrete Chebyshev series can be defined. In a discrete transform, the fundamental representation of a function $u(x)$ on the interval $(-1, 1)$ is in terms of the values at discrete points. Derivatives are approximated by analytic derivatives of the interpolating polynomial.

In a discrete Chebyshev transform we seek discrete polynomial coefficients $\tilde{u}_k$ and a set of collocation points x_j such that

$$u(x_j) = \sum_{k=0}^{N} \tilde{u}_k T_k(x_j); \qquad j = 0, 1, \cdots, N. \tag{10.42}$$

One can readily verify that if we introduce as the discrete points those known as the Gauss–Lobatto points,

$$x_j = \cos \frac{\pi j}{N}; \qquad j = 0, 1, \cdots, N, \tag{10.43}$$

we then have the discrete transform pair,

$$\tilde{u}_k = \frac{2}{N \bar{c}_k} \sum_{j=0}^{N} \frac{1}{\bar{c}_j} u(x_j) T_k(x_j); \qquad k = 0, 1, \cdots, N, \tag{10.44}$$

$$= \frac{2}{N \bar{c}_k} \sum_{j=0}^{N} \frac{1}{\bar{c}_j} u(x_j) \cos \frac{\pi j k}{N}; \qquad k = 0, 1, \cdots, N,$$

$$u(x_j) = \sum_{k=0}^{N} \tilde{u}_k T_k(x_j); \qquad j = 0, 1, \cdots, N, \tag{10.45}$$

$$= \sum_{k=0}^{N} \tilde{u}_k \cos \frac{\pi j k}{N}; \qquad j = 0, 1, \cdots, N.$$

Here,

$$\bar{c}_j = \begin{cases} 2 & \text{if} \quad j = 0, N, \\ 1 & \text{if} \quad 1 \leq j \leq N - 1. \end{cases}$$

These can be evaluated using a variant of the fast Fourier transform algorithm [217]. Equation (10.42) can be regarded as an interpolating formula for x values that are away from the discrete points x_j. For any value of x in the interval, we thus have as the interpolating formula:

$$u(x) = \sum_{k=0}^{N} \tilde{u}_k T_k(x). \tag{10.46}$$

As with the case of Fourier series, a relation exists between the finite and discrete Chebyshev coefficients,

$$\tilde{u}_k = \hat{u}_k + \sum_{\substack{j = 2mN \pm k \\ j > N}} \hat{u}_j, \tag{10.47}$$

which exhibits the form of the aliasing error.

10.3   Non-Linear Problems

Here we consider the application of spectral methods to the non-linear Burgers's equation

$$\frac{\partial u}{\partial t} + u\frac{\partial u}{\partial x} - \nu\frac{\partial^2 u}{\partial x^2} = 0, \qquad (10.48)$$

where ν is a positive constant, and initial conditions $u(x,0)$ and boundary conditions are assumed to be given. We will consider several different methods for discretizing Eq. (10.48) corresponding to different choices of trial functions and of test functions.

10.3.1   Fourier Galerkin

Let us look for a solution which is periodic on the interval $(0, 2\pi)$. We take as the trial space the set of all trigonometric polynomials of degree $\leq N/2$. The approximation function u^N is represented as the truncated Fourier series

$$u^N(x,t) = \sum_{k=-N/2}^{N/2-1} \hat{u}_k(t)e^{ikx}. \qquad (10.49)$$

The fundamental unknowns are the coefficients $\hat{u}_k(t)$, $k = -N/2, \cdots, N/2-1$. Using the complex conjugate of the trial functions as test functions, a set of ODE's for $\hat{u}_k$ are obtained by requiring that the residual of Eq. (10.48) be orthogonal to all the test functions, i.e.,

$$\frac{1}{2\pi}\int_0^{2\pi}\left(\frac{\partial u^N}{\partial t} + u^N\frac{\partial u^N}{\partial x} - \nu\frac{\partial^2 u^N}{\partial x^2}\right)e^{-ikx}dx = 0;$$
$$k = -\frac{N}{2}, \cdots, \frac{N}{2} - 1. \qquad (10.50)$$

Due to the orthogonality property of the test and trial functions, we have

$$\frac{\partial \hat{u}_k}{\partial t} + \left(\widehat{u^N\frac{\partial u^N}{\partial x}}\right)_k + k^2\nu\hat{u}_k = 0 ; \quad k = -\frac{N}{2}, \cdots, \frac{N}{2} - 1, \quad (10.51)$$

where the convolution sum is defined by

$$\left(\widehat{u^N\frac{\partial u^N}{\partial x}}\right)_k = \frac{1}{2\pi}\int_0^{2\pi}u^N\frac{\partial u^N}{\partial x}e^{-ikx}dx. \qquad (10.52)$$

Equation (10.52) is a special case of the more general quadratic non-linear term

$$(\widehat{uv})_k = \frac{1}{2\pi} \int_0^{2\pi} uv e^{-ikx} dx,$$ (10.53)

where u and v denote generic trigonometric polynomials of degree $N/2$. When their expansions are inserted into Eq. (10.53) and the orthogonality property is invoked, the expression

$$(\widehat{uv})_k = \sum_{p+q=k} \hat{u}_p \hat{v}_q$$ (10.54)

results. This is a convolution sum. In the "pure spectral" method, the higher harmonics generated in Eq. (10.54) with $|k| > N/2$ are neglected.

When fast transform methods are applied to evaluate the convolution sums such as indicated in Eq. (10.53), it is known as the *pseudo-spectral* method. The procedure is to use the FFT to transform $\hat{u}_m$ and $\hat{v}_n$ to physical space, perform the multiplication, and then use the FFT to transform back again. There is a subtle point, however, in that the result one gets in doing this, which we will call $(u_j v_j)_k$, differs from the pure Galerkin spectral result given in Eq. (10.54) as follows:

$$(u_j v_j)_k = (\widehat{uv})_k + \sum_{p+q=k\pm N} \hat{u}_p \hat{v}_q.$$ (10.55)

The second term on the right-hand side is the *aliasing error*. It arises from the fact that higher harmonics are generated when performing the product, and since the representation is not capable of accurately representing these higher harmonics, they manifest themselves as modifications to the lower-order harmonics.

There are several techniques for getting rid of an aliasing error. The most popular is to use a discrete transform with M rather than N points, where $M \geq 3/2N$, and to pad $\hat{u}_k$ and $\hat{v}_k$ with zeros for all the higher harmonics with $|k| > N/2$. This will cause one of the factors in the aliasing error product to be zero for all $|k| < N/2$ which are kept in the final product.

There is some debate on how important it is to remove the aliasing error. It appears to be more important to do so in problems that are marginally resolved. There is some support for the claim that, for any given problem, an aliased calculation will yield just as acceptable an answer as a de-aliased one, once sufficient resolution has been achieved.

10.3.2 Fourier Collocation

Here we again take periodicity on $(0, 2\pi)$ and take the same trial functions as in Eq. (10.49), but use delta functions as test functions. We now think of the approximate solution u^N as represented by its values at the grid points

$x_j = 2\pi j/N; j = 0, \cdots, N-1$. The delta function test functions require that the equation be satisfied exactly at the grid points.

$$\frac{\partial u^N}{\partial t} + u^N \frac{\partial u^N}{\partial x} - \nu \frac{\partial^2 u^n}{\partial x^2} \bigg|_{x=x_j} = 0; \qquad j = 0, 1, \cdots, N-1. \qquad (10.56)$$

Initial conditions for this are simply

$$u^N(x_j, 0) = u_0(x_j). \qquad (10.57)$$

The derivative terms are evaluated as in Eq. (10.21), which involves performing a discrete transform, differentiating, and transforming back to real space. The non-linear term is evaluated in real space as the pointwise product of two vectors $u^N(x_j)$ and $\partial u^N/\partial x|_{x_j}$. The Fourier collocation method is very similar to, and in many cases identical to, the Fourier Galerkin method when pseudo-spectral techniques are used to evaluate the convolutions of the latter.

10.3.3   Chebyshev Tau

Let us now consider solving Eq. (10.48) on the interval $(-1, 1)$ subject to the Dirichlet boundary conditions

$$u(-1, t) = u(1, t) = 0. \qquad (10.58)$$

We expand the solution as a Chebyshev series

$$u^N(x, t) = \sum_{k=0}^{N} \hat{u}_k T_k(x), \qquad (10.59)$$

with the Chebyshev coefficients comprising the fundamental representation of the approximation. Using the $T_k(x)$ as test functions, we enforce Eq. (10.48) as follows

$$\frac{2}{\pi c_k} \int_{-1}^{1} \left(\frac{\partial u^N}{\partial t} + u^N \frac{\partial u^N}{\partial x} - \nu \frac{\partial^2 u^N}{\partial x^2} \right) T_k(x) \left(1 - x^2\right)^{-\frac{1}{2}} dx = 0;$$
$$k = 0, \cdots, N-2. \qquad (10.60)$$

The boundary conditions impose the additional constraints

$$u^N(-1, t) = u^N(1, t) = 0.$$

Using the orthogonality relations, Eq. (10.60) reduces to

$$\frac{\partial \hat{u}_k}{\partial t} + \left(\widehat{u^N \frac{\partial u^N}{\partial x}} \right)_k - \nu \hat{u}_k^{(2)} = 0;$$
$$k = 0, 1, \cdots, N-2, \qquad (10.61)$$

where, from Eq. (10.41),

$$\hat{u}_k^{(2)} = \frac{1}{c_k} \sum_{\substack{p=k+2 \\ p+k \ even}}^{N} p\left(p^2 - k^2\right)\hat{u}_p,$$

and, from Eq. (10.60),

$$\left(\widehat{u^N \frac{\partial u^N}{\partial x}}\right)_k = \frac{2}{\pi c_k} \int_{-1}^{1} \left(u^N \frac{\partial u^N}{\partial x}\right) T_k(x)\left(1 - x^2\right)^{-\frac{1}{2}} dx.$$

In terms of the Chebyshev coefficients, the boundary conditions become

$$\sum_{k=0}^{N} \hat{u}_k = 0, \tag{10.62}$$

$$\sum_{k=0}^{N} (-1)^k \hat{u}_k = 0. \tag{10.63}$$

The initial conditions are

$$\hat{u}_k(0) = \frac{2}{\pi c_k} \int_{-1}^{1} u_0(x) T_k(x)(1 - x^2)^{-\frac{1}{2}} dx \quad ; \quad k = 0, 1, \cdots, N. \tag{10.64}$$

Equations (10.61)–(10.64) provide a complete set of ODEs for this approximation. An implicit time advance results in a upper triangular matrix that needs to be solved each time step. This normally requires $\mathcal{O}(N^2)$ operations. However, in many cases the recursion relations can be used to simplify this matrix into a near tridiagonal matrix that only requires $\mathcal{O}(N)$ operations per time step [217].

10.4 Time Discretization

Here we consider how to evolve the spectral amplitudes in time. Say that the original partial differential equation is an initial value problem given by

$$\frac{\partial \mathbf{u}}{\partial t} = \mathbf{f}(\mathbf{u}, t). \tag{10.65}$$

We expand in the set of trial functions and apply a projection operator $\mathbf{Q}_N$ which corresponds to multiplying by test functions and integrating over all space. We can then write the discrete time-dependent problem as

$$\mathbf{Q}_N \frac{d\mathbf{u}^N}{dt} = \mathbf{Q}_N \mathbf{F}_N \left(\mathbf{u}^N, t\right), \tag{10.66}$$

where $\mathbf{u}^N$ is the spectral approximation to $\mathbf{u}$, and $\mathbf{F}_N$ is the spectral approximation to $\mathbf{F}$. Now let $\mathbf{U}(t) = \mathbf{Q}_N \mathbf{u}^N(t)$ be either the grid values for a collocation method or the expansion amplitudes for a Galerkin method, so that it is enough to consider

$$\frac{d\mathbf{U}}{dt} = \mathbf{F}(\mathbf{U}), \qquad (10.67)$$

or, in the linearized version

$$\frac{d\mathbf{U}}{dt} = \mathbf{A} \cdot \mathbf{U}, \qquad (10.68)$$

where $\mathbf{A}$ is the matrix resulting from the linearization of Eq. (10.67).

The definition of stability of a spectral method is largely the same as that discussed in Section 2.4, except that now we look for a condition on the maximum time step as a function of N rather than of δx. If the maximum allowable time step, δt_{max}, is independent of N, the method is unconditionally stable.

The time advancement equation (10.68) is a system of ordinary differential equations (ODEs). In this form it is sometimes called *the method of lines*. There is an extensive literature on numerical methods for systems of ODEs which examine their stability. Some excellent reference books are [217, 218, 219, 220]. The crucial property of $\mathbf{A}$ is its spectrum, for this will determine the stability of the time discretization.

The spatial eigenfunctions for Fourier approximations are e^{ikx}. The corresponding eigenvalues for basic convection or diffusion equations are obvious.

$$\frac{\partial}{\partial x} \quad \rightarrow \quad ik,$$
$$\frac{\partial^2}{\partial x^2} \quad \rightarrow \quad -k^2.$$

For Chebyshev and other approximations, the eigenvalues can be computed, but are more complicated [217].

In most applications of spectral methods to partial differential equations, the temporal discretization uses conventional finite differences. As an example, consider the leapfrog method applied to the spectral formulation of a convection problem as given by Eq. (10.6),

$$u_k^{n+1} = u_k^{n-1} - 2ik\,\delta t\,v u_k^n.$$

The amplification factor r satisfies the quadratic equation

$$r^2 + 2ibr - 1 = 0,$$

with $b = k\,\delta t\,v$. The stability condition $|r| \le 1$ is equivalent to $|b| \le 1$, which translates into the time step restriction

$$\delta t \quad \leq \quad \frac{1}{v k_{max}},$$

$$\leq \quad \frac{2}{v N}.$$

This is the spectral equivalent of the CFL condition.

Many of the same considerations that enter into choosing a time advance scheme for a system of finite difference equations also enter into that for a spectral system. Parabolic and hyperbolic terms often need to be treated by different methods. However, because of the orthogonality of the eigenfunctions, there is typically no penalty for using an implicit method for the linear terms, even if explicit methods are used for the non-linear terms.

10.5 Implicit Example: Gyrofluid Magnetic Reconnection

As as example of the power of the spectral representation, we will consider the normalized set of equations solved by Loureiro and Hammett [221] in their two-dimensional two-field magnetic reconnection studies,

$$\frac{\partial n_e}{\partial t} + [\phi, n_e] \quad = \quad [\psi, \nabla_\perp^2 \psi] + \nu \nabla_\perp^2 n_e, \tag{10.69}$$

$$\frac{\partial \psi}{\partial t} + [\phi, \psi] \quad = \quad \rho_s^2 [n_e, \psi] + \eta \nabla_\perp^2 \psi, \tag{10.70}$$

$$n_e \quad = \quad \frac{1}{\rho_i^2} \left[\hat{\Gamma}_0(b) - 1 \right] \phi. \tag{10.71}$$

Here, n_e is the perturbed electron density, the in-plane velocity is represented as $\mathbf{u}_\perp = \hat{\mathbf{z}} \times \nabla \phi$, the in-plane magnetic field is given by $\mathbf{B}_\perp = \hat{\mathbf{z}} \times \nabla \psi$, ρ_i and ρ_s are the ion and ion-sound Larmor radius, assumed to be constants, ν is the (constant) viscosity, and η is the (constant) resistivity. We have also adopted the notation that $\nabla_\perp^2 \equiv \partial_x^2 + \partial_y^2$, and introduced the Poisson bracket between two scalar fields A and B as $[A, B] \equiv \partial_x A \partial_y B - \partial_y A \partial_x B$.

Equation (10.71) is known as the gyrokinetic Poisson equation [222]. In this approximation and normalization, the electric potential is the same as the velocity stream function ϕ. The integral operator $\hat{\Gamma}_0(b)$ expresses the average of the electrostatic potential over rings of radius ρ_i. In Fourier space, this operator takes a particularly simple form,

$$\hat{\Gamma}_0(b) = e^{-b} I_0(b), \tag{10.72}$$

where $b = k_\perp^2 \rho_i^2 \equiv (k_x^2 + k_y^2)\rho_i^2$ and I_0 is the modified Bessel function of zeroth order. When implemented numerically in real space, a Padé approximation to the operator is normally used [223],

$$\hat{\Gamma}_0(b) - 1 \approx -\frac{b}{1+b}.$$

Equation (10.71) then takes the real space form

$$\left(1 - \rho_i^2 \nabla_\perp^2\right) n_e = \nabla_\perp^2 \phi. \tag{10.73}$$

There is a clear advantage in treating this term in Fourier space where Eq. (10.71) becomes just an algebraic equation, as opposed to Eq. (10.73), which is a two-dimensional elliptic partial differential equation.

Linearizing Eqs. (10.69)–(10.71) about an equilibrium with $\mathbf{B}_\perp = B_0 \hat{\mathbf{y}}$, with B_0 a constant, and with the equilibrium perturbed density n_e and stream function ϕ zero, and ignoring the diffusive terms for now, one can combine the equations to give a single second-order (in time) equation of the form

$$\frac{\partial^2 \psi}{\partial t^2} = \mathcal{L}(\psi), \tag{10.74}$$

where $\mathcal{L}$ is just a linear operator that becomes a multiplicative operator in k-space, i.e., $\mathcal{L}(\psi) = -\hat{\omega}^2 \psi$, with

$$\hat{\omega}^2 = k_\perp^2 \left(\rho_s^2 - \frac{\rho_i^2}{\hat{\Gamma}_0(b) - 1}\right) k_y^2 B_0^2. \tag{10.75}$$

It is seen that Eq. (10.74) is of the same form as Eq. (9.83) and thus the method of differential approximation can be applied. In [221] they have found that a good approximation to Eq. (10.75) that remains valid even when the magnetic field has some spatial variation is given by

$$\hat{\omega}^2 = k_\perp^4 \left(\rho_s^2 - \frac{\rho_i^2}{\hat{\Gamma}_0(b) - 1}\right) B_{\perp,max}^2, \tag{10.76}$$

where $B_{\perp,max}^2$ is the maximum value of $B_\perp$ on the grid.

To solve Eqs. (10.69)–(10.71), we expand ψ in Fourier basis functions,

$$\psi(x, y, t) = \sum_{k_x=-N/2}^{N/2-1} \sum_{k_y=-N/2}^{N/2-1} \psi_\mathbf{k} e^{i(k_x x + k_y y)}, \tag{10.77}$$

and similarly for ϕ, and n_e. We multiply each equation by a test function of the form $e^{-i(k_x' x + k_y' y)}$, and integrate over one period of the periodic domain. Equation (10.71) becomes just an algebraic equation relating $n_{e\mathbf{k}}$ and $\phi_\mathbf{k}$. We can use this to eliminate $n_{e\mathbf{k}}$ from the other equations, which can then be written in the form

$$\frac{\partial \psi_\mathbf{k}}{\partial t} = \widehat{F(\phi, \psi)}_\mathbf{k} - \eta k^2 \psi_\mathbf{k}, \tag{10.78}$$

$$\frac{\partial \phi_\mathbf{k}}{\partial t} = \widehat{G(\phi, \psi)}_\mathbf{k} - \nu k^2 \phi_\mathbf{k}, \tag{10.79}$$

where we have written the combined non-linear convolution terms as $\widehat{F(\phi, \psi)}_{\mathbf{k}}$ and $\widehat{G(\phi, \psi)}_{\mathbf{k}}$ for notational simplicity.

It is noted that the diffusion terms can be integrated analytically. If we define $D_\eta = k^2 \eta$ and $D_\nu = k^2 \nu$, then we can make the variable transformations

$$\psi_{\mathbf{k}}(t) = e^{-D_\eta t} \tilde{\psi}_{\mathbf{k}}(t), \qquad \phi_{\mathbf{k}}(t) = e^{-D_\nu t} \tilde{\phi}_{\mathbf{k}}(t). \qquad (10.80)$$

This allows Eqs. (10.78) and (10.79) to be written in the form

$$\frac{\partial \tilde{\psi}_{\mathbf{k}}}{\partial t} = e^{D_\eta t} \widehat{F(\phi, \psi)}_{\mathbf{k}}, \qquad (10.81)$$

$$\frac{\partial \tilde{\phi}_{\mathbf{k}}}{\partial t} = e^{D_\nu t} \widehat{G(\phi, \psi)}_{\mathbf{k}}. \qquad (10.82)$$

The following two-step method has been used to advance Eqs. (10.81) and (10.82) from time $t^n = n\delta t$ to time $t^{n+1} = (n+1)\delta t$ [221],

$$\psi_{\mathbf{k}}^* = e^{-D_\eta \delta t} \psi_{\mathbf{k}}^n + \frac{\delta t}{2} \left(1 + e^{-D_\eta \delta t}\right) \widehat{F(\phi^n, \psi^n)}_{\mathbf{k}}, \qquad (10.83)$$

$$\phi_{\mathbf{k}}^* = e^{-D_\nu \delta t} \phi_{\mathbf{k}}^n + \frac{\delta t}{2} \left(1 + e^{-D_\nu \delta t}\right) \widehat{G(\phi^n, \psi^n)}_{\mathbf{k}}, \qquad (10.84)$$

$$\psi_{\mathbf{k}}^{n+1} = e^{-D_\eta \delta t} \psi_{\mathbf{k}}^n + \frac{\delta t}{2} e^{-D_\eta \delta t} \widehat{F(\phi^n, \psi^n)}_{\mathbf{k}}$$
$$+ \frac{\delta t}{2} \widehat{F(\phi^*, \psi^*)}_{\mathbf{k}} - \frac{\hat{\omega}^2 \delta t^2}{4} \left(\psi_{\mathbf{k}}^{n+1} - \psi_{\mathbf{k}}^n\right), \qquad (10.85)$$

$$\phi_{\mathbf{k}}^{n+1} = e^{-D_\nu \delta t} \phi_{\mathbf{k}}^n + \frac{\delta t}{2} e^{-D_\nu \delta t} \widehat{G(\phi^n, \psi^n)}_{\mathbf{k}}$$
$$+ \frac{\delta t}{2} \widehat{G(\phi^*, \psi^{n+1})}_{\mathbf{k}}. \qquad (10.86)$$

Equation 10.85 is solved algebraically for $\psi_{\mathbf{k}}^{n+1}$, which is inserted into Eq. 10.86 to solve for $\psi_{\mathbf{k}}^{n+1}$ It is seen that in the absence of diffusion ($D_\eta = D_\nu = 0$), the method described by Eqs. (10.83)–(10.86) is very similar to the method of differential approximation described by Eqs. (9.85) and (9.86), except that the time centering is achieved by a predictor-corrector method rather than a leapfrog time staggering. It is shown in Problem 10.5 that this method gives unconditional stability. The use of the spectral representation has not only greatly simplified the gyrokinetic Poisson equation, but it has converted the operator in the method of differential approximation from a differential operator to a simple multiplicative operator, thus greatly simplifying the solution procedure.

10.6   Summary

Spectral methods use global orthogonal functions as basis functions. They are especially powerful and efficient in problems with periodic boundary conditions so that Fourier series can be used, and in situations where a small number of basis functions do a reasonable job of representing the true solution. For non-linear problems, the fast Fourier transform (FFT) algorithm allows computation of non-linear products, or convolutions, in a time that scales like $N \log N$ instead of N^2, making the calculation of non-linear terms feasible for large problems. In problems without periodic boundary conditions, orthogonal polynomials such as the Chebyshev polynomials can be used and have many of the desirable properties of trigonometric functions. Considerations for time advance of spectral methods are very similar to those for finite difference methods. It often happens that one uses a mixed representation, which treats one or more (usually periodic) coordinates with the spectral representation, and one or more with either finite difference or finite element.

Problems

10.1: Derive the relation between the discrete Fourier coefficients $\tilde{u}_k$ and the exact Fourier coefficients $\hat{u}_k$ as given by Eq. 10.20.

10.2: Derive the orthonormalization condition for the Chebyshev polynomials, Eq. (10.28), by using the relations in Eq. (10.26) and trigonometric identities.

10.3: By differentiating Eq. (10.33), equating it to Eq. (10.35), and making use of the recurrence relation in Eq. (10.36), derive the relation

$$2k\hat{u}_k = c_{k-1}\hat{u}_{k-1}^{(1)} - \hat{u}_{k+1}^{(1)}; \qquad k \geq 1. \tag{10.87}$$

Show that this implies

$$\hat{u}_k^{(1)} = \frac{2}{c_k} \sum_{\substack{p=k+1 \\ p+k \text{ odd}}}^{\infty} p\hat{u}_p, \tag{10.88}$$

10.6: Verify that the discrete transform pair given by Eqs. (10.44) and (10.45) is an exact transform pair.

10.5: Consider the time advance described by Eqs. (10.83)–(10.86). In the approximation where $F(\widehat{\phi^n, \psi^n})_{\mathbf{k}} = f\phi_{\mathbf{k}}^n$ and $G(\widehat{\phi^n, \psi^n})_{\mathbf{k}} = g\phi_{\mathbf{k}}^n$, and there is no diffusion so that $D_\eta = D_\nu = 0$, derive an expression for the amplification factor and a condition for numerical stability.

Chapter 11

The Finite Element Method

11.1 Introduction

The finite element method has emerged as an extremely powerful technique for solving systems of partial differential equations. As described in the now-classic textbook by Strang and Fix [224], the method was originally developed on intuitive grounds by structural engineers. It was later put on a sound theoretical basis by numerical analysts and mathematicians once it was realized that it was, in fact, an instance of the Rayleigh–Ritz technique. It is similar to the finite difference method in that it is basically a technique for generating sparse matrix equations that describe a system. However, it provides a prescription for doing this that is in many ways more straightforward than for finite differences, especially when going to high-order approximations. We start with an example using the Ritz method in one dimension with low-order elements. In subsequent sections this is extended to higher-order elements, the Galerkin method, and multiple dimensions.

11.2 Ritz Method in One Dimension

Suppose that the problem to be solved is in a variational form, i.e., find a function $U(x)$ which minimizes a given functional $I(U)$. The basic idea of the Ritz method is to choose a finite number of trial functions $\phi_0(x), \cdots, \phi_N(x)$, and among all their linear combinations,

$$u(x) = \sum_{j=0}^{N} q_j \phi_j(x), \tag{11.1}$$

find the one that is minimizing.

The unknown weights q_j appearing in Eq. (11.1) are determined by solving a system of N discrete algebraic equations. The minimizing process then automatically seeks out the combination which is closest to $U(x)$. The trial functions $\phi_j(x)$ must be convenient to compute with, and also general enough to closely approximate the unknown solution $U(x)$. In the finite element method, they are piecewise low-order polynomials.

11.2.1   An Example

We wish to find the function $v(x)$ that minimizes the quadratic functional

$$I(v) = \int_0^1 \left[p(x)[v'(x)]^2 + q(x)[v(x)]^2 - 2f(x)v(x) \right] dx, \tag{11.2}$$

subject to the boundary conditions $v(0) = 0$ and $v'(1) = 0$. (We have denoted derivatives with respect to x by $()'$.) Here, $p(x)$, $q(x)$, and $f(x)$ are prescribed functions which satisfy the positivity constraints $p(x) \geq 0$, $q(x) > 0$.

Note that the associated Euler equation for this variational problem is

$$\left[-\frac{d}{dx}\left(p(x)\frac{d}{dx} \right) + q(x) \right] v(x) = f(x), \tag{11.3}$$

where we have used the boundary conditions to set the boundary term $pvv'|_0^1$ to zero. Mathematically, we want to minimize $I(v)$ over the space of all functions $v(x)$ such that $I(v)$ is finite, i.e., we require that

$$\int (\frac{dv}{dx})^2 dx \quad \text{and} \quad \int v^2 dx$$

be finite. This is equivalent to the requirement that v belongs to H^1, which is the infinite dimensional Hilbert space of functions whose first derivatives have finite energy.

The Ritz method is to replace H^1 by a finite dimensional subspace S contained in H^1. The elements of S are the trial functions. The minimization of I leads to the solution of a system of simultaneous linear equations, the number of equations coinciding with the dimension of S. The Ritz approximation to the solution is the function u which minimizes I over the space S, so that

$$I(u) \leq I(v) \tag{11.4}$$

for all v in S. We will consider how to determine $u(x)$ by using an example solution space, and also how to estimate the error, or the difference between $u(x)$ and the true minimizing solution in H^1, $U(x)$.

11.2.2   Linear Elements

We divide the interval $[0, 1]$ into N equally spaced intervals of spacing $h = 1/N$. The simplest choice for the subspace S is the space of functions that are linear over each interval $[(j-1)h, jh]$ and continuous at the nodes $x = jh$. For $j = 0, \cdots, N$ let ϕ_j be the function in S which equals one at the particular node $x = jh$, and vanishes at all the others (Figure 11.1). Every

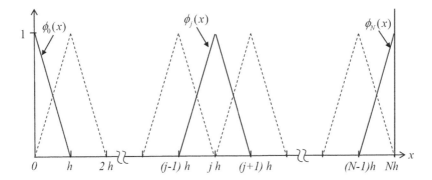

FIGURE 11.1: Linear finite elements ϕ_j equals 1 at node j, 0 at all others, and are linear in between. A half element is associated with the boundary nodes at $j = 0$ and $j = N$.

member of S can be written as a linear combination of the ϕ_j,

$$v(x) = \sum_{j=0}^{N} q_j \phi_j. \tag{11.5}$$

The "tent functions" ϕ_j are the simplest members of the more general set of piecewise polynomials. Note that the amplitudes q_j are the nodal values of the function and thus have direct physical significance. Also, note that the functions ϕ_j form a local basis in that they are orthogonal to all other elements except ϕ_{j+1} and ϕ_{j-1}. Only adjacent elements are directly coupled.

Let us now return to minimizing the functional $I(v)$, Eq. (11.2), using the linear elements so that we approximate $v(x)$ as in Eq. (11.5). For simplicity, we will now take the functions p and q to be constants. Computing one subinterval at a time, we find, as shown in Figure 11.2a,

$$\int_{jh}^{(j+1)h} v^2 dx = \int_{jh}^{(j+1)h} \left[q_{j+1} \left(\frac{x}{h} - j \right) + q_j \left(-\frac{x}{h} + (j+1) \right) \right]^2 dx$$

$$= \frac{h}{3} \left[q_j^2 + q_j q_{j+1} + q_{j+1}^2 \right].$$

Similarly, as shown in Figure 11.2b,

$$\int_{jh}^{(j+1)h} (v')^2 dx = \frac{1}{h} \left[q_j - q_{j+1} \right]^2.$$

We also need to compute the vector corresponding to the linear term, which corresponds to a load. This is given by

$$\int_0^1 f(x)v(x)dx = \sum_{j=1}^{N} q_j \int_0^1 f(x)\phi_j(x)dx. \tag{11.6}$$

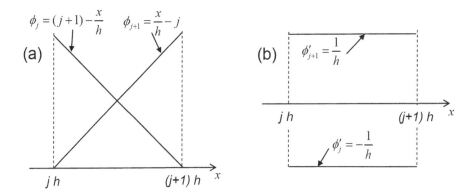

FIGURE 11.2: Expressions needed for overlap integrals for functions (a) and derivatives (b) for linear elements.

The usual method to evaluate Eq. (11.6) is to approximate f by linear interpolation at the nodes, i.e.,

$$f(x) = \sum_{k=0}^{N} f_k \phi_k(x), \tag{11.7}$$

where f_k is the value of f at the node points $x = kh$. We therefore approximate the integral

$$
\begin{aligned}
I &= \int_0^1 \left[p(v')^2 + qv^2 - 2fv \right] dx \\
&= \sum_{j=0}^{N-1} \left\{ \frac{p(q_j - q_{j+1})^2}{h} + \frac{qh \left(q_j^2 + q_j q_{j+1} + q_{j+1}^2 \right)}{3} \right. \\
&\quad \left. - 2 \left(\frac{h}{3} q_{j+1} f_{j+1} + \frac{h}{6} q_{j+1} f_j + \frac{h}{6} q_j f_{j+1} + \frac{h}{3} q_j f_j \right) \right\}. \tag{11.8}
\end{aligned}
$$

We next take note of the leftmost boundary condition to set $q_0 = 0$ and thus eliminate it from the system so that we only need to solve for the amplitudes $q_1, \cdots, q_N$. To extremize Eq. (11.8) with respect to the nodal values, we obtain a matrix equation to determine the q_j. Noting that each of the amplitudes away from the boundary appears in two terms in the sums, the conditions $\partial I / \partial q_i = 0$; for $i = 1, \cdots, N$ yields the matrix equation

$$\mathbf{K} \cdot \mathbf{q} = \mathbf{F}, \tag{11.9}$$

where $\mathbf{q}$ is the vector of unknown amplitudes

$$\mathbf{q} = [q_1, q_2, \cdots, q_N], \tag{11.10}$$

and the matrix $\mathbf{K}$ is given by the sum of two matrices, commonly called the "stiffness" matrix and the "mass" matrix.

$$
\mathbf{K} = \frac{p}{h}
\begin{bmatrix}
2 & -1 & & & & \\
-1 & 2 & -1 & & & \\
 & -1 & 2 & & \cdot & \\
 & & & \cdot & & \\
 & & & \cdot & \cdot & \cdot \\
 & & & -1 & 2 & -1 \\
 & & & & -1 & 1
\end{bmatrix}
+ \frac{qh}{6}
\begin{bmatrix}
4 & 1 & & & & \\
1 & 4 & 1 & & & \\
 & 1 & 4 & & \cdot & \\
 & & & \cdot & & \\
 & & & \cdot & \cdot & \cdot \\
 & & & 1 & 4 & 1 \\
 & & & & 1 & 2
\end{bmatrix}
$$

$$(11.11)$$

The load vector is given by

$$
\mathbf{F} = \frac{h}{6}
\begin{bmatrix}
f_0 + 4f_1 + f_2 \\
f_1 + 4f_2 + f_3 \\
\cdots \\
f_{j-1} + 4f_j + f_{j+1} \\
\cdots \\
f_{N-1} + 2f_N
\end{bmatrix}.
\qquad (11.12)
$$

Finally, the solution is obtained by solving the matrix equation in Eq. (11.8). We see that in this example $\mathbf{K}$ is a tridiagonal matrix (that is also diagonally dominant and symmetric) so that this matrix equation can be solved efficiently with methods discussed in Section 3.2.2.

We note that we did not need to explicitly apply the boundary condition $v'(1) = 0$ since this occurs naturally as a result of minimizing the functional I, Eq. (11.2). This can be understood because the minimizing function $v(x)$ must both satisfy Eq. (11.3) and make the boundary term vanish. One can verify that the last row of the matrix equation, Eq. (11.9), is consistent with the imposition of this boundary condition (see Problem 11.1).

11.2.3 Some Definitions

The finite element subspace is said to be of degree $k - 1$ if it contains in each element a complete polynomial of this degree. For linear elements, as considered in Section 11.2.2, the degree is one and $k = 2$.

The subspace S is said to possess continuity C^{q-1} if the qth derivative exists everywhere but is discontinuous at element boundaries. The linear elements have $q = 1$ and are therefore C^0. The Ritz method can be applied to a differential equation of order $2m$ if $q \geq m$. This is because the associated variational statement is only of order m, but the integrations by parts such as in going from Eq. (11.2) to Eq. (11.3) can increase the order of the differential equation over that of the variational integral by a factor of 2.

For the subsequent error analysis, we introduce norms of the square roots of the energies. For an interval $[a,b]$ and a function $u(x)$, we define the norms

$$\|u\|_0 \equiv \left[\int_a^b (u(x))^2\, dx\right]^{\frac{1}{2}},$$

$$\|u\|_2 \equiv \left[\int_a^b \left[(u''(x))^2 + (u'(x))^2 + (u(x))^2\right] dx\right]^{\frac{1}{2}},$$

etc.

11.2.4   Error with Ritz Method

Let us write the functional $I(v)$ in Eq. (11.2) as follows:

$$I(v) = a(v,v) - 2(f,v), \tag{11.13}$$

where the *energy integral* is given by

$$a(v,\omega) = \int_0^1 [p(x)v'(x)\omega'(x) + q(x)v(x)\omega(x)]\, dx. \tag{11.14}$$

We will show that the energy in the error, $U - u$, is a minimum.

The demonstration goes as follows: Let $u(x)$ be the computed function that minimizes I over the finite element function space S. It follows that for any ϵ and for any other function v belonging to S,

$$\begin{aligned}
I(u) &\leq I(u + \epsilon v) \\
&= a(u + \epsilon v, u + \epsilon v) - 2(f, u + \epsilon v) \\
&= I(u) + 2\epsilon[a(u,v) - (f,v)] + \epsilon^2 a(v,v).
\end{aligned}$$

It therefore follows that

$$0 \leq 2\epsilon\,[a(u,v) - (f,v)] + \epsilon^2 a(v,v).$$

Since ϵ may be of either sign, it follows that the quantity in brackets must vanish. We therefore have the result that

$$a(u,v) = (f,v),$$

for all functions v in S. Now, if U is the exact solution, it satisfies

$$a(U,v) = (f,v)$$

for all v. Subtracting gives

$$a(U,v) - a(u,v) = 0,$$

or

$$a(U - u, v) = 0. \tag{11.15}$$

Therefore, the error between the computed solution and the exact solution is orthogonal to all elements of v with respect to the energy inner product.

A detailed error analysis [224] shows that the energy integral of the error and the error norms will scale with the inter-element spacing h as

$$a(u - U, u - U) = \mathcal{O}\left(h^k, h^{2(k-m)}\right), \tag{11.16}$$

and

$$\|u - U\|_s = \mathcal{O}\left(h^{k-s}, h^{2(k-m)}\right), \tag{11.17}$$

where the approximation space is of order $k - 1$, and the variational problem is of order m. Applying this to the example in Section 11.2.2, where linear elements give $k = 2$ and the variational problem is first order so that $m = 1$, we find that both the energy integral and the $s = 0$ error norm will scale like $\mathcal{O}(h^2)$.

11.2.5 Hermite Cubic Elements

The linear elements discussed in Section 11.2.2 are continuous, but their first derivatives are not. (However, the first derivatives have finite energy, and hence they belong to H^1.) These elements are therefore classified as C^0. It is also possible to define a class of piecewise quadratic or piecewise cubic elements that are C^0 [224]. However, there is a special class of C^1 cubic elements, called *Hermite cubics*, that are constructed such that not only is the solution guaranteed to be continuous across element boundaries, but so is the first derivative. These can therefore be used on variational problems up to second order, or on differential equations up to fourth order by performing two integrations by parts.

There are two elements, or basis functions, associated with each node j that we will refer to as $\nu_{1,j}(x)$ and $\nu_{2,j}(x)$ (or sometimes just as ν_1 and ν_2 in context). The first, $\nu_{1,j}(x)$, has value unity at node j but its derivative $\nu'_{1,j}(x)$ vanishes there. The second basis function, $\nu_{2,j}(x)$, is zero at node j but its derivative, $\nu'_{2,j}(x)$, takes on the value of unity there. Both $\nu_{1,j}(x)$ and $\nu_{2,j}(x)$ and their derivatives $\nu'_{1,j}(x)$ and $\nu'_{2,j}(x)$ vanish at nodes $j+1$ and $j-1$. These elements are defined as follows and illustrated in Figure 11.3:

$$
\begin{aligned}
\nu_1(y) &= (|y| - 1)^2 \left(2|y| + 1\right), \\
\nu_2(y) &= y \left(|y| - 1\right)^2, \\
\nu_{1,j}(x) &= \nu_1\left(\frac{x - x_j}{h}\right), \\
\nu_{2,j}(x) &= h\,\nu_2\left(\frac{x - x_j}{h}\right).
\end{aligned}
$$

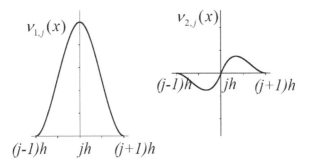

FIGURE 11.3: C^1 Hermite cubic functions enforce continuity of $v(x)$ and $v'(x)$. A linear combination of these two functions is associated with node j.

Since a cubic polynomial has four coefficients, it is completely determined in an interval by the value of the function and its derivative at the endpoints. The cubic polynomial in the interval between grid point j and $j+1$ can therefore be written

$$
\begin{aligned}
v(x) &= v_j\, \nu_{1,j}(x) + v'_j\, \nu_{2,j}(x) + v_{j+1}\, \nu_{1,j+1}(x) + v'_{j+1}\, \nu_{2,j+1}(x), \\
&= v_j + v'_j x + \left(-3v_j - 2hv'_j + 3v_{j+1} - hv'_{j+1}\right)\left(\frac{x}{h}\right)^2 \\
&\quad + \left(2v_j + hv'_j - 2v_{j+1} + hv'_{j+1}\right)\left(\frac{x}{h}\right)^3, \\
&= a_0 + a_1 x + a_2 x^2 + a_3 x^3.
\end{aligned}
$$

In matrix form, the four coefficients of the cubic polynomial in the interval $[x_j, x_{j+1}]$ can thereby be related to the values of the function and its derivative at the endpoints by

$$
\begin{bmatrix} a_0 \\[6pt] a_1 \\[6pt] a_2 \\[6pt] a_3 \end{bmatrix}
=
\begin{bmatrix}
1 & 0 & 0 & 0 \\[6pt]
0 & 1 & 0 & 0 \\[6pt]
-\frac{3}{h^2} & -\frac{2}{h} & \frac{3}{h^2} & -\frac{1}{h} \\[6pt]
\frac{2}{h^3} & \frac{1}{h^2} & -\frac{2}{h^3} & \frac{1}{h^2}
\end{bmatrix}
\cdot
\begin{bmatrix} v_j \\[6pt] v'_j \\[6pt] v_{j+1} \\[6pt] v'_{j+1} \end{bmatrix}.
\tag{11.18}
$$

If we call the above matrix $\mathbf{H}$, the vector containing the four polynomial coefficients $\mathbf{A} = (a_0, a_1, a_2, a_3)$, and the vector containing the four nodal parameters $\mathbf{q}_j = (v_j, v'_j, v_{j+1}, v'_{j+1})$, we can write this relation in the interval $[x_j, x_{j+1}]$ as $\mathbf{A} = \mathbf{H} \cdot \mathbf{q}_j$.

The integration of $v^2 = \left(a_0 + a_1 x + a_2 x^2 + a_3 x^3\right)^2$ is readily obtained:

$$\int_{x_j}^{x_{j+1}} v^2 dx = [a_0 \ a_1 \ a_2 \ a_3] \begin{bmatrix} h & \frac{h^2}{2} & \frac{h^3}{3} & \frac{h^4}{4} \\[4pt] \frac{h^2}{2} & \frac{h^3}{3} & \frac{h^4}{4} & \frac{h^5}{5} \\[4pt] \frac{h^3}{3} & \frac{h^4}{4} & \frac{h^5}{5} & \frac{h^6}{6} \\[4pt] \frac{h^4}{4} & \frac{h^5}{5} & \frac{h^6}{6} & \frac{h^7}{7} \end{bmatrix} \cdot \begin{bmatrix} a_0 \\ a_1 \\ a_2 \\ a_3 \end{bmatrix}. \tag{11.19}$$

If this matrix is denoted as $\mathbf{N_0}$, the result can be written

$$\int_{x_j}^{x_{j+1}} v^2 dx = \mathbf{A}^T \cdot \mathbf{N_0} \cdot \mathbf{A} = \mathbf{q}^T \cdot \mathbf{H}^T \cdot \mathbf{N_0} \cdot \mathbf{H} \cdot \mathbf{q}. \tag{11.20}$$

Therefore, the contribution to the mass matrix from this interval is given by

$$\mathbf{K_0} = \mathbf{H}^T \cdot \mathbf{N_0} \cdot \mathbf{H}. \tag{11.21}$$

If we add the contributions from interval $[x_{j-1}, x_j]$ to those of the interval $[x_j, x_{j+1}]$, multiply by $1/2$, and differentiate with respect to the nodal coefficients v_j and v_j', we obtain the rows of the mass matrix corresponding to node j for application to the differential equation given in Eq. (11.3). For $\mathbf{K_0} \cdot \mathbf{q}$, we have a matrix vector product that takes the form

$$\frac{qh}{420} \begin{bmatrix} 8h^2 & 13h & -3h^2 & 0 & \cdots & & & \\ 13h & 312 & 0 & 54 & -13h & 0 & \cdots & \\ -3h^2 & 0 & 8h^2 & 13h & -3h^2 & 0 & \cdots & \\ 0 & 54 & 13h & 312 & 0 & 54 & -13h & 0 \\ 0 & -13h & -3h^2 & 0 & 8h^2 & 13h & -3h^2 & 0 \\ & \cdots & 0 & 54 & 13h & 312 & 0 & 54 \\ & \cdots & 0 & -13h & -3h^2 & 0 & 8h^2 & 13h \\ & & & \cdots & & 0 & 54 & 13h & 312 \end{bmatrix} \cdot \begin{bmatrix} \vdots \\ v_{j-1} \\ v_{j-1}' \\ v_j \\ v_j' \\ v_{j+1} \\ v_{j+1}' \\ \vdots \end{bmatrix}.$$

The non-zero entries for the two rows corresponding to v_j and v_j' are complete as shown. The others (partially shown) are obtained by recursively shifting these two patterns over by 2 and down by 2. This matrix is the analogue of the second matrix in Eq. (11.11) calculated for the linear elements.

Similarly, we can integrate $(v')^2 = \left(a_1 + 2a_2 x + 3a_3 x^2\right)^2$:

$$\int_{x_j}^{x_{j+1}} (v')^2 dx = [a_0 \ a_1 \ a_2 \ a_3] \begin{bmatrix} 0 & 0 & 0 & 0 \\[4pt] 0 & h & h^2 & h^3 \\[4pt] 0 & h^2 & \frac{4h^3}{3} & \frac{3h^4}{2} \\[4pt] 0 & h^3 & \frac{3h^4}{2} & \frac{9h^5}{5} \end{bmatrix} \cdot \begin{bmatrix} a_0 \\ a_1 \\ a_2 \\ a_3 \end{bmatrix}. \tag{11.22}$$

If this matrix is denoted as $\mathbf{N}_1$, the contribution to the stiffness matrix from this interval is given by

$$\mathbf{K}_1 = \mathbf{H}^T \cdot \mathbf{N}_1 \cdot \mathbf{H}. \tag{11.23}$$

Again, we add the contributions from interval $[x_{j-1}, x_j]$ to those of the interval $[x_j, x_{j+1}]$, multiply by $1/2$, and differentiate with respect to the nodal coefficients v_j and v'_j to obtain the rows of the mass matrix corresponding to node j for application to the differential equation given in Eq. (11.3). The matrix vector product $\mathbf{K}_1 \cdot \mathbf{q}$ then takes the form:

$$\frac{p}{30h} \begin{bmatrix} 8h^2 & -3h & -h^2 & 0 & \cdots & & & \\ -3h & 72 & 0 & -36 & 3h & 0 & \cdots & \\ -h^2 & 0 & 8h^2 & -3h & -h^2 & 0 & \cdots & \\ 0 & -36 & -3h & 72 & 0 & -36 & 3h & 0 \\ 0 & 3h & -h^2 & 0 & 8h^2 & -3h & -h^2 & 0 \\ & \cdots & 0 & -36 & -3h & 72 & 0 & -36 \\ & \cdots & 0 & 3h & -h^2 & 0 & 8h^2 & -3h \\ & & \cdots & & 0 & -36 & -3h & 72 \end{bmatrix} \cdot \begin{bmatrix} \vdots \\ v_{j-1} \\ v'_{j-1} \\ v_j \\ v'_j \\ v_{j+1} \\ v'_{j+1} \\ \vdots \end{bmatrix}$$

This is the analogue of the first matrix in Eq. (11.11) calculated for the linear elements.

The error $\|u - U\|_0$ in these elements when applied to a second-order differential equation such as Eq. (11.3) is seen from Eq. (11.17) to be $\mathcal{O}(h^4)$. This is also what we would infer from the fact that a local piecewise cubic expansion can match the terms in a Taylor series expansion through x^3 and so we would expect the local error would be of order $x^4 \approx h^4$. Boundary conditions at node J are readily incorporated into these matrices by replacing the row corresponding to either v_J or v'_J by a row with a 1 on the diagonal and the other elements zero, and with the boundary value on the right side.

11.2.6 Cubic B-Splines

The basic cubic B-spline is shown in Figure 11.4. It is a piecewise continuous cubic polynomial which extends over four subintervals. The coefficients of the polynomial in each subinterval are such that $\phi(x), \phi'(x)$, and $\phi''(x)$ are continuous across subinterval boundaries, making it a C^2 element that could potentially be used for differential equations up to sixth order. For the uniform mesh spacing case shown in Figure 11.4, these are given in the four intervals

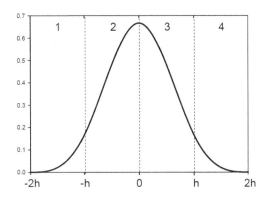

FIGURE 11.4: The basic cubic B-spline is a piecewise cubic with continuous value, first, and second derivatives.

as follows:

$$\phi^1 = \tfrac{4}{3} + 2y + y^2 + \tfrac{y^3}{6}; \qquad -2 \leq y < -1,$$

$$\phi^2 = \tfrac{2}{3} - y^2 - \tfrac{y^3}{2}; \qquad -1 \leq y < 0,$$

$$\phi^3 = \tfrac{2}{3} - y^2 + \tfrac{y^3}{2}; \qquad 0 \leq y < 1, \qquad (11.24)$$

$$\phi^4 = \tfrac{4}{3} - 2y + y^2 - \tfrac{y^3}{6}; \qquad 1 \leq y \leq 2.$$

These basic functions are translated so that

$$\phi_j(x) = \phi(\frac{x}{h} - j).$$

In contrast to the Hermite cubic, there is only one basis function per node, but the basis functions extend over two intervals on either side so that a given basis function will overlap with itself and six neighbors as shown in Figure 11.5. Because the second derivative is forced to be continuous across node boundaries, the function space represented by the cubic B-spline is a subspace of that represented by the Hermite cubic.

Consider the interval $jh < x < (j+1)h$. The function v will be represented by the sum of contributions from the four neighboring elements. Assuming this interval is sufficiently away from the boundary, the representation in that interval will be

$$v(x) = q_{j-1}\phi^4\left(\frac{x}{h} - j + 1\right) + q_j\phi^3\left(\frac{x}{h} - j\right)$$
$$+ q_{j+1}\phi^2\left(\frac{x}{h} - j - 1\right) + q_{j+2}\phi^1\left(\frac{x}{h} - j - 2\right). \qquad (11.25)$$

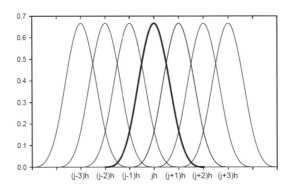

FIGURE 11.5: Overlapping elements for cubic B-spline. A given element overlaps with itself and six neighbors.

The derivative function v' is obtained by differentiating Eq. (11.25) with respect to x. We can then form v^2 and v'^2, integrate over the interval, and differentiate with respect to the amplitudes q_j to obtain the contribution to the matrix from that interval. Note that a given amplitude q_j will appear in four different interval integrals, corresponding to the four intervals in Eq. (11.24). This results in each amplitude being coupled to its six neighbors.

To incorporate the boundary conditions in a convenient manner, it is necessary to extend the sum over basis elements from $[0, N]$ to $[-1, N+1]$ and to modify slightly the form of the basis elements near the boundaries. For this purpose we introduce three new elements at each boundary, and use these to replace $\phi_{-1}(x), \phi_0(x), \phi_1(x), \phi_{N-1}(x), \phi_N(x), \phi_{N+1}(x)$. For the case of the left boundary, taken here to be $x = 0$, these boundary elements are defined as follows and illustrated in Figure 11.6. The element associated with $x = 2h$ is the basic B-spline shown in Figure 11.4. Defining $y \equiv x/h$ the element at $x = h$ (corresponding to $j = 1$) is defined over three intervals as follows:

$$
\begin{aligned}
\phi_1^1 &= \tfrac{1}{2}y^2 - \tfrac{11}{36}y^3; & 0 \le x < h, \\[4pt]
\phi_1^2 &= -\tfrac{1}{2} + \tfrac{3}{2}y - y^2 + \tfrac{7}{36}y^3; & h \le x < 2h, \qquad (11.26) \\[4pt]
\phi_1^3 &= \tfrac{3}{2} - \tfrac{3}{2}y + \tfrac{1}{2}y^2 - \tfrac{1}{18}y^3; & 2h \le x < 3h.
\end{aligned}
$$

The element corresponding to $j = 0$ is defined over only two intervals, and has a non-zero value of ϕ' at the origin:

$$
\begin{aligned}
\phi_0^1 &= y - \tfrac{3}{2}y^2 + \tfrac{7}{12}y^3; & 0 \le x < h, \\[4pt]
\phi_0^2 &= \tfrac{2}{3} - y + \tfrac{1}{2}y^2 - \tfrac{1}{12}y^3; & h \le x < 2h.
\end{aligned}
$$

$$(11.27)$$

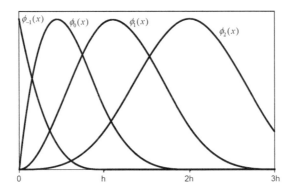

FIGURE 11.6: Special elements are utilized near boundary. Boundary conditions at $x = 0$ are set by specifying ϕ_{-1} (Dirichlet) or constraining a linear combination of ϕ_{-1} and ϕ_0 (Neumann).

Also required is an element associated with $j = -1$ that is defined over only a single interval, and which has a non-zero value at the origin:

$$\phi^1_{-1} \quad = \quad 1 - 3y + 3y^2 - y^3; \qquad 0 \le x < h \qquad (11.28)$$

Dirichlet boundary conditions for ϕ involve replacing the row corresponding to $j = -1$ with with a row with all zeros except a one on the diagonal, and with the boundary value specified on the right. Because both ϕ^1_0 and ϕ^1_{-1} have non-zero derivatives at the origin, Neumann boundary conditions (say $\phi' - 0$) would be implemented by constraining that $-3q_{-1} + q_0 = 0$. The special boundary elements at the right are just the reflections of these elements.

11.3 Galerkin Method in One Dimension

The Ritz technique applies only to problems of the classical variational type, in which a convex functional is minimized. The corresponding Euler differential equation is self-adjoint and elliptic. The Galerkin method is a means for extending the finite element technique to differential equations which are not necessarily the Euler equation for a corresponding variational statement.

Consider the differential equation

$$\mathcal{L}(U) = f, \qquad (11.29)$$

where $\mathcal{L}$ is a linear differential operator, $f(x)$ is a known function, and $U(x)$

is the function to be solved for. We multiply by a test function $V(x)$ and integrate over the domain,

$$(\mathcal{L}(U), V) = (f, V). \tag{11.30}$$

Here we have introduced the notation that for two functions f and g, and a one-dimensional domain $a \le x \le b$, we define the inner product to be

$$(f, g) \equiv \int_a^b f(x)g(x)dx. \tag{11.31}$$

If we introduce a solution space S to which U belongs, and require that Eq. (11.30) be satisfied for every test function $V(x)$ in that function space, it is called the *weak form* of Eq. (11.29). The Galerkin method is just the discretization of the weak form.

There is a close connection between the Galerkin method and the Ritz method if $\mathcal{L}$ is a *self-adjoint operator*. Let us return to our example in Section 11.2.1 and take as our operator that in Eq. (11.3),

$$\mathcal{L} = -\frac{d}{dx}\left(p(x)\frac{d}{dx}\right) + q(x),$$

and as our interval $[a, b] = [0, 1]$. After an integration by parts, we obtain

$$a(U, V) = (f, V), \tag{11.32}$$

where $a(U, V)$ is the energy integral defined in Eq. (11.14). Here, we assumed that the boundary conditions are such that the boundary terms in the integration by parts vanish. In general, if the order of $\mathcal{L}$ is $2m$, we can shift m derivatives to v by repeated integration by parts.

To discretize Eq. (11.30), let us take $\phi_1, \cdots, \phi_N$ as the basis for both the solution space and for the test function space. We represent the solution as a linear combination of the ϕ_j,

$$U = \sum_{j=1}^N q_j\phi_j,$$

where the amplitudes q_j are to be solved for. We then have the system

$$\sum_{j=1}^N q_j \left(L\phi_j, \phi_k\right) = (f, \phi_k); \qquad k = 1, \cdots, N,$$

or, in matrix form

$$\mathbf{G} \cdot \mathbf{Q} = \mathbf{F}, \tag{11.33}$$

where

$$G_{kj} = (L\phi_j, \phi_k),$$
$$F_k = (f, \phi_k).$$

Assuming that the solution and test space are polynomials with finite energy for the *mth* derivative, the elements in the matrix in Eq. (11.33) can be integrated by parts m times so that

$$(L\phi_j, \phi_k) \rightarrow a(\phi_j, \phi_k),$$

in which case it becomes identical with the discretization of the Ritz case.

The expected rate of convergence for the Galerkin method is [224]

$$\|U - u\|_0 = \mathcal{O}\left(h^k, h^{2(k-m)}\right),$$

where the finite element solution space and test space S is of degree $k - 1$, and the differential equation is order $2m$. This is in agreement with the result in Eq. (11.17) for the Ritz method.

Let us now consider an example problem in which we solve using the Galerkin method with cubic B-spline elements. The equation we will consider is

$$-p\frac{d^2}{dx^2}U + \tau\frac{d}{dx}U + qU = f(x), \tag{11.34}$$

where p, τ, and q are constants. Note that the presence of the term with τ causes the operator to be *non*-self-adjoint.

For the solution and projection function spaces we have

$$u(x) = \sum_{j=-1}^{N+1} q_j \phi_j(x), \tag{11.35}$$

$$v_j(x) = \phi_j(x); \qquad j = -1, \cdots, N+1. \tag{11.36}$$

Insertion of Eq. (11.35) into Eq. (11.34) and projecting with each element of Eq. (11.36), we obtain after integrating the first term by parts

$$\sum_{j=-1}^{N+1} \left[p\left(\phi'_k, \phi'_j\right) + \tau\left(\phi_k, \phi'_j\right) + q\left(\phi_k, \phi_j\right) \right] q_j = (f, \phi_k), \tag{11.37}$$

which can be written as a matrix equation

$$g_{kj}q_j = f_k. \tag{11.38}$$

To calculate the matrix g_{kj}, one must compute the overlap integrals shown in

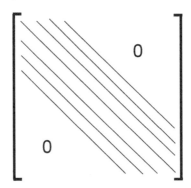

FIGURE 11.7: Band structure of the matrix g_{kj} that arises when the Galerkin method for a second-order differential equation is implemented using cubic B-spline elements.

Figure 11.5. A given element will overlap with itself and six neighbors. This will lead to a sparse matrix with seven non-zero diagonal bands as shown in Figure 11.7. The matrix elements are modified to include boundary conditions as discussed in Section 11.2.6. The matrix equation can then be solved using a LU decomposition algorithm for a banded matrix similar to that used in the tridiagonal matrix algorithm introduced in Section 3.2.2.

The presence of the first derivative term, with coefficient τ, in Eq. (11.34) spoils the self-adjointness property of the equation. It follows that the Galerkin method no longer corresponds to minimizing a positive definite functional, and therefore there is no guarantee that the solution found using the Galerkin method is a "good" approximation to the true solution in a meaningful sense. In practice, we find that as τ becomes larger so that the term $\tau u'$ dominates the second derivative term, large point-to-point oscillations can develop in the solution computed by the Galerkin method. In this case, finite difference methods, perhaps using one-sided (upstream) differences for this term, would yield a superior solution.

11.4   Finite Elements in Two Dimensions

In two dimensions, finite elements occupy a finite area rather than a finite interval. This area can be a rectangle or a triangle as shown in Figure 11.8. Rectangular elements have the advantage that they have a regular structure, which will simplify programming requirements, but they generally cannot be locally refined. Triangular elements have the advantage that they can more

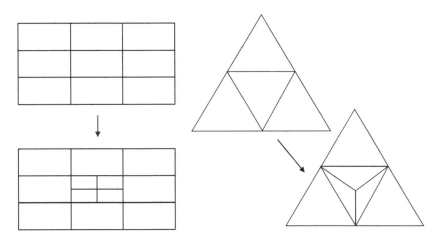

FIGURE 11.8: Rectangular elements cannot be locally refined without introducing hanging nodes. Triangular elements can be locally refined, but require unstructured mesh data structures in programming.

easily fit complex shapes and can be locally refined, but they generally require unstructured mesh logic which can make the programming more complex.

Elements are also characterized by the order of the polynomial that is defined within them. If an element with typical size h contains a complete polynomial of order M, then the error will be at most of order h^{M+1}. This follows directly from a local Taylor series expansion:

$$\phi(x,y) = \sum_{k=0}^{M} \sum_{l=0}^{k} \frac{1}{l!(k-l)!} \left[\frac{\partial^k \phi}{\partial x^k \partial y^{k-1}} \right]_{x_0,y_0} (x-x_0)^l (y-y_0)^{k-l} + \mathcal{O}(h^{M+1}).$$

Linear elements will therefore have an error $\mathcal{O}(h^2)$, quadratic elements will have an error $\mathcal{O}(h^3)$, etc.

Another property that characterizes elements is their interelement continuity. As discussed in Section 11.2.3, a finite element with continuity C^{q-1} belongs to Hilbert space H^q, and hence can be used for differential operators with order up to $2q$. This applicability is made possible using the corresponding variational statement of the problem (Ritz method) or by performing integration by parts in the Galerkin method, and thereby shifting derivatives from the unknown to the trial function.

11.4.1   High-Order Nodal Elements in a Quadrilateral

Suppose we have a structured mesh of quadrilateral elements, each of arbitrary shape as shown in Figure 11.9. We can map each element into the unit

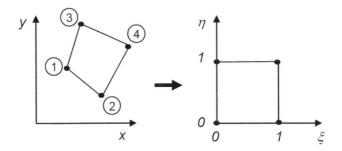

FIGURE 11.9: Each arbitrary quadrilateral element in the Cartesian space (x, y) is mapped into the unit square in the logical space for that element, (ξ, η).

square $(0 \leq \xi \leq 1)$, $(0 \leq \eta \leq 1)$ by means of the coordinate change

$$
\begin{aligned}
x(\xi, \eta) &= x_1 + (x_2 - x_1)\xi + (x_3 - x_1)\eta + (x_4 - x_3 - x_2 + x_1)\xi\eta, \\
y(\xi, \eta) &= y_1 + (y_2 - y_1)\xi + (y_3 - y_1)\eta + (y_4 - y_3 - y_2 + y_1)\xi\eta.
\end{aligned}
\tag{11.39}
$$

This mapping is invertible if the Jacobian is everywhere non-vanishing inside the element, which can be shown to be the case if the original quadrilateral in (x, y) space is convex: All of its angles must be less than π [224]. It can also be shown that the order of convergence is maintained by the mapping so that, for example, if the trial function space contains all biquadratics or bicubics in ξ and η, then it contains all quadratics or cubics in x and y and thus the error for a properly posed problem should scale like h^k with $k = 3$ or 4, respectively.

Let (m, n) be the indices of a particular quadrilateral element in the global space which includes all elements. For definiteness, let $m = 1, \cdots, M$ and $n = 1, \cdots, N$ so that there are $M \times N$ global 2D elements. For each element, one can then form a tensor product in the local logical coordinates ξ and η, so that within global element (m, n) we have the local expansion of a scalar field

$$
U(x, y) = \sum_{j=0}^{P} \sum_{k=0}^{P} \hat{U}_{j,k}^{m,n} h_j(\xi) h_k(\eta).
\tag{11.40}
$$

In a C^0 *nodal expansion*, we take the one-dimensional functions h_j to be Lagrange polynomials associated with a set of nodal points within the global element. The nodal points must include the ends of the domain and hence be shared with the neighboring domain in order for the solution to be continuous. There is a freedom, however, in how to choose the location of the interior points. This is discussed further in Section 11.4.2, but here we will take them to

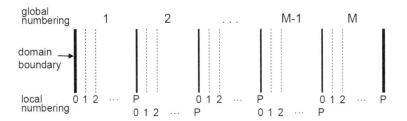

FIGURE 11.10: Relation between global numbering of elements and local numbering within an element in one dimension. In order for the solution to be continuous, the variable with (global number, local number) $= (m, P)$ must be the same as that with $(m+1, 0)$ for $m = 1, \cdots, M-1$.

be equally spaced. Figure 11.10 illustrates the relation between the global and local numbering in one dimension. In order for the solution to be continuous, we require $\hat{U}_{P,k}^{m,n} = \hat{U}_{0,k}^{m+1,n}$, etc. In this two-dimensional example there would therefore be a total of $(M \times P + 1) \times (N \times P + 1)$ unknowns (or degrees of freedom).

For both ξ and η, if we are given a set of $(P+1)$ nodal points within the element, which, for example, we denote by ξ_q, $0 \le q \le P$, the Lagrange polynomial $h_p(\xi)$ is the unique polynomial of order P which has a unit value at ξ_p and is zero at ξ_q $(q \ne p)$, i.e.,

$$h_q(\xi_q) = \delta_{pq},$$

where δ_{pq} is the Kronecker delta. The Lagrange polynomial can be written in product form as

$$h_p(\xi) = \frac{\displaystyle\prod_{q=0, q\ne p}^{P} (\xi - \xi_q)}{\displaystyle\prod_{q=0, q\ne p}^{P} (\xi_p - \xi_q)}. \tag{11.41}$$

The Lagrange approximation of a one-dimensional function $f(\xi)$ that passes through the $(P+1)$ nodal points ξ_j can thus be written

$$f(\xi) \cong \sum_{j=0}^{P} f(\xi_j) h_j(\xi) \equiv \sum_{j=0}^{P} \hat{f}_j h_j(\xi). \tag{11.42}$$

If $f(\xi)$ is a polynomial of order P then the relationship is exact.

As an example, we consider solving the Helmholtz equation in an arbitrary global domain that can be approximated by straight boundary segments. For a given function $\rho(x, y)$ and real constant λ, the 2D Helmholtz equation to determine $U(x, y)$ is given by

$$\nabla^2 U - \lambda^2 U = \rho. \qquad (11.43)$$

We assume the values of U on the domain boundary are specified (Dirichlet boundary conditions). Equation (11.43) corresponds to minimizing the functional

$$
\begin{aligned}
I(u) &= \int\int \left[\frac{1}{2}|\nabla U|^2 + \frac{1}{2}\lambda^2 U^2 + U\rho\right] dxdy, \\
&= \sum_{m=0}^{M}\sum_{n=0}^{N} I_{m,n}, \qquad (11.44)
\end{aligned}
$$

where $I_{m,n}$ involves an integral over a single global element. Using the non-orthogonal coordinate techniques of Chapter 5, we can use the local coordinate transformation given by Eq. (11.39) (different for each global element) to write

$$
\begin{aligned}
I_{m,n} &= \int_0^1\int_0^1 \left\{\frac{1}{2}\left[U_\xi^2|\nabla\xi|^2 + 2U_\xi U_\eta(\nabla\xi\cdot\nabla\eta) + U_\eta^2|\nabla\eta|^2\right.\right. \\
&\qquad\qquad \left.\left. + \lambda^2 U^2\right] + U\rho\right\} Jd\xi d\eta. \qquad (11.45)
\end{aligned}
$$

Here, the metric coefficients and Jacobian are given in terms of ξ and η derivatives of the transformations for x and y given in Eq. (11.39), using the same techniques as in Section 5.2.5. We have

$$
\begin{aligned}
|\nabla\xi|^2 &= J^{-2}\left(x_\eta^2 + y_\eta^2\right), & (\nabla\xi\cdot\nabla\eta) &= -J^{-2}\left(x_\xi x_\eta + y_\xi y_\eta\right), \\
|\nabla\eta|^2 &= J^{-2}\left(x_\xi^2 + y_\xi^2\right), & J &= x_\xi y_\eta - x_\eta y_\xi.
\end{aligned} \qquad (11.46)
$$

We next substitute the expansion for U from Eq. (11.40) into Eq. (11.45), noting that

$$
U_\xi = \sum_{j=0}^{P}\sum_{k=0}^{P} \hat{U}_{j,k}^{m,n} h_j'(\xi)h_k(\eta),
$$

$$
U_\eta = \sum_{j=0}^{P}\sum_{k=0}^{P} \hat{U}_{j,k}^{m,n} h_j(\xi)h_k'(\eta).
$$

We proceed to minimize $I_{m,n}$ with respect to each of the $\hat{U}_{j,k}^{m,n}$. In the global element (m, n), we have the following contribution to the matrix equation $\mathbf{A}\cdot\mathbf{x} = \mathbf{b}$:

$$
\begin{aligned}
a_{jk;pq} &= \int_0^1\int_0^1 \left[h_j'(\xi)h_k(\eta)h_p'(\xi)h_q(\eta)|\nabla\xi|^2\right. \\
&\quad + \left(h_j'(\xi)h_k(\eta)h_p(\xi)h_q'(\eta) + h_j(\xi)h_k'(\eta)h_p'(\xi)h_q(\eta)\right)(\nabla\xi\cdot\nabla\eta) \\
&\quad + \left. h_j(\xi)h_k'(\eta)h_p(\xi)h_q'(\eta)|\nabla\eta|^2 + \lambda^2 h_j(\xi)h_k(\eta)h_p(\xi)h_q(\eta)\right] Jd\xi d\eta, \\
b_{pq} &= \int_0^1\int_0^1 h_p(\xi)h_q(\eta)\rho\left(x(\xi,\eta), y(\xi,\eta)\right) Jd\xi d\eta. \qquad (11.47)
\end{aligned}
$$

The matrix is assembled by adding the contributions from each global element, taking into account the identity of points with (global, local) indices of (m, P) and $(m + 1, 0)$, etc., as discussed above and illustrated in Figure 11.10.

Once the matrix is assembled, it is modified in several ways. The Dirichlet boundary conditions are implemented simply by replacing the row and column in the matrix **A** corresponding to the boundary element by a row and column of zeros, except for a 1 on the diagonal and the value of the boundary condition in the corresponding location in the vector **b**. In the example being considered here, this would be the case for (global element, local element): $(m = 1, p = 0)$, $(m = M, p = P)$, $(n = 1, q = 0)$, $(n = N, q = P)$.

The matrix **A** can then be reduced in size substantially through *static condensation*. This takes note of the fact that the interior Lagrange nodes for each global rectangle, corresponding to those with both local numbering $p = 1, \cdots , P-1$ and $q = 1, \cdots , P-1$, are coupled only to the other Lagrange nodes of that rectangle. For each global rectangle, we can invert a $(P - 1) \times (P - 1)$ matrix to give a matrix equation for the values at the interior points in terms of the values at the boundary points corresponding to $p = 0, P$ or $q = 0, P$. These matrix relations can then be used to remove the interior values from the large matrix without introducing any new non-zero entries.

11.4.2 Spectral Elements

The *spectral element method* combines aspects of the spectral method described in Chapter 10 with the high-order nodal finite element method discussed in the last section. In fact, several of the most popular spectral element methods are just high-order nodal finite element methods, but with the spacing of the nodes non-uniform, as determined by the zeros of a polynomial or trigonometric function. A Lagrange polynomial is then put through these nodes. The set of special Lagrange polynomials so obtained have a number of desirable properties. They are more nearly orthogonal than are the set of Lagrange polynomials through uniformly spaced nodes. This leads to better conditioned matrices for most applications, and also to an expansion that exhibits improved convergence (spectral convergence) [225].

Let us first consider the Chebyshev spectral element method in one dimension [226]. The domain is broken up into discrete elements, with the mth element being of length h and defined in the interval $[a, b]$, where $b = a + h$. Within the mth global element, we represent the function $U(x)$ as the Lagrangian interpolant through the $P + 1$ Chebyshev–Gauss–Lobatto points

$$\bar{x}_j = \cos \frac{\pi j}{P}, \qquad j = 0, 1, \cdots , P. \tag{11.48}$$

The overbar represents the local element coordinate system defined by

$$\bar{x} = \frac{2}{h}(x - a) - 1, \tag{11.49}$$

so that $-1 \leq \bar{x} \leq 1$ for x within $a \leq x \leq b$. The Chebyshev interpolant of a function $u(x)$ within this interval is written as

$$U(x) = \sum_{j=0}^{P} U_j h_j(\bar{x}), \tag{11.50}$$

where the $h_j(\bar{x})$ are Lagrange interpolation functions defined within the element that satisfy

$$h_j(\bar{x}_k) = \delta_{jk}. \tag{11.51}$$

The coefficients U_j in Eq. (11.50) are thus the values of $U(x)$ at the Gauss–Lobatto points defined in Eq. (11.48). (Here $= \delta_{jk}$ is the Kronecker delta.) Using the relations from Section 10.2.4, the interpolation functions can be expressed as

$$h_j(\bar{x}) = \frac{2}{P} \sum_{k=0}^{P} \frac{1}{\bar{c}_j \bar{c}_k} T_k(\bar{x}_j) T_k(\bar{x}), \tag{11.52}$$

where if follows from Eqs. (10.26) and (11.48) that $T_k(\bar{x}_j) = \cos \frac{\pi k j}{P}$. If the solution is sufficiently differentiable, the arguments of Chapter 10 can be used to show that the error will decay faster than any power of P and thereby exhibit "spectral convergence."

The methods of the last section can be used to map an arbitrary shaped quadrilateral to the unit square in (ξ, η) space which can then be simply transformed by $\bar{\xi} = 2\xi - 1$ and $\bar{\eta} = 2\eta - 1$ to the intervals $[-1, 1]$. The Chebyshev interpolant of a variable $U(x, y)$ can be written for the $(m, n)^{th}$ element as a tensor product

$$U(x, y) = \sum_{j=0}^{P} \sum_{k=0}^{P} U_{j,k}^{m,n} h_j(\bar{\xi}) h_k(\bar{\eta}). \tag{11.53}$$

The expansion in Eq. (11.53) is seen to be identical to that used in Eq. (11.40), except that the interpolation functions are to be defined by Eq. (11.52) rather than Eq. (11.41), and thus the results of that section can be used.

A similar procedure could be followed by using the Lagrange polynomial through the Gauss–Lobatto–Legendre points which are defined by the roots of the polynomial $(1 - x^2)L'_P(x)$. This leads to the Legendre spectral element method.

11.4.3   Triangular Elements with C^1 Continuity

A particularly useful triangular finite element in two dimensions with C^1 continuity is known as the *reduced quintic* [7, 227, 228, 229] and is depicted in Figure 11.11. In each triangular element, the unknown function $\phi(x, y)$ is

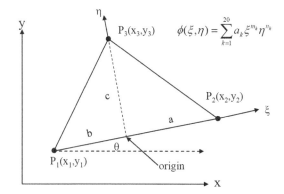

FIGURE 11.11: Reduced quintic finite element is defined by the four geometric parameters a, b, c, θ. A local (ξ, η) Cartesian system is used. The function and first two derivatives are constrained at the three vertex points, and C^1 continuity is imposed at the edges. Exponents m_i and n_i are given in Table 11.1.

written as a general polynomial of 5th degree in the local Cartesian coordinates ξ and η: $\phi(\xi, \eta) = \sum_{i=1}^{20} a_i \xi^{m_i} \eta^{n_i}$ (where the exponents m_i and n_i are given in Table 11.1), which would have 21 coefficients were not there additional constraints. Eighteen of the coefficients are determined from specifying the values of the function and its first five derivatives: $\phi, \phi_x, \phi_y, \phi_{xx}, \phi_{xy}, \phi_{yy}$ at each of the three vertices, thus guaranteeing that globally all first and second derivatives will be continuous at each vertex. Since the one-dimensional quintic polynomial along each edge is completely determined by these values specified at the endpoints, it is guaranteed that the expansion is continuous between elements.

The remaining three constraints come from the requirement that the normal derivative of ϕ at each edge, ϕ_n, reduce to a one-dimensional cubic polynomial along that edge. This implies that the two sets of nodal values completely determine ϕ_n everywhere on each edge, guaranteeing its continuity from one triangle to the next so that the element is C^1. One of these three constraints is trivial and has been used to reduce the number of terms from 21 to 20 in the sum.

In imposing these continuity constraints, the expansion is no longer a complete quintic, but it does contain a complete quartic with additional constrained quintic coefficients to enforce C^1 continuity between elements. Thus, the name "reduced quintic." If the characteristic size of the element is h, then it follows from a local Taylor's series analysis that the approximation error in the unknown function, $\phi - \phi^h$, will be of order h^5.

Another advantage of this element is that all of the unknowns (or degrees of freedom, DOF) appear at the vertices, and are thus shared with all the

TABLE 11.1: Exponents of ξ and η for the reduced quintic expansion $\phi(\xi,\eta) = \sum_{i=1}^{20} a_i \xi^{m_i} \eta^{n_i}$.

k	m_k	n_k	k	m_k	n_k	k	m_k	n_k	k	m_k	n_k
1	0	0	6	0	2	11	4	0	16	5	0
2	1	0	7	3	0	12	3	1	17	3	2
3	0	1	8	2	1	13	2	2	18	2	3
4	2	0	9	1	2	14	1	3	19	1	4
5	1	1	10	0	3	15	0	4	20	0	5

triangular elements that connect to this vertex. This leads to a very compact representation with relatively small matrices compared to other representations. Suppose that we are approximating a square domain by partitioning it into n^2 squares or $2n^2$ triangles. The reduced quintic will asymptotically have $N = 6n^2$ unknowns, or three unknowns for each triangle. This scaling can be verified by the fact that if we introduce a new point into any triangle and connect it to the three nearby points, we will have generated two new triangles and introduced six new unknowns. We contrast this with a Lagrange quartic element which has the same formal order of accuracy but asymptotically has eight unknowns per triangle [224].

For a given triangle, if we locally number the unknowns ϕ, ϕ_x, ϕ_y, ϕ_{xx}, ϕ_{xy}, ϕ_{yy} at vertex P_1 as $\Phi_1 - \Phi_6$, at P_2 as $\Phi_7 - \Phi_{12}$, and at P_3 as $\Phi_{13} - \Phi_{18}$, then we seek a relation between the polynomial coefficients a_i and the Φ_j. This is done in two parts. We first define the 20×20 matrix in each triangle that relates the 18 local derivatives and two constraints to the 20 polynomial coefficients. We call this matrix $\mathbf{T}$. The 20 rows of $\mathbf{T}$ are given by:

$$\phi^1 = a_1 - ba_2 + b^2 a_4 - b^3 a_7 + b^4 a_{11} - b^5 a_{16}$$
$$\phi^1_\xi = a_2 - 2ba_4 + 3b^2 a_7 - 4b^3 a_{11} + 5b^4 a_{16}$$
$$\phi^1_\eta = a_3 - ba_5 + b^2 a_8 - b^3 a_{12}$$
$$\phi^1_{\xi\xi} = 2a_4 - 6ba_7 + 12b^2 a_{11} - 20b^3 a_{16}$$
$$\phi^1_{\xi\eta} = a_5 - 2ba_8 + 3b^2 a_{12}$$
$$\phi^1_{\eta\eta} = 2a_6 - 2ba_9 + 2b^2 a_{13} - 2b^3 a_{17}$$

$$\phi^2 = a_1 + aa_2 + a^2a_4 + a^3a_7 + a^4a_{11} + a^5a_{16}$$
$$\phi_\xi^2 = a_2 + 2aa_4 + 3a^2a_7 + 4a^3a_{11} + 5a^4a_{16}$$
$$\phi_\eta^2 = a_3 + aa_5 + a^2a_8 + a^3a_{12}$$
$$\phi_{\xi\xi}^2 = 2a_4 + 6aa_7 + 12a^2a_{11} + 20a^3a_{16}$$
$$\phi_{\xi\eta}^2 = a_5 + 2aa_8 + 3a^2a_{12}$$
$$\phi_{\eta\eta}^2 = 2a_6 + 2aa_9 + 2a^2a_{13} + 2a^3a_{17}$$

$$\phi^3 = a_1 + ca_3 + c^2a_6 + c^3a_{10} + c^4a_{15} + c^5a_{20}$$
$$\phi_\xi^3 = a_2 + ca_5 + c^2a_9 + c_{14}^3 + c^4a_{19}$$
$$\phi_\eta^3 = a_3 + 2ca_6 + 3c^2a_{10} + 4c^3a_{15} + 5c^4a_{20}$$
$$\phi_{\xi\xi}^3 = 2a_4 + 2ca_8 + 2c^2a_{13} + 2c^3a_{18}$$
$$\phi_{\xi\eta}^3 = a_5 + 2ca_9 + 3c^2a_{14} + 4c^3a_{19}$$
$$\phi_{\eta\eta}^3 = 2c_6 + 6ca_{10} + 12c^2a_{15} + 20c^3a_{20}$$

$$0 = 5b^4ca_{16} + (3b^2c^3 - 2b^4c)a_{17} + (2bc^4 - 3b^3c^2)a_{18}$$
$$\quad + (c^5 - 4b^2c^3)a_{19} - 5bc^4a_{20}$$
$$0 = 5a^4ca_{16} + (3a^2c^3 - 2a^4c)a_{17} + (-2ac^4 - 3a^3c^2)a_{18}$$
$$\quad + (c^5 - 4a^2c^3)a_{19} - 5ac^4a_{20}.$$

This satisfies $\mathbf{\Phi}' = \mathbf{T}\,\mathbf{A}$, where $\mathbf{\Phi}'$ denotes the vector of length 20 produced by stringing together the 18 values of the function and derivatives with respect to the local Cartesian coordinates ξ and η at the three vertices, and with the final two elements zero, and $\mathbf{A}$ is the vector produced by the 20 polynomial coefficients. This can be solved for the coefficient matrix by inverting $\mathbf{T}$, thus $\mathbf{A} = \mathbf{T}^{-1}\mathbf{\Phi}'$. A useful check is to verify that the numerically evaluated determinant of $\mathbf{T}$ has the value $-64(a+b)^{17}c^{20}(a^2+c^2)(b^2+c^2)$. Since the final two elements of $\mathbf{\Phi}'$ are zero, we can replace $\mathbf{T}^{-1}$ by the 20×18 matrix $\mathbf{T}_2$ which consists of the first 18 columns of $\mathbf{T}^{-1}$.

To get the coefficient matrix $\mathbf{A}$ in terms of the vector containing the actual derivatives with respect to (x, y), we have to apply the rotation matrix $\mathbf{R}$. This is compactly defined in terms of the angle θ appearing in Figure 11.11 by

$$\mathbf{R} = \begin{bmatrix} \mathbf{R}_1 & & \\ & \mathbf{R}_1 & \\ & & \mathbf{R}_1 \end{bmatrix}, \tag{11.54}$$

where

$$\mathbf{R}_1 = \begin{bmatrix} 1 & 0 & 0 & 0 & 0 & 0 \\ 0 & \cos\theta & \sin\theta & 0 & 0 & 0 \\ 0 & -\sin\theta & \cos\theta & 0 & 0 & 0 \\ 0 & 0 & 0 & \cos^2\theta & 2\sin\theta\cos\theta & \sin^2\theta \\ 0 & 0 & 0 & -\sin\theta\cos\theta & \cos^2\theta - \sin^2\theta & \sin\theta\cos\theta \\ 0 & 0 & 0 & \sin^2\theta & -2\sin\theta\cos\theta & \cos^2\theta \end{bmatrix}.$$

If we then define the matrix $\mathbf{G} = \mathbf{T}_2\mathbf{R}$, this relates the coefficient matrix directly to the unknown vector consisting of the function and derivatives with respect to (x, y), i.e., $\mathbf{A} = \mathbf{G}\boldsymbol{\Phi}$, or in component notation: $a_i = \sum_{j=1}^{18} g_{i,j}\Phi_j$ for $i = 1, 20$. The 20×18 matrix $g_{i,j}$ depends only on the shape and orientation of the individual triangle, and in general will be different for each triangle. The general expression for the unknown function ϕ in a given triangle is

$$\phi(\xi, \eta) = \sum_{i=1}^{20} a_i \xi^{m_i} \eta^{n_i} = \sum_{i=1}^{20}\sum_{j=1}^{20} g_{i,j}\phi_j \xi^{m_i}\eta^{n_i},$$

$$= \sum_{j=1}^{18} \nu_j \phi_j, \tag{11.55}$$

where we have defined the basis functions as

$$\nu_j \equiv \sum_{i=1}^{20} g_{i,j} \xi^{m_i}\eta^{n_i}; \qquad j = 1, 18. \tag{11.56}$$

The 18 basis functions for each triangle, as defined by Eq. (11.56), have the property that they have a unit value for either the function or one of its first or second derivatives at one vertex and and zero for the other quantities at this and the other nodes. They also have the C^1 property embedded.

All of the integrals that need to be done to define the matrices that occur in the Galerkin method are of the form of 2D integrals of polynomials in ξ or η over the triangles. These can either be evaluated by numerical integration [230], or by making use of the analytic formula:

$$F(m, n) \equiv \int\int_{triangle} \xi^m \eta^n d\xi d\eta = c^{n+1}\frac{\left[a^{m+1} - (-b)^{m+1}\right]m!n!}{(m+n+2)!}. \tag{11.57}$$

For example, to evaluate the mass matrix, we would have

$$\int\int \nu_i(\xi, \eta)\phi(\xi, \eta)d\xi d\eta = \sum_{k=1}^{18}\left[\sum_{i=1}^{20}\sum_{l=1}^{20} g_{i,j}g_{l,k}F(m_i + m_l, n_i + n_l)\right]\Phi_k$$

$$\equiv \sum_{k=1}^{18} M_{jk}\Phi_k. \tag{11.58}$$

The implementation of boundary conditions requires some discussion [231]. Recall that for each scalar variable being solved for, there are six unknowns at each node corresponding to the function and all its first and second derivatives. If Dirichlet or Neumann boundary conditions are being applied at a boundary that is aligned with the $\hat{x}$ or $\hat{y}$ axis, the imposition of boundary conditions is straightforward. Consider homogeneous Dirichlet boundary conditions being applied at a boundary node where the boundary lies on the x axis. We then replace the rows of the matrix that correspond to the trial functions ν_i having non-zero values of ϕ, ϕ_x and ϕ_{xx} with a row from the identity matrix of all zeros but a one on the diagonal, and with zero on the right side of the equation. This is equivalent to removing these basis functions from the system. If Neumann conditions were being applied, we would similarly replace the rows corresponding to ϕ_y and ϕ_{xy}. If we define the operator

$$\mathcal{L}(\phi) = [\phi, \phi_x, \phi_y, \phi_{xx}, \phi_{xy}, \phi_{yy}], \qquad (11.59)$$

then these conditions are all of the form $\mathcal{L}_i(\phi) = C$. This is easily implemented because the basis functions ν_j at the boundary point satisfy the orthogonality conditions

$$\mathcal{L}_i(\nu_j) = \delta_{ij}. \qquad (11.60)$$

But what if the boundaries are not aligned with the x or y axis? In this case, if we define the vector corresponding to normal and tangential derivatives of the solution as

$$\mathcal{L}'(\phi) = [\phi, \phi_n, \phi_t, \phi_{nn}, \phi_{nt}, \phi_{tt}], \qquad (11.61)$$

then the boundary conditions we need to impose are of the form $\mathcal{L}'_i(\phi) = C$. Here n is the coordinate locally normal to the boundary, and t is locally tangent to the boundary. We also define the local curvature as $\kappa = \hat{n} \cdot d\hat{t}/ds$ where ds is the arc length along the boundary curve. The derivatives of ϕ with respect to the normal and tangential coordinates (n, t) are related to those with respect to the global coordinates (x, y) at a given boundary point by the transformation

$$\mathcal{L}'_i(\phi) = M_{ij}\mathcal{L}_j(\phi), \qquad (11.62)$$

where M is the unimodular matrix

$$\mathbf{M} = \begin{bmatrix} 1 & 0 & 0 & 0 & 0 & 0 \\ 0 & n_x & n_y & 0 & 0 & 0 \\ 0 & -n_y & n_x & 0 & 0 & 0 \\ 0 & 0 & 0 & n_x^2 & 2n_xn_y & n_y^2 \\ 0 & -\kappa n_y & \kappa n_x & -n_xn_y & n_x^2 - n_y^2 & n_xn_y \\ 0 & -\kappa n_x & -\kappa n_y & n_y^2 & -2n_xn_y & n_x^2 \end{bmatrix}. \qquad (11.63)$$

In order to impose a boundary condition on a curved boundary (or even a straight boundary not aligned with the $\hat{x}$ or $\hat{y}$ axis) and for the residual to be constrained as much as possible, we want to define a set of trial functions μ_i at each boundary point that satisfy a similar orthogonality condition to that in the axis-aligned boundary case, i.e.,

$$\mathcal{L}_i'(\mu_j) = \delta_{ij}. \tag{11.64}$$

The new trial functions must be a linear combination of the old basis functions since the latter span the solution space, and thus we can write for some matrix **N**,

$$\mu_i = N_{ij}\nu_j. \tag{11.65}$$

It follows from Eqs. (11.60), (11.62), (11.64), and (11.65) that $N_{ij} = (M_{ji})^{-1}$,

$$\mathbf{N} = \begin{bmatrix} 1 & 0 & 0 & 0 & 0 & 0 \\ 0 & n_x & n_y & \kappa n_y^2 & -\kappa n_x n_y & \kappa n_x^2 \\ 0 & -n_y & n_x & 2\kappa n_x n_y & -\kappa(n_x^2 - n_y^2) & -2\kappa n_x n_y \\ 0 & 0 & 0 & n_x^2 & n_x n_y & n_y^2 \\ 0 & 0 & 0 & -2n_x n_y & n_x^2 - n_y^2 & 2n_x n_y \\ 0 & 0 & 0 & n_y^2 & -n_x n_y & n_x^2 \end{bmatrix}. \tag{11.66}$$

For each point that lies on the boundary, we therefore substitute the six trial functions μ_i for the original six trial functions ν_i. These boundary vertex trial functions obey an orthogonality condition on their derivatives with respect to a local coordinate system (n, t) defined relative to the local boundary orientation. We then substitute the appropriate rows for rows imposing the boundary conditions as in the case of the coordinate aligned boundaries. Using these linearly transformed trial functions to impose the boundary conditions is optimal in the sense that the components of the residual associated with the boundary vertex are maximally constrained.

11.5   Eigenvalue Problems

The finite element method is natural for solving eigenvalue problems of the form

$$\mathcal{L}(U) = \lambda U, \tag{11.67}$$

where $\mathcal{L}$ is a self-adjoint operator, λ is the scalar eigenvalue, and U is the unknown function. This is exactly the form of the linear ideal MHD equation of motion that was presented in Chapter 8. The basic idea is to formulate the problem as the minimization of the Rayleigh quotient. Using piecewise polynomials as trial functions leads to a matrix eigenvalue problem to obtain the solution.

For definiteness, let us take for the operator in Eq. (11.67) the same linear operator considered in Section 11.3,

$$\mathcal{L}(U) = -\frac{d}{dx}\left(p(x)\frac{dU}{dx}\right) + q(x)U. \tag{11.68}$$

The Raleigh-Ritz method is to form the Raleigh quotient and then substitute a finite dimensional subspace S for the full admissible function space H^1. Choosing a basis $\phi_1(x), \cdots, \phi_N(x)$ that spans S, we represent the solution as the linear combination

$$U = \sum_{j=1}^{N} a_j \phi_j. \tag{11.69}$$

The Rayleigh quotient is then defined as

$$R(U) = \frac{a(U,U)}{(U,U)} = \frac{\int [p(x)U'^2 + q(x)U^2]dx}{\int U^2 dx}. \tag{11.70}$$

It is readily verified that the condition that $R(U)$ be stationary is equivalent to Eq. (11.67), with $\lambda = R(U)$.

Insertion of Eq. (11.69) into Eq. (11.70) yields

$$R(a_1, \cdots, a_N) = \frac{\sum_{j=1}^{N} \sum_{k=1}^{N} a_j a_k \int (p(x)\phi_j'\phi_k' + q(x)\phi_j\phi_k)dx}{\sum_{j=1}^{N} \sum_{k=1}^{N} a_j a_k \int \phi_j\phi_k dx}. \tag{11.71}$$

The critical points of R are the solution to the matrix eigenvalue problem

$$K_{jk}a_k = \lambda^h M_{jk}a_k, \tag{11.72}$$

where

$$K_{jk} = \int (p(x)\phi_j'\phi_k' + q(x)\phi_j\phi_k)dx,$$

$$M_{jk} = \int (\phi_j\phi_k)dx.$$

Solving Eq. (11.72) using matrix eigenvalue techniques will yield a sequence of eigenvalues $\lambda_1 \leq \lambda_2 \leq \cdots \leq \lambda_N$. The fundamental frequency λ_1 is often the quantity of greatest significance. It is an important property of the finite element methods that the λ_1^h computed from Eq. (11.72) always lies above the true result λ_1,

$$\lambda_1^h \geq \lambda_1.$$

This follows since λ_1^h is the minimum value of $R(u)$ over the subspace S, and

λ_1 is the minimum over the entire admissible space H^1. For the MHD problem, this says that the discrete approximation is always more stable than the true system.

It is worth noting that the eigenvalue equation, Eq. (11.67), can also be solved using the Galerkin method. In this case, we multiply Eq. (11.67) by a trial function v and integrate by parts. If we expand U as in Eq. (11.69) and require that the weak form hold for all functions v in the subspace S, we obtain a matrix eigenvalue problem that is equivalent to Eq. (11.72).

11.5.1   Spectral Pollution

When using the finite element method to calculate the unstable eigenmodes of a highly magnetized plasma, special care must be given to the representation of the displacement field. When the ideal MHD stability problem is cast in the form of Eq. (11.67), interest normally lies in calculating the most unstable modes, or the eigenmodes with the smallest (or most negative) eigenvalues $\lambda = \omega^2$.

However, as is illustrated in Figure 8.4, a wide range of other modes are present with large positive eigenvalues corresponding to stable oscillations of the plasma column. It has been recognized for some time that the existence of these stable modes can "pollute" the unstable spectrum, making it practically impossible to obtain meaningful solutions for the unstable modes.

The root cause of this phenomena can be seen by consideration of the variational form of the potential energy as given in Eq.(8.49). The first three terms in that equation are all positive definite, and thus stabilizing. The numerical representation of the displacement field must be such as to allow the discrete form of those terms to vanish to a high degree if the effect of the smaller, potentially negative terms is to be computed.

In the development of the ideal MHD finite element eigenvalue code PEST, the authors chose to represent the plasma displacement $\boldsymbol{\xi}$ in terms of three scalar fields (ζ, δ, τ) defined by the relation

$$\boldsymbol{\xi} = \frac{R^2}{g} \left[i\mathbf{B} \times \nabla_\perp \zeta + \delta \nabla \theta \times \mathbf{B} \right] + i\tau \mathbf{B}. \qquad (11.73)$$

Here, $\mathbf{B} = \nabla \Psi \times \nabla \phi + g \nabla \phi$ is the equilibrium axisymmetric magnetic field, and $\nabla_\perp$ means "the gradient in the plane perpendicular to $\nabla \phi$". The angle θ that appears in Eq. (11.73) is the poloidal angle in a straight field line coordinate system with $J \sim R^2$; see Table 5.1. The factors of $i \equiv \sqrt{-1}$ are included so that when a spectral representation is used for the θ and ϕ coordinates and a dependence $\equiv e^{i(m\theta - n\phi)}$ is used with m and n integers, the scalar fields will all be real. In the strong toroidal field limit, the magnetic field is dominated by the term $\mathbf{B} \sim g \nabla \phi$, and we see that the large stabilizing second term in Eq. (8.49) is completely independent of the scalar fields ζ and τ. This means that these fields will be used to minimize the smaller terms associated with the Alfvén and slow magnetoacoustic branches without significant interference from the fast magnetoacoustic branch. Also, the scalar field τ is predominantly associated with the slow wave as seen from Eq. (1.96) and the discussion that follows it, and the scalar field δ is the dominant one that is involved in perpendicular compressibility and hence the fast wave. Thus, this representation of the displacement vector field helps to avoid one part of the spectrum polluting the others.

In straight circular cylindrical geometry as discussed in Section 8.4, it is common to use a spectral representation for the angle θ and the longitudinal coordinate z and to use finite elements for the radial coordinate r. Assuming θ and z dependence as as $e^{i(m\theta - kz)}$, Appert et al. [232] introduced the following displacement projection

$$\boldsymbol{\xi} = \xi_1(r)\hat{r} - \frac{i}{m}\left(r\xi_2(r) - \xi_1(r)\right)\hat{\theta} + \frac{i}{k}\xi_3(r)\hat{z}, \tag{11.74}$$

where $\xi_i(r); i = 1 \cdots 3$ are functions of r to be solved for. This gives

$$\nabla \cdot \boldsymbol{\xi} = \xi_1'(r) + \xi_2(r) + \xi_3(r). \tag{11.75}$$

They reason that in order to avoid pollution of the rest of the spectrum by the fast wave, they should choose a finite element basis for the three functions $\xi_1(r), \xi_2(r)$, and $\xi_3(r)$ that will allow $\nabla \cdot \boldsymbol{\xi}$ to be arbitrarily small over a given grid interval in r. From the form of Eq. (11.75), we see that this will follow if the order of the finite elements used to expand $\xi_1(r)$ is 1 higher than the order used to expand $\xi_2(r)$ and $\xi_3(r)$. They chose linear (tent) functions to expand $\xi_1(r)$ and constant (hat) functions to expand $\xi_2(r)$ and $\xi_3(r)$. Another choice that has been used [233] is cubic B-spline for $\xi_1(r)$ and the derivative of a cubic B-spline for $\xi_2(r)$ and $\xi_3(r)$.

This set of cylindrical projections was extended by Chance et al. [234] to one that decouples the parallel from the perpendicular displacements and allows for both $\nabla \cdot \boldsymbol{\xi}_\perp = 0$ and $\nabla \cdot \boldsymbol{\xi} = 0$. The displacement field is represented as

$$\boldsymbol{\xi} = \frac{1}{r}\left(\xi_1 + m\xi_2\right)\hat{r} + i\left(\xi_2' + B_\theta \xi_3\right)\hat{\theta} + i\left(-\frac{B_\theta}{B_z}\xi_2' + \xi_3 B_z\right)\hat{z}, \tag{11.76}$$

or

$$\boldsymbol{\xi} = \frac{\xi_1}{r}\hat{r} + i\frac{\mathbf{B}}{B_z} \times \nabla_\perp \xi_2 + i\xi_3 \mathbf{B}, \tag{11.77}$$

where we have again used $\nabla_\perp$ to indicate the gradient in the plane perpendicular to $\hat{z}$. For finite elements, they used linear elements for $\xi_2(r)$, constant elements for $\xi_3(r)$, and for $\xi_1(r)$ they used a special finite element $\phi_{1j}(r)$ defined by

$$\phi_{1j} = \begin{cases} \dfrac{\int_{r_{j-1}}^{r} B_\theta r / B_z \, dr}{\int_{r_{j-1}}^{r_j} B_\theta r / B_z \, dr} & \text{if } r_{j-1} < r < r_j, \\[3mm] \dfrac{\int_{r}^{r_{j+1}} B_\theta r / B_z \, dr}{\int_{r_j}^{r_{j+1}} B_\theta r / B_z \, dr} & \text{if } r_j < r < r_{j+1}, \\[3mm] 0 & \text{otherwise.} \end{cases} \tag{11.78}$$

This form for ϕ_{1j} was chosen to allow

$$\nabla \cdot \boldsymbol{\xi}_\perp = \frac{1}{r}\frac{d\xi_1}{dr} + k\frac{B_\theta}{B_z}\frac{d\xi_2}{dr} = 0, \tag{11.79}$$

to machine accuracy.

11.5.2 Ideal MHD Stability of a Plasma Column

Here we consider the problem of determining the stability of a large aspect ratio toroidal plasma surrounded by a vacuum region, which is in turn surrounded by a perfectly conducting wall. A complete representation for the plasma displacement and magnetic vector potential in a torus that is designed to both minimize spectral pollution and be general enough to be used in a non-linear calculation is given by [30]

$$\boldsymbol{\xi} = R^2 \nabla U \times \nabla\phi + \omega R^2 \nabla\phi + R^{-2}\nabla_\perp \chi, \tag{11.80}$$
$$\mathbf{A} = R^2 \nabla\phi \times \nabla f + \psi\nabla\phi - F_0 \ln R\hat{Z}. \tag{11.81}$$

Here (R, ϕ, Z) form a cylindrical coordinate system, the subscript $\perp$ indicates "perpendicular to $\nabla\phi$," i.e., in the (R, Z) plane, and F_0 is a constant which represents an externally imposed toroidal magnetic field. The displacement is thus determined by the three scalar functions (U, ω, χ). The magnetic field is given by $\mathbf{B} = \nabla \times \mathbf{A}$, where the magnetic vector potential is defined by the two scalar functions (f, ψ), the constant F_0, and the gauge condition $\nabla_\perp \cdot R^{-2}\mathbf{A} = 0$.

Let a be a typical minor radius and R_0 be a typical major radius. We take the large aspect ratio limit $a/R_0 \sim \epsilon \to 0$. Gradients in the (R, Z) plane are assumed ordered as $\nabla_\perp \sim 1/a$. The standard tokamak ordering corresponds to ordering the quantities that appear in the magnetic field and pressure as $f \sim R_0\epsilon^4 \tilde{f}$, $\psi \sim R_0^2\epsilon^2\tilde{\psi}$ $p \sim \epsilon^2\tilde{p}$, $F_0 \sim R_0$, $\nabla R \sim 1$. To the lowest non-trivial order in ϵ, it is sufficient to consider the straight cylindrical equilibrium geometry with circular cross section as shown in Figure 8.5. We then use a (r, θ, z) cylindrical coordinate system where now $\hat{z}$ is a unit vector replacing $R\nabla\phi$. In this geometry, the equilibrium only depends on the coordinate r. We take the plasma region to be $0 < r < a$ and the vacuum region to be $a < r < b$.

Dropping the ordering parameter ϵ for notational simplicity, and similarly dropping the tilde, and normalizing lengths such that $R_0 = 1$, we can write the equilibrium magnetic field and current density as

$$\mathbf{B} = \nabla\psi_0 \times \hat{z} + F_0\hat{z}, \qquad \mathbf{J}_0 = -\nabla^2\psi_0\hat{z}. \tag{11.82}$$

Letting primes denote derivatives with respect to r, we have for the equilibrium poloidal magnetic field and the safety factor

$$B_0^\theta = -\psi_0', \qquad q(r) = rF_0/B_0^\theta. \tag{11.83}$$

To lowest order, the only component of the displacement and perturbed magnetic field that enter are

$$\boldsymbol{\xi} = \nabla U \times \hat{z}, \qquad \mathbf{Q} = \nabla \times (\boldsymbol{\xi} \times \mathbf{B}_0) = \nabla \psi \times \hat{z}. \qquad (11.84)$$

This is linearized reduced MHD in a cylinder. We adopt a spectral representation in the periodic variables θ and z, and only keep a single mode so that for the perturbed variables, $\psi, U = \psi(r), U(r) \exp i(m\theta - kz)$. It then follows that

$$\mathbf{B} \cdot \nabla U = i\frac{B_0^\theta}{r}(m - kq(r))U \equiv iF(r)U. \qquad (11.85)$$

To determine the ideal stability of this system, we vary the Rayleigh quotient, Eq. (8.28), which is equivalent to varying the Lagrangian

$$L = \rho_0\omega^2 T - \delta W_f - \delta W_V. \qquad (11.86)$$

To evaluate the plasma energy term, we start from Eq. (8.49). Applying the above orderings, we find that to lowest order

$$
\begin{aligned}
\delta W_f &= \int_P d\tau \left\{ |\mathbf{Q}|^2 + \frac{\mathbf{J}_0 \cdot \mathbf{B}_0}{B_0^2}\mathbf{B}_0 \times \boldsymbol{\xi}^* \cdot \mathbf{Q} \right\}, \\
&= \int_P d\tau \left\{ |\nabla\psi|^2 - i\nabla^2\psi_0 \left[U^*, \psi\right] \right\}, \\
&= \int_P d\tau \left\{ |\nabla(FU)|^2 - i\nabla^2\psi_0 \left[U^*, FU\right] \right\}. \qquad (11.87)
\end{aligned}
$$

We can make use of Eq. (8.68), derived in Section 8.4.2, to write the vacuum contribution as

$$2\delta W_v = 2\pi m F^2 \frac{1 + (a/b)^{2m}}{1 - (a/b)^{2m}}|U(a)|^2. \qquad (11.88)$$

The kinetic energy term is simply

$$2T = \int_P d\tau \left\{ |\nabla U|^2 \right\}. \qquad (11.89)$$

Combining these, we obtain

$$
\begin{aligned}
\frac{1}{\pi}L &= \int_P rdr \left\{ \rho_0\omega^2 \left(U'^2 + \frac{m^2}{r^2}U^2 \right) - \left((FU)'^2 + \frac{m^2}{r^2}(FU)^2 \right) \right. \\
&\quad \left. - \nabla^2\psi_0 \frac{m}{r}\left(2UU'F + U^2F' \right) \right\} \\
&\quad - mF^2\frac{1 + (a/b)^{2m}}{1 - (a/b)^{2m}}|U(a)|^2. \qquad (11.90)
\end{aligned}
$$

To apply the finite element method, we expand the $U(r)$ in basis functions,

$$U(r) = \sum_{i=1}^{N} a_i \phi_i(r). \tag{11.91}$$

We proceed to substitute this expansion into Eq. (11.90) and take derivatives with respect to each of the a_i^s. We define the matrices

$$
\begin{aligned}
A_{ij} &= \rho_0 \int_0^a r\, dr \phi_i' \phi_j' + \rho_0 m^2 \int_0^a dr \frac{1}{r} \phi_i \phi_j, \\
B_{ij} &= \int_0^a r\, dr (F\phi_i)'(F\phi_j)' + m^2 \int_0^a dr \frac{1}{r} F^2 \phi_i \phi_j \\
&\quad + m \int_0^a dr \nabla^2 \psi_0 \left[F'\phi_i \phi_j + F\phi_i' \phi_j + F\phi_i \phi_j' \right] \\
&\quad + m F^2 \left[\frac{1 + (a/b)^{2m}}{1 - (a/b)^{2m}} \right] \delta_{NN}.
\end{aligned}
\tag{11.92}
$$

If we define the vector of unknown amplitudes as $\mathbf{x} = [a_0, a_1, \cdots, a_N]$, then the eigenvalue problem takes the form

$$\omega^2 \mathbf{A} \cdot \mathbf{x} = \mathbf{B} \cdot \mathbf{x}. \tag{11.93}$$

As a check, we observe that if the equilibrium current density is constant, $\nabla^2 \psi_0 = J_0$, then $F(r)$ becomes constant, so we can set $F = F_0$. This has an analytic solution [235] in that there is only one unstable mode given by

$$\rho_0 \gamma^2 = J_0 F - \frac{2F^2}{1 - (a/b)^{2m}}. \tag{11.94}$$

A generalization of this calculation to the full ideal MHD equations in an elliptical plasma column is given in [234], and to a circular cross-section ideal MHD plasma with a vacuum region and a resistive wall in [233].

11.5.3 Accuracy of Eigenvalue Solution

The errors in the Ritz *eigenfunctions*,

$$\|U_\ell - u_\ell\|,$$

are of the same order as the errors in the approximate solution to the steady-state problem $Lu = f$ considered earlier. However, the *eigenvalues* are as accurate as the energies in the eigenfunctions. It follows that for an approximation space of order $k - 1$,

$$\|\lambda_\ell^h - \lambda_\ell\| < \|U_\ell - u_\ell\|_m^2 = O[h^{k-m}]^2,$$

for $2m^{th}$ order problems. The reason, as we have seen before, is that the Rayleigh quotient is flat near a critical point. Moderately accurate trial functions yield very accurate eigenvalues.

11.5.4 Matrix Eigenvalue Problem

There exist standard techniques for solving the matrix eigenvalue problem, and an excellent reference for this is the book by Gourlay and Watson [236]. One of the most widely used methods is called *inverse iteration*. In its simplest form, for the eigenvalue problem $Ax = \lambda x$, inverse iteration proceeds by solving a linear system at each step: $y_{n+1} = A^{-1}x_n$. Then the approximation to λ is $\lambda_{n+1} = 1/\|y_{n+1}\|$, and the new approximation to x is the normalized vector $x_{n+1} = \lambda_{n+1}y_{n+1}$. If we imagine that the starting vector x_0 is expanded in terms of the true eigenvectors v_j, $x_0 = \sum c_j v_j$, then the effect of n inverse iterations is to amplify each component by $(\lambda_j)^{-n}$. If λ_1 is distinctly smaller than the other eigenvalues, then the first component will become dominant, and x_n will approach the unit eigenvector v_1. The convergence is like that of a geometric series, depending on the fraction λ_1/λ_2, although the matrix A can be replaced by $A - \lambda_0 I$ to shift all eigenvalues uniformly by λ_0.

There is another method for solving the generalized band eigenvalue problem,

$$KQ = \lambda MQ, \tag{11.95}$$

that is due to a matrix theorem due to Sylvester [236]: The number of eigenvalues less than a given λ_0 can be determined just by counting the number of negative pivots when Gauss elimination is applied to $K - \lambda_0 M$. Thus, for example, an algorithm based on bisection can be used. Suppose it is determined that there are n_0 eigenvalues below the first guess λ_0. Then the Gaussian pivots for $K - \lambda_0/2M$ reveal the number n_1 which is below $\lambda_0/2$; and the remaining $n_0 - n_1$ eigenvalues must lie between $\lambda_0/2$ and λ_0. Repeated bisection will isolate any eigenvalue. In other variations, this method can be sped up by using the *values* of the pivots rather than just their signs, or it can be supplemented with inverse iteration once λ_K is known approximately.

11.6 Summary

The finite element method is a natural choice if the problem to be solved is of the classical variational type: Find a function $U(\mathbf{x})$ that minimizes a functional $I(U)$. The Galerkin approach extends the finite element method to any system of equations. You simply select a class of trial elements, multiply the equation by every test element in the class, and integrate over the domain. This implies that the residual will be orthogonal to all elements in the class. Integration by parts is used to shift derivatives from the original equation to the test element. It is essential that the elements have sufficient continuity to be consistent with the final form of the equations. Because of the integration by parts, C^0 elements can be used for equations with up to second derivatives and C^1 elements can be used for equations with up to fourth derivatives. If the family of finite elements used is such that a complete Taylor series can be represented to some order, then the error in the approximation will scale like the error in truncating the Taylor series expansion. By using non-uniform node spacing interior to a global finite element, spectral C^0 elements can be defined that have some advantages over high-order elements based on uniform node spacing. If a linear stability problem can be put into variational form, it is natural to use the finite element method to convert it to a matrix eigenvalue problem which can be readily solved using standard mathematical libraries. The eigenvalue will be more accurate than the eigenfunction if variational principles are followed.

Problems

11.1: Verify that the last row of the matrix equation, Eq. (11.9), is consistent with the imposition of the boundary condition $v'(1) = 0$.

11.2: Consider the analogue of the matrix equation, Eq. (11.9), for the Hermite cubic elements of Section 11.2.5 in which $\mathbf{K} = \mathbf{K}_0 + \mathbf{K}_1$ as given by the matrices following Eqs. (11.21) and (11.23). Using Taylor series expansion, show that both the even and odd rows of the of the matrix equation are consistent with the differential equation, Eq. (11.3) for the case $p = q = $ const.

11.3: Calculate the "bending matrix" that corresponds to minimizing $(v'')^2$ for the Hermite cubic finite elements of Section 11.2.5.

11.4: Consider a 2D rectangular region with $M \times N$ global elements as in Section 11.4.1. Each global element has $(P+1) \times (P+1)$ local Lagrange points defined (including points on the inter-element boundaries) so that there are $(PM + 2) \times (PN + 2)$ unknowns. How many unknowns would remain after static condensation was applied?

11.5: Calculate the matrix elements one obtains when using the Chebyshev spectral elements of Section 11.4.2 applied to the 2D Helmholtz equation given by Eq. (11.43). Assume that the domain and elements are rectangular so that the coordinate transformation given by Eq. (11.39) is just a renormalization. Make explicit use of the integrals given in Eqs. (10.32) and (10.30) in expressing the result.

11.6: Using the linear finite elements of Section 11.2.2, write a computer program to solve Eq. (11.67) in the domain $0 \le x \le 1$ for the operator given by Eq. (11.68) for $p(x) = k^2 = $ constant, $q(x) = 0$. Compare the solution with the analytic solution $\lambda_n = k^2 n^2 \pi$, and show that the lowest-order mode, λ_1, converges as h^2 as the element spacing $h \to 0$.

Bibliography

[1] Braginski, S. I. 1966. Transport processes in a plasma. In *Review of Plasma Physics, Vol. I*, ed. M.A. Leontovich, 205-311. New York: Consultants Bureau.

[2] Ramos, J. J. 2005. General expression of the gyroviscous force. *Phys. Plasmas* **12**:112301.

[3] Simakov, A. N. and P. J. Catto. 2005. Evaluation of the neoclassical radial electric field in a collisional tokamak. *Phys. Plasmas* **12**:012105.

[4] Chang, Z. and J. D. Callen. 1992. Generalized gyroviscous force and its effect on the momentum balance equation. *Phys. Fluids B* **4**:1766-71.

[5] Ferraro, N. and S. C. Jardin. 2006. Finite element implementation of Braginskii's gyroviscous stress with application to the gravitational instability. *Phys. Plasmas* **13**:092101.

[6] Breslau, J. A. and S. C. Jardin. 2003. Global extended magnetohydrodynamic studies of fast magnetic reconnection. *Phys. Plasmas* **10**:1291-1298.

[7] Jardin, S. C., J. Breslau, and N. Ferraro. 2007. A high-order implicit finite element method for integrating the two-fluid magnetohydrodynamic equations in two dimensions. *J. Comput. Phys.* **226**:2146-2174.

[8] Jardin, S. C., A. Janos, and M. Yamada. 1986. The effect of a column inductive transformer on the S-1 spheromak. *Nuclear Fusion* **26**:647-55.

[9] Ward, D. J. and S. C. Jardin. 1989. Modeling the effects of the sawtooth instability in tokamaks using a current viscosity term. *Nuclear Fusion* **29**:905-14.

[10] Hubba, J. D. 2007. NRL Plasma Formulary, NRL/PU/6790-07-500.

[11] Furth, H., J. Killeen, and M. Rosenbluth. 1960. Finite-resistivity instabilities of a sheet pinch. *Phys. Fluids* **6**:459-84.

[12] Coppi, B., J. M. Greene, and J. L. Johnson. 1966. Resistive instabilities in a diffuse linear pinch. *Nucl. Fusion* **6**:101-17.

[13] Glasser, A. H., S. C. Jardin, and G. Tesauro. 1984. Numerical solution of the resistive magnetohydrodynamic boundary-layer equations. *Phys. Fluids* **27**:1225-1242.

[14] Freidberg, J. P. 1987. *Ideal Magnetohydrodynamics*. New York: Plenum Press.

[15] Ramos, J. J. 2005. Fluid formalism for collisionless magnetized plasmas. *Phys. Plasmas* **12**:052102.

[16] Chew, G. F., M. L. Goldberger, and F. E. Low. 1956. The Boltzmann equation and the one-fluid hydromagnetic equations in the absence of particle collisions. *Proc. R. Soc. London, Ser. A* **236**:112-18.

[17] Schmalz, R. 1981. Reduced, 3-dimensional, non-linear equations for high-beta plasmas including toroidal effects. *Phys. Letters A* **82**:14-17.

[18] Strauss, H. R. 1976. Nonlinear 3-dimensional magnetohydrodynamics of noncircular tokamaks. *Phys. Fluids* **19**:134.

[19] Friedrichs, K. O. 1955. Nonlinear wave motion in magnetohydrodynamics. LAMS–2105, Los Alamos Scientific Laboratory.

[20] Friedrichs, K. O. and H. Kranzer. 1958. Notes on Magnetohydrodynamics VIII – Nonlinear Wave Motion. New York University NYO–6486.

[21] Ramos, J. J. private communication. 2009.

[22] Richtmyer, R. D. and Y. W. Morton. 1967. *Difference Methods for Initial Value Problems*. New York: Wiley & Sons.

[23] Roache, P. J. 1972. *Computational Fluid Dynamics*. Albequerque, NM: Hermosa Publishers.

[24] Roache, P. J. 1998. *Fundamentals of Computational Fluid Dynamics*. Albequerque, NM: Hermosa Publishers.

[25] Strikwerda, J. C. 2004. *Finite Difference Schemes and Partial Differential Equations*. Philadelphia: SIAM.

[26] Garabedian, P. R. 1964. *Partial Differential Equations*. New York: Wiley.

[27] Brigham, E. O. 1974. *The Fast Fourier Transform*. Englewood Cliffs, NJ: Prentice-Hall.

[28] Hirt, C. W. 1986. Heuristic stability theory for finite-difference equations. *J. Comput. Phys.* **2**:339-355.

[29] Evans, L. C. 1998. *Partial Differential Equations*. Providence: American Mathematical Society.

[30] Breslau, J., N. Ferraro, S. C. Jardin. 2009. Some properties of the M3D-C^1 form of the 3D magnetohydrodynamics equations. *Phys. Plasmas* **16**:092503.

[31] Acton, F. S. 1990. *Numerical Methods That Work*. Trumbull, CT: Spectrum Publishing.

[32] Molchanov, I. N. and L. D. Nikolenko. 1972. On an approach to integrating boundary problems with a non-unique solution. *Inform. Process. Lett.* **1**:168-172.

[33] Duff, I. S. 1998. Direct Methods. Technical Report RAL-98-056. Didcot, UK: Rutherford Appleton Laboratory.

[34] Dongarra, J., I. Duff, D. Sorensen, and H. van der Vorst. 1998. *Numerical Linear Algebra for High-Performance Computers*. Philadelphia, PA: SIAM.

[35] Heath, M., E. Ng, and B. Peyton. 1991. Parallel algorithms for sparse linear systems. *SIAM Review*, **33**:420-460.

[36] Demmel, J. W., J. R. Gilbert, and X. S. Li. 1999. An asynchronous parallel supernodal algorithm for sparse Gaussian elimination. *SIAM J. Matrix Anal. Appl.* **20**:915-952.

[37] Amestoy, P. R., I. S. Duff, and J. Y. L'Excellent. 2000. Multifrontal parallel distributed symmetric and unsymmetric solvers. *Computer Methods in Appl. Mechanics and Engineering*. **184**:501-20.

[38] Varga, R. S. 1962. *Matrix Iterative Analysis*. Englewood Cliffs, NJ: Prentice-Hall.

[39] Dahlquist, G. and N. Anderson. 1974. *Numerical Methods*. Englewood Cliffs, NJ: Prentice-Hall.

[40] Martucci, S. A. 1994. Symmetric convolution and the discrete sine and cosine transforms. *IEEE Trans. Sig. Processing*. **42**:1038-51.

[41] Briggs, W. L. 1987. *A Multigrid Tutorial*. Philadelphia: SIAM.

[42] Hackbush, W. and U. Trottenberg. 1982. *Multigrid Methods*. Berlin: Springer-Verlag.

[43] Brandt, A., S. F. McCormick, and J. W. Ruge. 1984. Algebraic multigrid (AMG) for sparse matrix equations. In *Sparsity and Its Applications*, ed. D. J. Evans. Cambridge: Cambridge University Press.

[44] Ruge, J. W. and K. Stuben. 1987. Algebraic multigrid (AMG). In *Multigrid Methods*, ed. S. F. McCormick. Vol. 3 of Frontiers in Applied Mathematics. Philadelphia: SIAM.

[45] Kershaw, D. 1978. The incomplete Cholesky-conjugate gradient method for the iterative solution of systems of linear equations. *J. Comput. Phys.* **26**:43-65.

[46] Barrett, R., M. Berry, T. Chan, et al., 1967. *Templates for the Solution of Linear Systems: Building Blocks for Iterative Methods*, (2nd edition). Philadelphia: SIAM.

[47] Golub, G. H. and C. F. Van Loan. 1996. *Matrix Computations. Third Edition.* Baltimore: The Johns Hopkins University Press.

[48] Smith, B., P. Bjorstat, W. Gropp, 1996. *Domain Decomposition, Parallel Multilevel Methods for Elliptic Partial Differential Equations*, Cambridge: Cambridge University Press.

[49] Toselli, A. and O. Widlund, *Domain Decomposition Methods-Algorithms and Theory*, Springer Series in Computational Mathematics, Vol. 34

[50] Axelsson, O. 1977. Solution of Linear Systems of Equations. *Lecture Notes in Math No. 572.* Berlin: Springer-Verlag.

[51] Feng, Y. T. 2006. On the discrete dynamic nature of the conjugate gradient method. *J. Comput. Phys.* **211**:91-98.

[52] Saad, Y. and M. H. Schultz. 1986. GMRES: A generalized minimal residual algorithm for solving nonsymmetric linear systems. *SIAM J. Sci. Stat. Comput.* **7**:856.

[53] Arnoldi, W. E. 1951. The principle of minimized iterations in the solution of the matrix eigenvalue problem. *Quarterly of Applied Mathematics.* **9**:17-29.

[54] Gould, N. I. M. and J. A. Scott. 1998. Sparse approximate-inverse preconditioners using norm-minimization techniques. *SIAM J. Sci. Comp.* **19**:605.

[55] Greenbaum, A. 1997. *Iterative Methods for Solving Linear Systems.* Philadelphia: SIAM.

[56] Cooley, J. W. and J. W. Tukey. 1965. Algorithm for machine calculation of complex Fourier series. *Math. Computation* **19**:297-301.

[57] Hockney, R. W. 1970. The potential calculation and some applications. In *Methods in Computational Physics, Vol. 9*, ed. B. Alder, S. Fernbach, and M. Rotenberg. New York: Academic Press.

[58] Grad H. and H. Rubin. 1958. Hydromagnetic equilibria and force-free fields. *Proceedings of the 2nd UN Conf. on the Peaceful Uses of Atomic Energy.* **31**:190 Beneva: IAEA.

[59] Shafranov, V. D. 1958. On magnetohydrodynamical equilibrium configurations. *Sov. Phys.-JETP* **6**:545-54.

[60] Lust, R. and A. Schlüter. 1957. Axisymmetric magnetohydrodynamic equilibrium configurations. *Z. Naturforsch.* **12a**:850.

[61] Strauss, H. R. 1973. Toroidal magnetohydrodynamic equilibrium with toroidal flow. *Phys. Fluids* **16**:1377.

[62] Lao, L., H. St. John, Q. Peng, et. al. 2005. MHD equilibrium reconstruction in the DIII-D tokamak. *Fus. Sci. and Tech.* **48**:968.

[63] Semenzato, S., R. Gruber, and H. Zehrfeld. 1984. Computation of symmetric ideal MHD flow equilibria. *Computer Phys. Reports* **1**:389.

[64] Guazzotto, L. and R. Betti. 2005. Magnetohydrodynamics equilibria with toroidal and poloidal flow. *Phys. Plasmas* **12**:056107.

[65] Belien, A. J., M. A. Botchev, J. P. Goedbloed, et al. 2005. FINESSE: Axisymmetric MHD equilibria with flow. *J. Comput. Phys.* **182**:91-117.

[66] Stix, T. 1973. Decay of poloidal rotation in a tokamak plasma. *Phys. Fluids* **16**:1260.

[67] Hassam, A. and R. M. Kulsrud. 1978. Time evolution of mass flows in a collisional tokamak. *Phys. Fluids* **21**:2271.

[68] Ferraro, N. M. and S. C. Jardin. 2009. Calculations of two-fluid magneto-hydrodynamic axisymmetric steady-states. *J. Comput. Phys* **228**:7742-70.

[69] Aydemir, A. Y. 2009. An intrinsic source of radial electric field and edge flows in tokamaks. *Nuclear Fusion* **49**:065001.

[70] Grad, H. 1967. Toroidal containment of a plasma. *Phys. Fluids* **10**:137.

[71] Cooper, A. et al. 1980. Beam-induced tensor pressure tokamak equilibria. *Nucl. Fus.* **20**:985.

[72] Shafranov, V. D. 1966. Plasma equilibrium in a magnetic field. In *Reviews of Plasma Physics, Vol. II.* Ed. M. A. Leontovich. New York: Consultants Bureau.

[73] Jardin, S. C. 1978. Stabilization of axisymmetric instability in poloidal divertor experiment. *Phys. Fluids* **21**:1851-5.

[74] Jardin, S. C. and D. A. Larrabee. 1982. Feedback stabilization of rigid axisymmetric modes in tokamaks. *Phys. Fluids* **22**:1095-8.

[75] Zakharov, L. E. and V. D. Shafranov. 1973. Equilibrium of a toroidal plasma with noncircular cross section. *Sov. Phys. Tech. Phys.* **18**:151.

[76] Reusch, M. F. and G. H. Neilson. 1986. Toroidally symmetric polynomial multipole solutions of the vector Laplace equation. *J. Comput. Phys.* **64**:416.

[77] Solov'ev, L. S. 1967. The theory of hydromagnetic stability of toroidal plasma configurations. *Sov. Phys. JETP* **26**:400.

[78] Chance, M., J. Greene, R. Grimm, et al. 1978. Comparative numerical studies of ideal magnetohydrodynamic instabilities. *J. Comput. Phys.* **28**:1.

[79] Zheng, S. B., A. J. Wootton, and E. R. Solano. 1996. Analytical tokamak equilibrium for shaped plasmas. *Phys. Plasmas* **3**:1176-9

[80] Weening, R. H. 2000. Analytic spherical torus plasma equilibrium model. *Phys. Plasmas* **7**:3654-62

[81] Shi, B. 2005. Analytic description of high poloidal beta equilibrium with a natural inboard poloidal field null. *Phys. Plasmas* **12**:122504.

[82] Cerfon, A. J. and J. P. Freidberg. 2010. "One size fits all" analytic solutions to the Grad–Shafranov equation. *Phys. Plasmas* **17**:032502.

[83] Atanasiu, C. V., S. Gunter, K. Lackner, et al. 2004. Analytical solutions to the Grad-Shafranov equation. *Phys. Plasmas* **11**:3510.

[84] Guazzotto, L. and J. P. Freidberg. 2007. A family of analytic equilibrium solutions for the Grad–Shafranov equation. *Phys. Plasmas* **14**:112508.

[85] Lao, L. L., S. P. Hirshman, and R. M. Wieland. 1981. Variational moment solutions to the Grad–Shafranov equation. *Phys. Fluids* **24**:1431-1441.

[86] Ling, K. M and S. C. Jardin. 1985. The Princeton spectral equilibrium code- PSEC. *J. Comput. Phys.* **58**:300-35.

[87] Johnson, J. L, H. E. Dalhed, J. M. Greene, et al. 1979. Numerical determination of axisymmetric toroidal magnetohydrodynamic equilibria. *J. Comput. Phys.* **32**:212-234.

[88] Jackson, J. D. 1975 *Classical Electrodynamics, Second Edition.* New York: Wiley.

[89] Lackner, K. 1976. Computation of ideal MHD equilibria. *Comput. Phys. Comm.* **12**:33-44.

[90] Akima, H. 1974. A method of bivariate interpolation and smooth surface fitting based on local procedures. *Comm. of the ACM* **17**:18.

[91] Lao, L., H. St. John, R. Stambaugh, et al. 1985. Reconstruction of current profile parameters and plasma shapes in tokamaks. *Nucl. Fus.* **25**:1611.

[92] Zakharov, L., E. L. Foley, F. M. Levinton, et al. 2008. Reconstruction of the q and p profiles in ITER from external and internal measurements. *Plasma Physics Reports* **34**:173.

[93] Zwingmann, W., L. Eriksson, and P. Stubberfield. 2001. Equilibrium analysis of tokamak discharges with anisotropic pressure. *Plasma Phys. Control. Fusion* **43**:1441-1456.

[94] Bateman, G., and Y.-K. M. Peng. 1977. Magnetohydrodynamic stability of flux-conserving tokamak equilibria. *Phys. Rev. Lett.* **38**:829-31.

[95] Kruskal M. and R. Kulsrud. 1958. Equilibrium of a magnetic confined plasma in a toroid. *Phys. Fluids* **1**:265.

[96] Hamada S. 1959. Notes on magnetohydrodynamic equilibrium. *Progr. Theoret. Phys.* **22**:145-6.

[97] Greene, J. M. and J. L. Johnson. 1962. Stability criterion for arbitrary hydromagnetic equilibria. *Phys. Fluids* **12**:510-7.

[98] Glasser, A. H. 1978. *Lecture Notes on Ballooning Modes.* Princeton, NJ.: Princeton University Plasma Physics Laboratory.

[99] Grimm, R. C., R. L. Dewar, J. Manickam. 1983. Ideal MHD stability calculations in aisymmetric toroidal coordinate systems. *J. Comput. Phys.* **49**:94-117.

[100] Boozer, A. H. 1980. Guiding center drift equations. *Phys. Fluids* **23**:904-8

[101] Boozer, A. H. 1981. Plasma equilibrium with rational magnetic surfaces. *Phys. Fluids* **24**:1999-2003

[102] White, R. B. and M. S. Chance. 1984. Hamiltonian guiding center drift orbit calculation for plasmas of arbitrary cross section. *Phys. Fluids* **27**:2455-67

[103] DeLucia, J., S. C. Jardin, A. M. M. Todd. 1980. An iterative metric method for solving the inverse tokamak equilibrium problem. *J. Comput. Phys.* **37**:183-204.

[104] Andre, R. Private communication. 2009.

[105] LoDestro, L. and L. D. Pearlstein. 1994. On the Grad–Shafranov equation as a eigenvalue problem, with implications for q-solvers. *Phys. Plasmas* **1**:90-95.

[106] Greene, J. M., J. L. Johnson, K. Weimer. 1971. Tokamak Equilibrium. *Phys. Fluids* **14**:671.

[107] Hirshman, S. P. and J. C. Whitson. 1983. Steepest-descent moment method for three-dimensional magnetohydrodynamic equilibria. *Phys. Fluids* **28**:3553.

[108] Johnson, J., C. R. Oberman, R. M. Kulsrud, et al. 1958. Some stable hydromagnetic equilibria. *Phys. Fluids* **1**:281.

[109] Kadomtsev, B. B. 1960. Equilibrium of a plasma with helical symmetry. *Soviet Physics JETP* **37**:962.

[110] Grad H. and J. Hogan. 1970. Classical diffusion in a tokamak. *Phys. Rev. Lett.* **24**:1337.

[111] Pao, Y. P. 1976. Classical diffusion in toroidal plasmas. *Phys. Fluids* **19**:1177-82.

[112] Hirshman, S. P. and S. C. Jardin. 1979. Two-dimensional transport of tokamak plasmas. *Phys. Fluids* **22**:731.

[113] Hazeltine, R. D. and F. L. Hinton. 1973. Collision-dominated plasma transport in toroidal confinement systems. *Phys. Fluids.* **16**:1883-9.

[114] Hirshman, S. P. 1977. Transport of a multiple-ion species plasma in Pfirsch–Schluter regime. *Phys. Fluids* **20**:589.

[115] Hirshman, S. P. 1978. Moment equation approach to neoclassical transport-theory. *Phys. Fluids* **21**:224.

[116] Hirshman, S. P. and D. J. Sigmar. 1981. Neoclassical transport of impurities in tokamak plasmas. *Nuc. Fusion* **21**:1079-1201.

[117] Hirshman, S. P. 1988. Finite-aspect-ratio effects on the bootstrap current in tokamaks. *Phys. Fluids* **31**:3150-52.

[118] Sauter, O., C. Angioni, and Y. R. Lin-Liu. 1999. Neoclassical conductivity and bootstrap current formulas for general axisymmetric equilibria and arbitrary collisionality regime. *Phys. Plasmas* **6**:2834.

[119] Helander, P. and D. J. Sigmar. 2002. *Collisional Transport in Magnetized Plasmas*. Cambridge: Cambridge University Press.

[120] Kinsey, J. E., G. M. Staebler, and R. E. Waltz. 2005. Predicting core and edge transport barriers in tokamaks using the GLF23 drift-wave transport model. *Phys Plasmas* **12**:052503.

[121] Bateman, G., A. H. Kritz, and J. E. Kinsey, et. al. 1998. Predicting temperature and density profiles in tokamaks. *Phys. Plasmas* **5**:1793.

[122] Weiland, J. 1999. *Collective Modes in Inhomogeneous Plasmas and Advanced Fluid Theory.* Bristol: Institute of Physics.

[123] Erba, M., T. Aniel, and V. Basiuk. 1998. Validation of a new mixed Bohm/gyro-Bohm model for electron and ion heat transport against the ITER, Tore Supra and START database discharges. *Nucl. Fusion* **38**:1013.

[124] Jardin, S. C., M. G. Bell, and N. Pomphrey. 1993. TSC simulation of Ohmic discharges in TFTR. *Nuclear Fusion* **33**:371-82.

[125] Blum J. and J. LeFoll. 1984. Plasma equilibrium evolution at the resistive diffusion timescale. *J. Comp. Phys. Reports* **1**:465-494.

[126] Taylor, J. B., Private communication. 1975.

[127] Jardin, S. C. 1981. Self-consistent solutions of the plasma transport equations in an axisymmetric toroidal system. *J. Comput. Phys.* **43**:31-60.

[128] Jardin, S. C. and W. Park. 1981. Two-dimensional modeling of the formation of spheromak configurations. *Phys. Fluids* **24**:679-88.

[129] Jardin, S. C., N. Pomphrey, and J. DeLucia. 1986. Dynamic modeling of transport and positional control of tokamaks. *J. Comput. Phys.* **66**:481-507.

[130] Hazeltine, R. D., F. L. Hinton, and M. N. Rosenbluth. 1973. Plasma transport in a torus of arbitrary aspect ratio. *Phys. Fluids* **16**:1645-53.

[131] Ascher, U. M., S. J. Ruuth, and B. T. R. Wetton. 1995. Implicit explicit methods for time-dependent partial-differential equations. *SIAM J. Numer. Anal.* **32**:797-823.

[132] DuFort, E. C., and S. P. Frankel. 1953. Stability conditions in the numerical treatment of parabolic differential equations. *Math. Tables and Other Aids to Computation* **7**: 135-52.

[133] Jardin, S. C., G. Bateman, G. W. Hammett, et al. 2008. On 1D diffusion problems with a gradient-dependent diffusion coefficient. *J. Comput. Phys.* **227**:8769-8775.

[134] Shestakov, A. I., R. H. Cohen, J.A.Crotinger, et al. 2003. Self-consistent modeling of turbulence and transport. *J. Comput. Phys.* **185**: 399-426.

[135] Yanenko, M. N. 1970. *The Method of Fractional Steps.* New York: Springer-Verlag,

[136] Douglas J. and J. Gunn. 1965. A General Formulation Alternating Direction Methods, I. *Number. Math.* **6**:428.

[137] Günter, S. G., Q. Yu, J. Krüger, and K. Lackner. 2005. Modeling of heat transport in magnetised plasmas using non-aligned coordinates. *J. Comput. Phys.* **209**:354-370.

[138] Sharma, P. and G. Hammett. 2007. Preserving monotonicity in anisotropic diffusion. *J. Comput. Phys.* **227**:123-142.

[139] Yuan, G. and F. Zuo. 2003. Parallel difference schemes for heat conduction equations. *Intern. J. Computer Math.* **80**:993-997.

[140] Günter, S., K. Lackner. 2009. A mixed implicit-explicit finite difference scheme for heat transport in magnetised plasmas. *J. Comput. Phys.* **228**:282-293.

[141] Bernstein, I. B., E. A. Frieman, M. D. Kruskal, and R. M. Kulsrud. 1958. An energy principle for hydromagnetic stability problems. *Proc. Royal Soc. London Ser. A* **244**:17.

[142] Kadomtsev, B. B. 1966. Hydromagnetic stability of a plasma. *Reviews of Plasma Physics.* **2**:153-199. New York: Consultants Bureau.

[143] Goedbloed, J. P. 1975. The spectrum of ideal magnetohydrodynamics of axisymmetric toroidal systems. *Phys. Fluids* **18**:1258-68.

[144] Hameiri, E. 1985. On the essential spectrum of ideal magnetohydrodynamics. *Commun. Pure Appl. Math.* **38**:43.

[145] Lifshitz, A. E. 1989. *Magnetohydrodynamics and Spectral Theory.* London: Kluwer Academic.

[146] Friedman, B. 1956. *Principles and Techniques of Applied Mathematics.* New York: Wiley & Sons.

[147] Appert, K., R. Gruber, J. Vaclavik. 1974. Continuous spectra of a cylindrical magnetohydrodynamic equilibrium. *Phys. Fluids* **17**:1471-2.

[148] Goedbloed, J. P. 1983. Lecture Notes on Ideal Magnetohydrodynamics. *Rijnhuizen Report 83-145.* FOM-Instituut voor Plasmapysica. Rijnhuizen, The Netherlands.

[149] Cheng, C. Z. and M. S. Chance. 1986. Low-n shear Alfvén spectra in axisymmetrical toroidal plasmas. *Phys. Fluids B* **29**:3695-701.

[150] Nührenberg, C. 1999. Compressional ideal magnetohydrodynamics: Unstable global modes, stable spectra, and Alfvén eigenmodes in Wendelstein 7–X-type equilibria. *Phys. Plasmas* **6**:137-47.

[151] Li, Y. M., S. M. Mahajan, D. W. Ross. 1987. Destabilization of global Alfvén eigenmodes and kinetic Alfvén waves by alpha-particles in tokamak plasma. *Phys. Fluids* **30**:1466-84.

[152] Fu, G. Y. and J. W. VanDam. 1989. Stability of the global Alfvén eigenmode in the presence of fusion alpha-particles in an ignited tokamak plasma. *Phys. Fluids B* **12**:2404-13.

[153] Cheng, C. Z. 1990. Alpha-particle effects on low-n magnetohydrodynamic modes. *Fusion Technology* **18**:443-54.

[154] Smith, S. P. 2009. Magnetohydrodynamic stability spectrum with flow and a resistive wall. Ph.D. diss., Princeton University, Princeton NJ.

[155] Frieman, E., and M. Rotenberg. 1960. On hydromagnetic stability of stationary equilibria. *Rev. Mod. Phys.* **32**:898-902.

[156] Guazzotto, L., J. P. Freidberg, and R. Betti. 2008. A general formulation of magnetohydrodynamic stability including flow and a resistive wall. *Phys. Plasmas* **15**:072503.

[157] Smith, S. P., S. C. Jardin, J. P. Freidberg. et al. 2009. Numerical calculations demonstrating complete stabilizaition of the ideal MHD resistive wall mode by longitudinal flow. *Phys. Plasmas* **16**:084505.

[158] Laval, G., C. Mercier, and R. M. Pellat. 1965. Necessity of energy principles for magnetostatic stability. *Nucl. Fusion* **5**:156.

[159] Freidberg, J. P. 1982. Ideal magnetohydrodynamic theory of magnetic fusion systems. *Rev. Mod. Phys.* **54**:801-902.

[160] Grimm, R. C., J. M. Greene, and J. L. Johnson. 1976. Computation of the magnetohydrodynamic spectrum in axisymmetric toroidal confinement systems. In *Methods in Computational Physics Vol. 16*, ed. B. Alder, S. Fernbach, M. Rotenberg, J. Killeen, 253-281. New York: Academic Press.

[161] Gruber R., F. Troyon, D. Berger, et al. 1981. ERATO stability code. *Computer Phys. Comm.* **21**:323-71.

[162] Goedbloed, J. P. and P. H. Sakanaka. 1974. New approach to magnetohydrodynamic stability. *Phys. Fluids* **17**:908-29.

[163] Bateman, G. 1978. *MHD Instabilities*. Cambridge, MA. MIT Press.

[164] Newcomb, W. A. 1960. The hydromagnetic stability of a diffuse linear pinch. *Ann. Phys.* **10**:232.

[165] Cheng, C. Z. and M. S. Chance. 1987. NOVA: A nonvariational code for solving the MHD stability of axisymmetric toroidal plasmas. *J. Comput. Phys.* **71**:124-146.

[166] Harley, T. R., C. Z. Cheng, and S. C. Jardin. 1992. The computation of resistive MHD instabilities in axisymmetrical toroidal plasmas. *J. Comput. Phys.* **103**:43-62.

[167] Ward, D., S. C. Jardin, and C. Z. Cheng. 1993. Calculations of axisymmetrical stability of tokamak plasmas with active and passive feedback. *J. Comput. Phys.* **104**:221-40.

[168] Cheng, C. Z. 1992. Kinetic extensions of magnetohydrodynamics for axisymmetrical toroidal plasmas. *Physics Reports-Review Section of Physics Letters* **211**:1-51.

[169] Lüst, R. and E. Martensen. 1960. The many-valued character of the scalar magnetic potential in the plasma hydromagnetic stability problem. *Z. Naturforsch. A* **15**:706-13.

[170] Chance, M. S. 1997. Vacuum calculations in azimuthally symmetric geometry. *Phys. Plasmas* **4**:2161.

[171] Bineau, M. 1962. Hydromagnetic stability of a toroidal plasma-variational study of the integral of energy. *Nucl. Fusion* **2**:130-47.

[172] Dobrott, D., D. Nelson, J. Greene, et al. 1977. The theory of ballooning modes in tokamaks with finite shear. *Phys. Rev. Lett.* **39**:943.

[173] Connor, J. W., R. J. Hastie, J. B. Taylor. 1978. Shear, periodicity, and plasma ballooning modes. *Phys. Rev. Lett.* **40**:396-99.

[174] Greene, J. M. and J. L. Johnson. 1968. Interchange instabilities in ideal hydromagnetic theory. *Plasma Physics* **10**:729.

[175] Mercier, C. 1962. Stability criterion of a hydromagnetic toroidal system with scalar pressure. *Nucl. Fusion* **Suppl. 2**:801-08.

[176] Shafranov, V. and Y. Yurchenko. 1968. Condition for flute instability of a toroidal-geometry plasma. *Sov. Phys. JETP* **26**:682.

[177] Wilson, H. R., P. B. Snyder, G. Huysmans, et al. 2002. Numerical studies of edge localized instabilities in tokamaks. *Phys. Plasmas* **9**:1277-1286.

[178] Connor, J. W., R. J. Hastie, J. B. Taylor. 1979. High mode number stability of an axisymmetric toroidal plasma. *Proc. R. Soc. London, Ser. A* **A365**:1-17.

[179] Hirsch, C. 1990. *Numerical Computation of Internal and External Flows, Volumes I and II.* New York: John Wiley & Sons.

[180] Leveque, R. J. 2002. *Finite Volume Methods for Hyperbolic Problems.* Cambridge: Cambridge University Press.

[181] Kreiss, H. O. 1964. On difference approximations of the dissipative type for hyperbolic differential equations. *Comm. Pure Appl. Math.* **17**:335.

[182] Lax, P. D. 1965. Weak solutions of nonlinear hyperbolic equations and their numerical computation. *Comm. Pure Appl. Math.* **7**:159.

[183] Courant, R., K. O. Friedrichs, and H. Lewy. 1928. Über die partiellen Differenzengleichungen der mathematischen Physik. *Math. Ann.* **100**:32-74.

[184] Courant, R., K. O. Friedrichs, and H. Lewy. 1967. On the partial differential equations of mathematical physics. *IBM J.* **11**:215-234.

[185] Lax, P. D. and B. Wendroff. 1960. Systems of conservation laws. *Commun. Pure Appl. Math.* **13**:217-237

[186] MacCormack, R. W. 1969. The effects of viscosity in hypervelocity impact cratering. *AIAA Paper* **69**:354.

[187] Zalesak, S. T. 1979. Fully multidimensional flux-corrected transport algorithms for fluids. *J. Comput. Phys.* **31**:35.

[188] Warming, R. F. and R. M. Beam. 1975. Upwind second-order difference schemes and applicaitons in unsteady aerodynamic flows. In *Proc. AIAA 2nd Computational Fluid Dynamics Conf.*, Hartford, CT.

[189] Courant, R., E. Isaacson, and M. Rees. 1952. On the solution of nonlinear hyperbolic differential equations by finite differences. *Commun. Pure Appl. Math.* **5**:243-55.

[190] Godunov, S. K. 1959. A difference scheme for numerical solution of discontinuous solution of hydrodynamic equations. *Math. Sbornik* **47**:271-306.

[191] Roe, P. L. 1981. Approximate Riemann solvers, parameter vectors, and difference schemes. *J. Comput. Phys.* **43**:357-72.

[192] Powell, K. G., P. L. Roe, T. J. Linde, et al. 1999. A solution-adaptive upwind scheme for ideal magnetohydrodynamics. *J. Comput. Phys.* **154**:284-309.

[193] Cargo, P. and G. Gallice. 1997. Roe matrices for ideal MHD and systematic construction of Roe matrices for systems of conservation laws. *J. Comput. Phys.* **136**:446-66.

[194] Aslan, N. 1996. Two-dimensional solutions of MHD equations with an adapted Roe method. *Int. J. Numer. Meth. in Fluids* **23**:1211.

[195] Sankaran, K., L. Martinelli, S. C. Jardin, and E. Y. Choueiri. 2002. A flux-limited numerical method for solving the MHD equations to simulate propulsive plasma flows. *Int. J. Numer. Methods in Eng.* **53**:1415-1432.

[196] Sankaran K., E. Y. Choueri, S. C. Jardin. 2005. Comparison of simulated magnetoplasmadynamic thruster flowfields to experimental measurements. *J. of Propulsion and Power* **21**:129-38.

[197] Boris, J. P. and D. L. Book. 1973. Flux corrected transport I, SHASTA, a fluid transport algorithm that works. *J. Comput. Phys.* **11**:38-69.

[198] Harten, A. and G. Zwas. 1972. Self-adjusting hybrid schemes for shock comjputations. *J. Comput. Phys.* **9**:568.

[199] Van Leer, B. 1973. Towards the ultimate conservative difference scheme I. The quest of monotonicity. *Springer Lecture Notes Phys.* **18**:163.

[200] Roe, P. L. 1985. Some contributions to the modeling of discontinuous flows. *Lect. Notes Appl. Math* **22**:163.

[201] Van Leer, B. 1974. Towards the ultimate conservative difference scheme II. Monotonicity and conservation combined in a second order scheme. *J. Comput. Phys.* **14**:361.

[202] Sweby, P. K. 1984. High resolution schemes using flux limiters for hyperbolic conservation laws. *SIAM J. Numer. Anal.* **21**:995-1011.

[203] Jardin, S. C. 1985. Multiple time-scale methods in tokamak magneto-hydrodynamics. In *Multiple Timescales*, ed. J. Brackbill and B. Cohen. New York: Academic Press.

[204] Beam, R. M. and R. F. Warming. 1976. An implicit finite-difference algorithm for hyperbolic systems in conservation-law form. *J. Comput. Phys.* **22**:87.

[205] Breslau, J. A. and S. C. Jardin. 2003. A parallel algorithm for global magnetic reconnection studies. *Computer Phys. Comm.* **151**:8-24.

[206] Arakawa, A. 1997. Computational design for long term numerical integration of the equations of fluid motion: Two-dimensional incompressible flow. Part I. *J. Comput. Phys.* **135**:103-114.

[207] Caramana, E. 1991. Derivation of implicit difference schemes by the method of differential approximation. *J. Comput. Phys.* **96**:484-493.

[208] Sovinec, C. R., A. H. Glasser, T. A. Gianakon, et al. 2004. Nonlinear magnetohydrodynamics simulation using high-order finite elements. *J. Comput. Phys.* **195**:355-386.

[209] Harned, D. S. and W. Kerner. 1986. Semi-implicit method for three-dimensional compressible magnetohydrodynamic simulation. *J. Comput. Phys.* **65**:57.

[210] Harned, D. S. and D. D. Schnack. 1986. Semi-implicit method for long time scale magnetohydrodynamic computations in three dimensions. *J. Comput. Phys.* **24**:57-70.

[211] Harned, D. S. and Z. Mikic. 1989. Accurate semi-implicit treatment of the Hall effect in magnetohydrodynamic calculations. *J. Comput. Phys.* **83**:1-15.

[212] Knoll, D. A. and D. E. Keyes. 2004. Jacobian-free Newton–Krylov methods: A survey of approaches and applications. *J. Comput. Phys.* **193**:357-397.

[213] Brown, P. N. and Y. Saad. 1990. Hybrid Krylov methods for nonlinear systems of equations. *SIAM J. Sci. Stat. Comput.* **11**:450-481.

[214] Chan, T. F. and K. R. Jackson. 1984. Nonlinearly preconditioned Krylov subspace methods for discrete Newton algorithms. *SIAM J. Sci. Stat. Comput.* **5**:533-542.

[215] Gottlieb D. and S. A. Orszag. 1978. *Numerical Analysis of Spectral Methods: Theory and Applications*. Philadelphia: SIAM-CBMS.

[216] Suetin, P. K. 2001. Chebyshev polynomials. In Hazewinkel, M. *Encyclopaedia of Mathematics*. Kluwer Academic Publishers.

[217] Canuto, C., M. Y. Hussaini, A. Quarteroni, T. A. Zang. 1987. *Spectral Methods in Fluid Dynamics*. New York: Springer-Verlag.

[218] Gear, C. W. 1971. *Numerical Initial Value Problems in Ordinary Differential Equations*. Englewood Cliffs, NJ: Prentice-Hall.

[219] Lambert, J. D. 1973. *Computational Methods in Ordinary Differential Equations*. New York: John Wiley & Sons.

[220] Shampine L. F., and M. K. Gordon. 1975. *Computer Solution of Ordinary Differential Equations: The Initial Value Problem*. New York: W. H. Freeman.

[221] Loureiro, N. F. and G. W. Hammett. 2008. An iterative semi-implicit scheme with robust damping. *J. Comput. Phys.* **227**:4518-4542.

[222] Lee, W. W. 1983. Gyrokinetic approach in particle simulation. *Phys. Fluids* **26**:556-562.

[223] Hammett, G. W., W. Dorland, and F. W. Perkins. 1992. Fluid models of phase mixing, Landau damping, and nonlinear gyrokinetic dynamics. *Phys. Fluids B* **7**:2052-2061.

[224] Strang C. and G. J. Fix. 1973. *An Analysis of the Finite Element Method*. Englewood Cliffs, NJ: Prentice-Hall.

[225] Karniadakis, G. E. and S. J. Sherwin. 1999. *Spectral/hp Element Methods for CFD*. New York: Oxford University Press.

[226] Patera, A. T. 1984. A spectral element method for fluid dynamics: Laminar flow in a channel expansion. *J. Comput. Phys.* **54**:468-488.

[227] Cowper, G. R., E. Kosko, G. Lindberg, et al. 1969. Static and dynamic applications of a high-precision triangular plate bending element. *AIAA J.* **7**:1957.

[228] Jardin, S. C. 2004. A triangular finite element with first-derivative continuity applied to fusion MHD applications. *J. Comput. Phys.* **200**:133-152.

[229] Jardin, S. C. and J. A. Breslau. 2005. Implicit solution of the four-field extended-magnetohydrodynamic equations using high-order high-continuity finite elements. *Phys. Plasmas* **12**:056101.

[230] Dunavant, D. A. 1985. High degree efficient symmetrical Gaussian quadrature rules for the triangle. *Int. J. Numer. Methods Eng.* **21**:1129-1148.

[231] Ferraro, N., S. C. Jardin, X. Luo. 2009. Boundary conditions with reduced quintic finite elements. *PPPL-4497* Princeton, NJ: Princeton University Plasma Physics Laboratory.

[232] Appert, K., D. Berger, R. Gruber, and J. Rappaz. 1975. New finite-element approach to normal mode analysis in magnetohydrodynamics. *J. Comput. Phys.* **18**:284.

[233] Smith, S. P. and S. C. Jardin. 2008. Ideal magnetohydrodynamic stability spectrum with a resistive wall. *Phys. Plasmas* **15**:080701.

[234] Chance, M. S., J. M. Greene, R. C. Grimm, and J. L. Johnson. 1977. Study of MHD spectrum of an elliptic plasma column. *Nucl. Fusion* **17**:65.

[235] Shafranov, V. D. 1970. Hydromagnetic stability of a current-carrying pinch in a strong longitudinal magnetic field. *Soviet Physics Technical Physics* **15**:241.

[236] Gourlay, A. R. and G. A. Watson. 1973. *Computational Methods for Matrix Eigenproblems.* New York: John Wiley & Sons.

Index

Accuracy
    and conservative differencing, 39
    order of, 39
Adiabatic index, 10, 145
Adiabatic variables, 158, 162
Alfvén wave, 19
    Alfvén time, 9
    Alfvén velocity, 9
Aliasing error, 272, 273, 278, 279
Anti-diffusion, 248
Arc length, 123
Arnoldi iteration, 76

Back substitution, 54, 55, 57, 79
Ballooning modes, 226
    balloon equation, 230
    ballooning representation, 229
    ordering, 227
Beam–Warming method, 244
Block-Tridiagonal, *see* tridiagonal
Boltzmann equation, 1
Boozer coordinates, 136
Boundary value problem, 46, 143
Burgers equation, 278

CFL condition
    definition, 38
    domain of dependence, 38
    implicit methods and, 249
    Lax–Friedrichs method, 236
    Lax–Wendroff method, 248
    partially implicit method, 250
    spectral equivalent, 283
Characteristics, 14
    characteristic curves, 14
    characteristic equation, 37
    domain of dependence, 16

    domain of influence, 16
    in ideal MHD, 16
    spacelike, 15
    timelike, 15
Charge density
    definition of, 2
    quasineutrality, 5
Chebyshev polynomials, 273
    discrete, 277
    expansion of function in, 275
Cholesky factorization
    incomplete, 81
CIR method, 245
Closure relations
    higher order, 10
    ideal MHD, 9
    need for, 5
    reduced MHD, 10
    resistive MHD, 8
    surface averaged MHD, 10
    two-fluid MHD, 6
Collision dominance, 8
Collision operator, 1
Collisional friction, 5
Condition number of a matrix, 76, 80
Conjugate gradient method, 73
    and dynamic relaxation, 76
    stencil for, 75
    with preconditioning, 82
Conservation form
    boundary conditions, 12
    definition, 10
    difficulties with, 12
    MHD equations, 11
Conservative differencing
    benefits of, 40
    Grad–Shafranov equation, 107

Neumann BC and, 51
non-linear diffusion, 176
upwind methods, 246
Consistency, 31
Continuous spectrum, 202
Convergence
    accelerated, 146
    and eigenvalues, 59
    anisotropic diffusion, 189
    criteria, 75, 82
    DuFort–Frankel method, 173
    finite difference solution, 31
    Grad–Shafranov equation, 108
    non-linear diffusion equation, 178
    rate of, 61, 62, 64, 66, 70, 76
Convolution sum, 278
Coordinate velocity, 152
Courant–Friedrichs–Lewy, *see* CFL
        condition
Courant–Isaacson–Rees, 245
Current density
    2D slab, 254
    asymptotic solution, 140
    axisymmetric equilibrium, 156
    axisymmetric geometry, 94
    central value, 109
    definition of, 2
    definition of in MHD, 4
    flux coordinates, 129
    force vector, 158
    linear, 203
    linearized, 197
    lowest order in cylinder, 320
    on axis, 213
    parallel, 129
        surface averaged, 131, 158
    relation to species velocity, 4
    surface integrated, 157

Debye length, 4
Diagonally dominant, 58, 293
Differential approximation method,
        257
Differential particle number, 156
Differential volume element, 121, 123

surface, 130, 145, 156
Diffusion equations
    one dimensional, 171
Diffusion time scales
    equilibration, 167
    flux diffusion, 165
    heat conduction, 167
    particle diffusion, 166
    skin time, 166
Direct solvers, 57
Discrete Chebyshev series, 277
Discrete transform, 270
Distribution function, 1
Domain decomposition methods, 81
Domain of dependence
    and characteristics, 16
    and domain of influence, 16
    and the CFL condition, 38
    explicit vs implicit, 30, 250
    finite difference, 29
DuFort–Frankel method, 184, 255
Dynamic relaxation, 65

Edge localized modes (ELMs), 231
Eigenvalue problems, 316
Electron heat flux, 157
Electron viscosity, 7
ELITE code, 232
Elliptic equations
    as steady state limit, 63
    finite difference methods for, 45
    for Ψ, 107
    for velocity potential, 256
    in vacuum, 102
    matrices associated with, 53
    matrix iterative methods, 58
    physical approach, 62
        dynamic relaxation, 65
        first order methods, 63
    transform methods, 86
Energetic particle modes, 203
Energy principle, 205
    comparison theorem, 211
    extended, 207
    physical significance, 210

proof, 206
reduction in a cylinder, 216
reduction in a torus, 225
Entropy
    conservation, 105, 106
    constraint in equilibrium, 165
    entropy disturbance, 19
    equation for, 10
    per unit mass, 10
    surface averaged density, 157
Equilibration, 7
Equilibrium equation
    and Grad–Shafranov equation, 93
    variational forms, 105
        Grad–Hirshman, 106
        ideal MHD, 105
Equilibrium identities
    curvature vector, 209
    divergence of $\boldsymbol{\xi}$, 209
    normal magnetic field, 210
Equilibrium surface current, 198
Equipartition term, 5, 150, 158
ERATO code, 206
Errors, 30
Eulerian velocity, 153
Explicit methods
    compared to implicit methods, 40
    definition of, 29
    hyperbolic equations
        centered space derivatives, 235
        limiter methods, 247
        one-sided space differences, 242
    parabolic equations
        DuFort–Frankel, 173
        FTCS, 172
        multiple dimensions, 183
Exponential convergence, 267
Exponential decay, 271
Extended magnetohydrodynamics, 4
External sources, 1

Fast wave, 19
Finite element method, 267, 289
    boundary conditions, 315

cubic B-spline, 300, 301
Hermite cubic, 298
linear elements, 291, 292
natural, or Neumann, 293
reduced quintic, 315
two dimensions, 309
continuity, 293, 305
cubic B-splines, 298
eigenvalue problems, 316
energy integral, 294
energy norms, 294
errors, 322
Galerkin method, 302
Hermite cubic, 295
linear elements, 290
mass matrix, 293
nodal expansion, 306
order, 305
reduced quintic, 310
Ritz method, 289
    error, 294
spectral elements, 309
    Chebyshev, 309
    Legendre, 310
    properties, 309
    spectral convergence, 310
static condensation, 309
stiffness matrix, 293
unknowns per triangle, 312
Finite Larmor radius (FLR), 6
Finite transform, 270
Finite volume method, 27, 41, 42
Fluid state vector, 37
Fluid velocity, 3
Flux coordinates, *see* Magnetic flux coordinates
Flux corrected transport, 248
Flux surface averaging, 154
Flux vector, 37
Force vector, 158
Fourier collocation, 279
Fourier expansion
    continuous, 270
    discrete, 272
Fourier Galerkin, 278

Fourier transform
  fast, 83, 184, 277, 279
    2D elliptic equations, 86
    based on prime numbers, 86
  finite, 33, 82
  real finite, 83
Friction force, 7
Fundamental variables, 5

Galerkin method, 301
Gauss elimination, 53
  pivoting, 53
Gauss–Lobatto points, 277, 310
Gauss–Seidel method, 60
Generalized Ohm's law, 5
GFL23, 178
Givens rotation, 79
Godunov method, 246
Grad–Hogan method, 163
Grad–Shafranov equation, 45, 93
  equilibrium reconstruction, 116
  exact solutions, 102
  free boundary, 106
    backaveraging, 108
    blending, 108
    boundary condition, 109
    critical points, 113
    free functions, 108
    Green's function, 110
    magnetic feedback systems, 114
    Picard iteration, 108
    von Hagenow method, 111
  helical, 148
  Solovév solution, 104
  vacuum solution, 102
  with adiabatic constraints, 164
  with poloidal flow, 97
  with tensor pressure, 97
  with toroidal flow, 95
Gram–Schmidt procedure, 72, 74, 76
Guiding center drift orbits, 136
Gyrofluid magnetic reconnection, 283
Gyrokinetic Poisson equation, 283, 285
Gyroviscosity, 6

gyroviscous cancellation, 6
gyroviscous stress, 6

Hall terms
  and dispersion relation, 24
  definition, 23
  semi-implicit operator, 261
Heat equation, *see* parabolic equations
Heat flux, 3, 5
Helmholtz equation, 307
Hessenburg matrix, 77
Hyper-resistivity, 7
Hyperbolic equations, 235
  θ-implicit method, 251
  alternating direction implicit, 252
  Beam–Warming method, 244
  Crank–Nicolson method, 251
  differential approximation, 257, 284
    Caramana method, 259
    split θ-implicit, 259
  FTCS, 235
  Lax–Friedrichs method, 236
  Lax–Wendroff methods, 237
  leapfrog method, 239, 282
  limiter methods, 247
  MacCormack differencing, 238
  Newton–Krylov, 262
  partially implicit, 253
  semi-implicit method, 260
    two-fluid MHD, 261
  trapezoidal leapfrog, 241
  upwind differencing, 242, 245

Ideal MHD, 9, 195
  boundary conditions, 198, 199
  discrete spectrum, 201
  equation of motion, 197, 200
  linear force operator, 197
    completeness, 201
    continuous spectrum, 202, 214, 220
    eigenvalues, 200
    orthogonality, 200

resolvent set, 202
self-adjointness, 200
spectral properties, 201
linear stability
cylindrical geometry, 213
initial value approach, 199
normal-modes approach, 199
Rayleigh principle, 204
variational forms, 204
with flow, 203
Ill posed, 36
Implicit methods
compared to explicit methods, 40
definition of, 29
hyperbolic equations, 249
parabolic equations
BTCS, 174
Crank–Nicolson, 175
multiple dimensions, 184
non-linear implicit, 175
θ-implicit, 174
with spectral methods, 283
Initial value problem, 46
Inverse equilibrium equation, 136
J-solver, 138
q-solver, 137
expansion solution, 139
iterative metric method, 136
steepest descent method, 144
variational solution, 140
Inverse iteration, 323
Ion cyclotron frequency, 23
Ion heat flux, 157
Ion magnetization velocity, 6
Ion skin depth, 23

Jacobi's method, 60, 64
as preconditioner, 80
rate of convergence, 61
Jacobian, 121, 132, 306, 308
Jacobian matrix, 37

Krylov space methods, 70
conjugate gradient method, 73
with preconditioning, 82

GMRES, 76
preconditioning, 76, 80
steepest descent, 73

Lagrange polynomial, 306, 307
Lagrangian velocity, 153
Larmor radius, 6
limit of zero, 8
Lax equivalence theorem, 32
Lax–Friedrichs method, 38
Lax–Wendroff method, 237, 238, 244, 247, 248
Leapfrog method, 239–241, 255
Limiter methods, 247
Local extremum diminishing, 243
Loop voltage, 138, 156
LU decomposition of a matrix
back substitution, 54
defined, 54
general direct solvers, 57
incomplete, 81
MUMPS, 57
relation to Gauss elimination, 54
SUPER_LU, 57
Lundquist number, 9, 150, 255

Machine error, *see* round-off error
Magnetic axis
calculation of, 113
exact solution, 104
flux change at, 101
Mercier criterion at, 231
normalized flux, 108, 117
poloidal flux at, 113, 163, 164
Shafranov shift of, 140
singularity at, 121
Magnetic field
axisymmetric, 94, 151
equilibrium, 129
Magnetic flux coordinates, 122
arc length, 129
axisymmetric, 122
basis vectors, 124
contravariant, 124
covariant, 124

Boozer coordinates, 136
Christoffel symbols, 148
constant arc length, 133
constant area, 133
constant volume, 133
constructing, 131
curl, 126
current density, 129
differential volume, 130
divergence, 126
gradient, 125
inverse metric tensor, 127
metric elements, 129
metric tensor, 127
parallel current, 129
poloidal current, 130
poloidal flux, 130
safety factor, 131
straight field line, 133, 135
surface average, 130
time dependent, 151, 154
    coordinate velocity, 152
    Jacobian, 152
    time derivatives, 152
toroidal current, 130
toroidal flux, 130, 155
volume enclosed, 130
Magnetic Lundquist number, *see* Lundquist
    number
Magnetic moment, 98
Magnetic vector potential, 320
Magnetization velocity, *see* ion mag-
    netization velocity
Magnetoacoustic wave, 19
Magnetohydrodynamics (MHD), 4
Matrix eigenvalue problem, 323
Matrix iterative approach, 57
    convergence, 59
    Gauss–Seidel method, 60
    Jacobi's method, 60, 61, 64
    successive over-relaxation, 61
Maxwell's equations, 2
Mercier criterion, 231
    at magnetic axis, 231
Method of lines, 282

Micro-instability, 162
Moment equations, 2
Monotonicity preserving, 249
Multigrid methods, 66
    algebraic multigrid, 70
    coarse grid correction, 68
    interpolation operator, 69
    nested iteration, 68
    restriction operator, 69
Mutual inductance
    between plasma and coil, 163
    between two coils, 163

Newton's method, 177
    Newton–Krylov, 262
Number density, 3
Numerical diffusion, 248
Numerical stability, 31
    definition of, 29
    relation to truncation error, 36
    Von Neumann analysis, 32

Ohm's law, 5, 150
Ordinary differential equation, 46,
    282

Parabolic equations, 171
    BTCS, 174
    ADI, 186
    alternating direction, 185
    anisotropic diffusion, 188
    boundary conditions, 179
        at coordinate origin, 179
    BTCS, 176
    Crank–Nicolson, 175, 184, 185
    Douglas–Gunn method, 187
    DuFort–Frankel, 173, 184
    fractional steps, 185
    FTCS, 172, 184
    hybrid DuFort–Frankel, 190
    modified Crank–Nicolson, 175
    non-linear implicit , 175
    one dimensional, 171
    semi-implicit method, 184
    splitting, 185
    θ-implicit, 174, 176, 184

vector forms, 180
Parseval identity, 271
Particle flux, 157
Permeability of free space, 2
Permittivity of free space, 2
PEST code, 206, 318
Picard iteration, 107
Plasma frequency, 4
Poisson bracket, 257, 283
    Arakawa differencing, 257
Poisson's equation, 45
    one dimensional, 46
    two dimensional, 48
        Neumann BC, 50
Poloidal angle, 122
Poloidal flux function, 94
    evolution equation, 151
    meaning of, 99
Poloidal magnetic flux, 99
Preconditioning, 80
    additive Schwarz, 81
    incomplete Cholesky, 81
    incomplete LU, 81
    Jabobi, 81
Predictor-corrector, 183
Pseudo-spectral, 279

QR decomposition, 79
Quasineutrality, 5

Radiation, 158
Rate of strain tensor, 6
Ratio of specific heats, 10
Ray surface diagram, 21
Reciprocal normal surface diagram, 20
Reduced MHD, 10, 256, 320
Relative transport, 158
Relaxation
    dynamic, 65
    Jacobi's method, 60, 64
    Richardson, 60
Residual
    conjugate gradient method, 74
    finite element Galerkin, 316, 323

Krylov space, 71
    minimum (GMRES), 76, 79
    multigrid, 67, 68
    spectral Galerkin, 278
    steepest descent, 73
Resistivity, 7
    resistive instabilities, 9, 23
    resistive MHD, 8
    resistive time scale, 150, 164, 180
Riemann problem, 246
Ritz method, 289
Rotational transform, 156
    evolution equation, 157
Round-off error, 31
Runge–Kutta method, 46

Safety factor, 131
Scalar pressure, 3
Schwarz method, additive, 81
Self adjoint, 302
Similarity transformation, 38
Slow wave, 19
Solvability constraint, 46, 50, 51
SOR, *see* successive over-relaxation
Sparse matrix
    definition, 47
    finite element method, 267, 289
    from 2D elliptic equation, 49
    general direct solver for, 57
    implicit hyperbolic, 253, 259, 260
    implicit parabolic, 184
    iterative solvers for, 58
    multivariate Newton iteration, 263
    sparseness pattern, 47
Spectral accuracy, 269, 271
Spectral methods, 267
    aliasing error, 279
    collocation, 269, 277, 279, 282
    convolution sum, 278
    exponential convergence, 267
    Galerkin, 269, 278–280, 282
    implicit, 283
    pseudo-spectral, 279
    spectral accuracy, 269, 271, 274

stability, 282
tau, 269, 280
time discretization, 281
Spectral pollution, 318
Spectral radius, 59
Speed of light, 2
Staggered locations
    anisotropic diffusion, 189
    conservation form, 42
    conservative upwind, 244
    for density and velocity, 40
    leapfrog method, 240
Steepest descent
    accelerated
        inverse equilibrium, 146
        plasma equilibrium, 165
    inverse equilibrium, 144
    Krylov space, 73
Stiffness
    hyperbolic equations, 257
    in ideal MHD, 23
    matrix, finite element, 293
    partially implicit MHD, 253
Straight field line coordinates, 133
Stress tensor, 3, 5
Sturm–Liouville problem, 274
Successive over-relaxation, 61
Surface average, 130
Surface averaged MHD, 10
Surface averaged transport, 156
Suydam's condition, 219
Sylvester's theorem, 323

TAE modes, 202
Taylor method, 164
Taylor series, 28
Temperature
    anisotropic diffusion, 188
    Boltzmann constant, 3
    constant on flux surface, 154
    current driven by, 161
    diffusion in 1D, 176
    equilibration, 7
    equilibrium with flow, 95, 97
    expansion at origin, 179

gradient, 176
profile, 168
resistive MHD, 9
species, 3
Tokamak, 99
    circuit equations, 163
    curvature index, 115
    equilibrium reconstruction, 116
    flux loops, 117
    internal inductance, 100
    loop voltage, 138, 156
    magnetic feedback system, 114
    OH solenoid, 99
    poloidal beta, 100
    Shafranov formula, 100
    shaping coils, 102
    vertical feedback system, 102, 115
    vertical field coils, 99
Toroidal angle, 122
    straight field line coordinates, 135
Toroidal angular momentum density
    evolution equation, 157
Toroidal elliptic operator, 94, 110, 137
Toroidal field function, 94
    evolution equation, 151
Toroidal flux, 122, 130
Total energy density
    conservation form and, 12
    fluid equations, 37
Total variation diminishing, 249
Transport fluxes, 159
Transport model, 158
    anomalous transport, 162
    banana regime, 160
    bootstrap current, 161
    external current drive, 159, 161
    Pfirsch–Schlüter regime, 159
Triangular form, 54, 55, 59
Tridiagonal
    algorithm, 47
    block-tridiagonal, 56, 143, 144, 183
    matrix, 47, 293

Truncation error
    ADI, 187
    and implicit methods, 249
    BTCS, 174
    Crank–Nicolson, 175, 251
    definition, 30
    differential approximation, 259
    Douglas–Gunn, 188
    DuFort–Frankel, 173
    for model problem, 31
    Lax–Freidrichs, 237
    Lax–Wendroff, 238
    leapfrog, 239
    order of, 40
    parabolic FTCS, 172
    relation to stability, 36
    semi-implicit method, 185
    time step splitting, 186
    upwind differencing, 242
Two-fluid
    closure model, 8
    dispersion relation, 23

Upwind differencing, 183, 242
    Beam–Warming method, 244
    system of equations, 245

Vacuum solution
    cylindrical geometry, 215
    multipolar expansion, 103
    toroidal geometry, 223
Vector diffusion equation, 182
Von Hagenow method, 111
Von Neumann stability analysis
    amplification factor, 33, 38
    fractional step method, 186
    higher order equations, 37
    introduction to, 32
    leapfrog method, 39
    local stability, 35
    multiple space dimensions, 39
    parabolic FTCS, 172
    semi-implicit method, 185

Weak form, 302

For Product Safety Concerns and Information please contact our EU representative GPSR@taylorandfrancis.com Taylor & Francis Verlag GmbH, Kaufingerstraße 24, 80331 München, Germany

T - #0012 - 160425 - C0 - 234/156/20 [22] - CB - 9781439810217 - Gloss Lamination